Shock Wave and High Pressure Phenomena

Founding Editor
R. A. Graham, USA

Honorary Editors
L. Davison, USA
Y. Horie, USA

Editorial Board
G. Ben-Dor, Israel
F. K. Lu, USA
N. Thadhani, USA

More information about this series at http://www.springer.com/series/1774

Shock Wave and High Pressure Phenomena

L.L. Altgilbers, M.D.J. Brown, I. Grishnaev, B.M. Novac, I.R. Smith, I. Tkach, and Y. Tkach: Magnetocumulative Generators

T. Antoun, D.R. Curran, G.I. Kanel, S.V. Razorenov, and A.V. Utkin: Spall Fracture

J. Asay and M. Shahinpoor (Eds.): High-Pressure Shock Compression of Solids

S.S. Batsanov: Effects of Explosion on Materials: Modification and Synthesis Under High-Pressure Shock Compression

G. Ben-Dor: Shock Wave Reflection Phenomena

L.C. Chhabildas, L. Davison, and Y. Horie (Eds.): High-Pressure Shock Compression of Solids VIII

L. Davison: Fundamentals of Shock Wave Propagation in Solids

L. Davison, Y. Horie, and T. Sekine (Eds.): High-Pressure Shock Compression of Solids

V.L. Davison and M. Shahinpoor (Eds.): High-Pressure Shock Compression of Solids III

R.P. Drake: High-Energy-Density Physics

A.N. Dremin: Toward Detonation Theory

J.W. Forbes: Shock Wave Compression of Condensed Matter

V.E. Fortov, L.V. Altshuler, R.F. Trunin, and A.I. Funtikov: High-Pressure Shock Compression of Solids VII

B.E. Gelfand, M.V. Silnikov, S.P. Medvedev, and S.V. Khomik: Thermo-Gas Dynamics of Hydrogen Combustion and Explosion

D. Grady: Fragmentation of Rings and Shells

Y. Horie, L. Davison, and N.N. Thadhani (Eds.): High-Pressure Shock Compression of Solids VI

J. N. Johnson and R. Cherét (Eds.): Classic Papers in Shock Compression Science

V.K. Kedrinskii: Hydrodynamics of Explosion

C.E. Needham: Blast Waves

V.F. Nesterenko: Dynamics of Heterogeneous Materials

S.M. Peiris and G.J. Piermarini (Eds.): Static Compression of Energetic Materials

M. Sućeska: Test Methods of Explosives

M.V. Zhernokletov and B.L. Glushak (Eds.): Material Properties under Intensive Dynamic Loading

J.A. Zukas and W.P. Walters (Eds.): Explosive Effects and Applications

Jiun-Ming Li • Chiang Juay Teo
Boo Cheong Khoo • Jian-Ping Wang
Cheng Wang
Editors

Detonation Control for Propulsion

Pulse Detonation and Rotating Detonation Engines

Editors
Jiun-Ming Li
Temasek Laboratories
National University of Singapore
Singapore, Singapore

Boo Cheong Khoo
Temasek Laboratories
National University of Singapore
Singapore, Singapore

Cheng Wang
State Key Laboratory of Explosion
Science and Technology
Beijing Institute of Technology
Beijing, China

Chiang Juay Teo
Mechanical Engineering Department
National University of Singapore
Singapore, Singapore

Jian-Ping Wang
Center for Combustion and Propulsion
CAPT & SKLTCS
Department of Mechanics and
Engineering Science
College of Engineering
Peking University
Beijing, China

ISSN 2197-9529　　　　　ISSN 2197-9537　(electronic)
Shock Wave and High Pressure Phenomena
ISBN 978-3-319-68905-0　　　ISBN 978-3-319-68906-7　(eBook)
https://doi.org/10.1007/978-3-319-68906-7

Library of Congress Control Number: 2017958815

© Springer International Publishing AG 2018
This work is subject to copyright. All rights are reserved by the Publisher, whether the whole or part of the material is concerned, specifically the rights of translation, reprinting, reuse of illustrations, recitation, broadcasting, reproduction on microfilms or in any other physical way, and transmission or information storage and retrieval, electronic adaptation, computer software, or by similar or dissimilar methodology now known or hereafter developed.
The use of general descriptive names, registered names, trademarks, service marks, etc. in this publication does not imply, even in the absence of a specific statement, that such names are exempt from the relevant protective laws and regulations and therefore free for general use.
The publisher, the authors and the editors are safe to assume that the advice and information in this book are believed to be true and accurate at the date of publication. Neither the publisher nor the authors or the editors give a warranty, express or implied, with respect to the material contained herein or for any errors or omissions that may have been made. The publisher remains neutral with regard to jurisdictional claims in published maps and institutional affiliations.

Printed on acid-free paper

This Springer imprint is published by Springer Nature
The registered company is Springer International Publishing AG
The registered company address is: Gewerbestrasse 11, 6330 Cham, Switzerland

Preface

For the past seventy-six years, since 1941 when Hoffman first proposed the use of detonation waves for propulsion applications, there has been sustained efforts by many researchers throughout the world to develop a real and practical propulsion system which employs detonation waves in a controllable manner. Especially noteworthy achievements are early efforts on rotating detonation engines by Nicholls at the University of Michigan and by Voitsekhovsky in Novosibirsk in the 1950s and 1960s, and continual research activities worldwide on pulse detonation engines that were initiated by Bussing's engine demonstration in the 1990s. Recently, through theoretical performance studies, experimental investigation, and analysis via numerical simulations, the technology development of rotating detonation engines has been revisited by international leading researchers and has proven to be capable of implementation in real applications for different purposes, such as in rocket motors, ramjets, and turbojets. These leading researchers are devoting to carry out examinations and verification of system-level rotating detonation engines currently to allow this technology to achieve a higher level of technology readiness level as can be seen in the effort by Piotr Wolański, Sergey Frolov, Kailas Kailasanath, Fred Schauer, Matthew Fotia, Christopher Brophy, Frank Lu, Kenneth Yu, Jiro Kasahara, Nobuyuki Tsuboi, Ken Matsuoka, Francois Falempin, Bruno Le Enaour, Jeong-Yeol Choi, Jian-Ping Wang, Boo Cheong Khoo, Chiang Juay Teo, and Jiun-Ming Li.

This book is a collection of state-of-the-art research contributions from these international leading experts who have presented their works at the 2016 International Workshop on Detonation for Propulsion (IWDP) held in Singapore and the 2015 IWDP held in Beijing. This workshop has successfully brought together scientists from various parts of the world for the exchange of cutting-edge technical knowledge and has served as a platform to promote more collaborative opportunities internationally since 2011 (2011, Bourges and Pusan; 2012, Tsukuba; 2013, Tainan; 2014, Warsaw). These forms of knowledge and experience sharing involving various aspects by the different groups constitute one of the main reasons why the rotating detonation engine has progressed remarkably rapidly in recent years.

The chapters in this book strive to meet the needs of scientists, young researchers, young engineers, students, and others in the field of shock waves, combustion, and aerospace propulsion. Each chapter manuscript was subjected to a peer-review process and subsequently revised by the chapter authors accordingly. A system-level design of the novel detonation engines (RDE & PDE), performance analysis, and investigation employing advanced experimental and numerical methods are presented. The world's first successful sled demonstration of a rocket RDE system and innovations in the development of a kilohertz PDE system are reported. It is hoped that readers will obtain in a straightforward manner an understanding of the RDE & PDE design, operation and testing approaches, and further specific integration schemes for diverse applications, e.g., rocket motor for space propulsion and turbojet/ramjet engines for air-breathing propulsion. A comprehensive survey of detonation-based engine technology is also introduced for the readers to have a broad overview and development to date. Finally, a fundamental research on basic detonation re-initiation phenomenon is included because such knowledge of detonation re-initiation mechanism serves as a useful guide for the detonation chamber design of RDE & PDE. In addition, the materials involving the use of optical diagnostics techniques in this study, such as Schlieren imaging, planar laser-induced fluorescence (PLIF), open shutter photography, and soot foil technique will be useful for researchers and engineers involved in the relevant fields. This book hopefully attracts the interests of young researchers/engineers to join in the ongoing revolution of new propulsion system. Lastly, this book can serve as a reference textbook for faculty and graduate students and for graduate-level courses in shock waves, combustion, and propulsion in universities. The following is a brief outline of what the readers can gleam for each chapter:

Chapter 1 Critical design of rotation detonation engine for air-breathing propulsion and performance evaluation process for rotation detonation engine.

Chapter 2 Strategic schemes to integrate turbojet and rotating detonation chamber and its direct performance measurement using gaseous hydrogen and liquid Jet-A as fuels.

Chapter 3 Investigations of detonation liquid rocket engine with natural gas for space propulsion and development of computational methods to simulate detailed flow structure within an operating continuous-detonation combustor with nozzle.

Chapter 4 Sled demonstration of rotation detonation rocket engine for thrust measurement and propellant injector shape effect on engine performance.

Chapter 5 Computational method development with robust-weighted compact nonlinear scheme for two-dimensional simulation of rotating detonation engine.

Chapter 6 Background, current progress, and challenge of rotating detonation engines.

Chapter 7 Progress and accomplishments of rotating detonation engine research at Peking University, China.

Chapter 8 Novel kilohertz pulse detonation chamber: concept, experiment, and one-dimensional numerical analysis.

Chapter 9 Fundamental detonation re-initiation phenomenon of stable and unstable detonation with multiple diagnostic methods: Schlieren photography, planar laser-induced fluorescence, open shutter photography, and soot foil technique.

The last section of this book constitutes the minutes of the Panel Discussion conducted at the 2016 International Workshop on Detonation for Propulsion (IWDP). Readers can gain further insights on the main concerns, critical research issues, and challenges of detonation-based engines for propulsion based on the dialogue of the leading researchers.

The editors would like to thank all chapter authors for contributing their latest valuable research results to promote these special collections for detonation engine development. Specifically, we wish to express our appreciation to Piotr Wolański and Sergey Frolov for their unwavering support. The editors would like to extend appreciation to the reviewers for their timely cooperation and expertise. During the early work on the manuscript, we have benefited from many colleagues who attended the 2015 IWDP and we would like to show our gratitude to the organizers of the 2015 IWDP (co-editors of this book): Jian-Ping Wang and Cheng Wang.

Finally, we are greatly thankful to Christopher Coughlin of Springer Publishers. His initiative and encouragement made this volume possible in the series on Shock Wave and High Pressure Phenomena. We would also like to express our appreciation to Springer editors Dominic Manoharan and HoYing Fan for their recommendation and close cooperation during the preparation of the book manuscript.

Singapore, Singapore
Jiun-Ming Li
Chiang Juay Teo
Boo Cheong Khoo

Contents

1. **Performance of Rotating Detonation Engines for Air Breathing Applications** 1
   Matthew L. Fotia, John Hoke, and Frederick Schauer

2. **Development of Gasturbine with Detonation Chamber** 23
   Piotr Wolański, Piotr Kalina, Włodzimierz Balicki, Artur Rowiński,
   Witold Perkowski, Michał Kawalec, and Borys Łukasik

3. **Flow Structure in Rotating Detonation Engine with Separate Supply of Fuel and Oxidizer: Experiment and CFD** 39
   Sergey M. Frolov, Viktor S. Aksenov, Vladislav S. Ivanov,
   Sergey N. Medvedev, and Igor O. Shamshin

4. **Application of Detonation Waves to Rocket Engine Chamber** 61
   Jiro Kasahara, Yuichi Kato, Kazuaki Ishihara, Keisuke Goto,
   Ken Matsuoka, Akiko Matsuo, Ikkoh Funaki, Hideki Moriai,
   Daisuke Nakata, Kazuyuki Higashino, and Nobuhiro Tanatsugu

5. **Numerical Simulation on Rotating Detonation Engine: Effects of Higher-Order Scheme** 77
   Nobuyuki Tsuboi, Makoto Asahara, Takayuki Kojima,
   and A. Koichi Hayashi

6. **Review on the Research Progresses in Rotating Detonation Engine** 109
   Mohammed Niyasdeen Nejaamtheen, Jung-Min Kim,
   and Jeong-Yeol Choi

7. **Continuous Detonation Engine Researches at Peking University** ... 161
   Jian-Ping Wang, Song-Bai Yao, and Xu-Dong Han

8. **Pulse Detonation Cycle at Kilohertz Frequency** 183
   Ken Matsuoka, Haruna Taki, Jiro Kasahara, Hiroaki Watanabe,
   Akiko Matsuo, and Takuma Endo

9 On the Investigation of Detonation Re-initiation Mechanisms
   and the Influences of the Geometry Confinements
   and Mixture Properties 199
   Lei Li, Jiun-Ming Li, Chiang Juay Teo, Po-Hsiung Chang,
   Van Bo Nguyen, and Boo Cheong Khoo

**2016 International Workshop on Detonation for Propulsion:
Panel Discussion** 237

Contributors

Viktor S. Aksenov Center for Pulsed Detonation Combustion, Semenov Institute of Chemical Physics (ICP), National Research Nuclear University MEPhI, Moscow, Russia

Makoto Asahara Department of Mechanical Engineering, Gifu University, Gifu, Japan

Włodzimierz Balicki Institute of Aviation, Warsaw, Poland

Po-Hsiung Chang Temasek Laboratories, National University of Singapore, Singapore, Singapore

Jeong-Yeol Choi Department of Aerospace Engineering, Pusan National University, Busan, Republic of Korea

Takuma Endo Department of Mechanical System Engineering, Hiroshima University, Higashi-Hiroshima, Hiroshima, Japan

Matthew L. Fotia Innovative Scientific Solutions Inc., Dayton, OH, USA

Sergey M. Frolov Center for Pulsed Detonation Combustion, Semenov Institute of Chemical Physics (ICP), National Research Nuclear University MEPhI, Moscow, Russia

Ikkoh Funaki Japan Space Exploration Agency, Sagamihara, Kanagawa, Japan

Keisuke Goto Nagoya University, Nagoya, Aichi, Japan

Xu-Dong Han Center for Combustion and Propulsion, CAPT & SKLTCS, Department of Mechanics and Engineering Science, College of Engineering, Peking University, Beijing, China

Kazuyuki Higashino Muroran Institute of Technology, Muroran, Hokkaido, Japan

John Hoke Innovative Scientific Solutions Inc., Dayton, OH, USA

Kazuaki Ishihara Nagoya University, Nagoya, Aichi, Japan

Vladislav S. Ivanov Center for Pulsed Detonation Combustion, Semenov Institute of Chemical Physics (ICP), National Research Nuclear University MEPhI, Moscow, Russia

Piotr Kalina Institute of Aviation, Warsaw, Poland

Jiro Kasahara Department of Aerospace Engineering, Nagoya University, Nagoya, Aichi, Japan

Yuichi Kato Nagoya University, Nagoya, Aichi, Japan

Michał Kawalec Institute of Aviation, Warsaw, Poland

Boo Cheong Khoo Temasek Laboratories, National University of Singapore, Singapore, Singapore

Jung-Min Kim Department of Aerospace Engineering, Pusan National University, Busan, Republic of Korea

A. Koichi Hayashi Department of Mechanical Engineering, Aoyama Gakuin University, Sagamihara, Kanagawa, Japan

Takayuki Kojima Chofu Aerospace Center, Japan Aerospace Exploration Agency, Chofu-shi, Tokyo, Japan

Jiun-Ming Li Temasek Laboratories, National University of Singapore, Singapore, Singapore

Lei Li Temasek Laboratories, National University of Singapore, Singapore, Singapore

Borys Łukasik Institute of Aviation, Warsaw, Poland

Akiko Matsuo Department of Mechanical Engineering, Keio University, Yokohama, Kanagawa, Japan

Ken Matsuoka Department of Aerospace Engineering, Nagoya University, Nagoya, Aichi, Japan

Sergey N. Medvedev Center for Pulsed Detonation Combustion, Semenov Institute of Chemical Physics (ICP), National Research Nuclear University MEPhI, Moscow, Russia

Hideki Moriai Mitsubishi Heavy Industries Ltd., Komaki, Aichi, Japan

Daisuke Nakata Muroran Institute of Technology, Muroran, Hokkaido, Japan

Van Bo Nguyen Temasek Laboratories, National University of Singapore, Singapore, Singapore

Mohammed Niyasdeen Nejaamtheen Department of Aerospace Engineering, Pusan National University, Busan, Republic of Korea

Witold Perkowski Institute of Aviation, Warsaw, Poland

Artur Rowiński Institute of Aviation, Warsaw, Poland

Frederick Schauer USAF Air Force Research Laboratory, Wright-Patterson AFB, OH, USA

Igor O. Shamshin Center for Pulsed Detonation Combustion, Semenov Institute of Chemical Physics (ICP), National Research Nuclear University MEPhI, Moscow, Russia

Haruna Taki Department of Aerospace Engineering, Nagoya University, Nagoya, Aichi, Japan

Nobuhiro Tanatsugu Muroran Institute of Technology, Muroran, Hokkaido, Japan

Chiang Juay Teo Mechanical Engineering Department, National University of Singapore, Singapore, Singapore

Nobuyuki Tsuboi Department of Mechanical and Control Engineering, Kyushu Institute of Technology, Kitakyushu, Fukuoka, Japan

Jian-Ping Wang Center for Combustion and Propulsion, CAPT & SKLTCS, Department of Mechanics and Engineering Science, College of Engineering, Peking University, Beijing, China

Hiroaki Watanabe Department of Mechanical Engineering, Keio University, Yokohama, Kanagawa, Japan

Piotr Wolański Institute of Aviation, Warsaw, Poland

Song-Bai Yao Center for Combustion and Propulsion, CAPT & SKLTCS, Department of Mechanics and Engineering Science, College of Engineering, Peking University, Beijing, China

Chapter 1
Performance of Rotating Detonation Engines for Air Breathing Applications

Matthew L. Fotia, John Hoke, and Frederick Schauer

Abstract The performance of air breathing rotating detonation engines is presented through the use of experimental measurements made at the Air Force Research Laboratory's Detonation Engine Research Facility. The performance scaling characteristics observed between various physical device configurations is discussed; included is the influence of mass flow rate, air injection area ratio and nozzle area ratio. The impact of geometry on the unsteady nature of rotating detonation engines and the combustors' efficient production of thrust is presented, along with an analysis of the pressure-gain properties of rotating detonation engines. Finally, the impact of transitioning from hydrogen fueled to hydrocarbon fueled devices is examined through the direct comparison of experimental measurements of operation on both gaseous hydrogen and ethylene fuels. Previous pulsed detonation engine work is used to provide a reference for this comparison.

1 Introduction

Rotating detonation engines have been shown to be viable pressure-gain combustion devices. The operability of such combustion systems and the manner in which the performance of these systems scale are still developing areas of investigation. The identification of the pertinent scaling laws to describe the performance of these systems and the performance of these devices when coupled to various internal nozzle configurations is the focus of the current chapter. In particular, the detonation channel scaling is examined while attempting to hold the effect of propellant mixing constant. The impact of operation of rotating detonation engines on hydrocarbon fuels is also presented. The successful scaling and transition to operation on more

M.L. Fotia (✉) • J. Hoke
Innovative Scientific Solutions Inc., Dayton, OH, USA
e-mail: matt.fotia@gmail.com

F. Schauer
USAF Air Force Research Laboratory, Wright-Patterson AFB, OH, USA

© Springer International Publishing AG 2018
J.-M. Li et al. (eds.), *Detonation Control for Propulsion*, Shock Wave and High Pressure Phenomena, https://doi.org/10.1007/978-3-319-68906-7_1

logistically supportable fuels is a key aspect in the future application of rotating detonation pressure-gain combustion technology to aerospace propulsion systems.

Operation of rotating detonation engines of different nominal detonation channel diameters and channel widths has been reported by Shank et al. (2012), Naples et al. (2013), Fotia et al. (2015, 2016, 2017), Russo et al. (2011) and Dyer et al. (2012). Shank et al. presented the initial experimental development effort on a nominal six-inch diameter rotating detonation engine with a focus on mapping the mass flow and equivalence ratio operating space of this new device. Naples et al. examined this device further through the use of a quartz outer-body and high-speed chemiluminescence imaging to provide basic data on the various angles present in the flow structure for use in the validation of modeling efforts.

Fotia et al. (2016) reported the effects of exhaust flow nozzling on the measured thrust production of a nominal 6-inch diameter rotating detonation engine. A discussion on appropriate stagnation states is used to identify the effects of pressure-gain in the system, while examining the comparative effect on performance of both bluff-body and plug nozzle exhaust schemes. The different regimes of ignition observed in this scale of rotating detonation engine are detailed by Fotia et al. (2015), where an air ejector configuration is used to provide independent backpressure to the detonation channel.

Kindracki et al. (2011a) examined the thrust production and evaluated the specific impulse developed by a rotating detonation engine with aerospike nozzles. The variation of pressures and velocities were presented by Kindracki et al. (2011b) for bluff-body arrangement of these devices.

Dyer et al. tested a larger diameter device with a twenty-inch detonation channel diameter and found that for rotating detonation devices there is a critical interaction between the fuel/air mixing and detonation propagation. Other recent experimental investigations include the study of the detonation wave/fuel plenum interaction and the potential coupling between the two by Fotia et al. (2014) where a two-dimensional test-section was used in a single-pass operation configuration. Frolov et al. (2015) demonstrated the operation of a large-scale hydrogen-air continuous detonation combustor, in which backpressuring obstacles had been placed in the combustion channel to enhance thrust production. It was found that the experimental results matched well with previous computational findings. The mode of detonation propagation within a rotating detonation engine has been examined by Lin et al. (2015) where single wave, single/dual-wave hybrid and dual-wave modes of operation have been observed in the same device by varying the injection conditions. As well, the self-ejection of oxidizer into the annular detonation channel has been studied by Bykovskii et al. (2010, 2011) for operation with both oxygen and air.

The modeling and simulation of rotating detonating engines have been more recently conducted by Schwer and Kailasanath (2010, 2011), Kaemming et al. (2017), Paxson (2014, 2016) and Davidenko et al. (2008). Schwer and Kailasanath (2010) developed a numerical procedure for modeling the flow field of rotation detonation engines which shows good structural agreement with experimental observations, while indicating the potential for losses in the combustion of fuel outside of the detonation wave and the post-detonation shocking of combustion products. These same authors (Schwer and Kailasanath, 2011) later used this algorithm to examine the effects of various engine size parameters that included the nominal diameter, length and depth of the detonation channel, as well as the area ratio of the propellant injection Scheme.

A reduced-order thermodynamic modeling approach was used by Kaemming et al. (2017) to capture the key physical mechanisms present in rotating detonation engines. This work considered four different thermodynamic pathways through the engine, and proposed coupled models for the injection of propellants and the exhaust of combustion products from the device. The results of this modeling agreed well with experimental performance measurements and provided confirmation that pre-combustion static pressure and deflagrative losses are first-order drivers of performance in these pressure-gain combustion systems. An experimental and analytical analysis of the detailed two-dimensional calculation of the structure of the rotating detonation and its resultant performance was presented by Hishida et al. (2009).

Paxson (2014) proposed a model that allowed for the reduction of the computational time and resources required to simulate a rotating detonation engine by considering a periodic two-dimensional computational space. This model was later used (2016) to investigate the impact of the exhaust throat area convergence on the performance of the combustor. The focus of this work was on the influence of the reflected wave mechanics between the nozzle throat and the air injector, showing that there exists a coupled relationship between the back-pressuring provided by the nozzle and the propellant fill Mach number provided by a given injection area.

A parametric study of flow field parameters was conducted by Davidenko et al., in which the injection total pressure and the spatial period of device operation are identified as scaling factors for the geometry and reactive flow pressure respectively. Average injection mass flux was found to be a factor in driving these two parameters. Rankin et al. (2014) showed through simulation that a reduction in exhaust plume unsteadiness in the rotating detonation engine is possible as it transits a converging-diverging nozzle arrangement. This natural reduction in flow turbulence has implications when considering the efficiency of energy conversion into thrust by the nozzle.

2 Experimental Setup

The data that will be discussed in this chapter was primarily collected through the use of a thrust stand installed at the Air Force Research Laboratory's Detonation Engine Research Facility at Wright-Patterson AFB. The testing was conducted on the six-inch diameter modular research rotation detonation engine test-section at flow rates from 0.61 to 1.82 ± 0.03 kg/s, and global equivalence ratios between 0.6 and 1.35 ± 0.02.

This rotating detonation engine design has been characterized by Naples et al. (2013) and Fotia et al. (2016, 2017) and is shown in Fig. 1.1, with the oxidizer and fuel feeds noted. Key features of this configuration include the radial outward flow

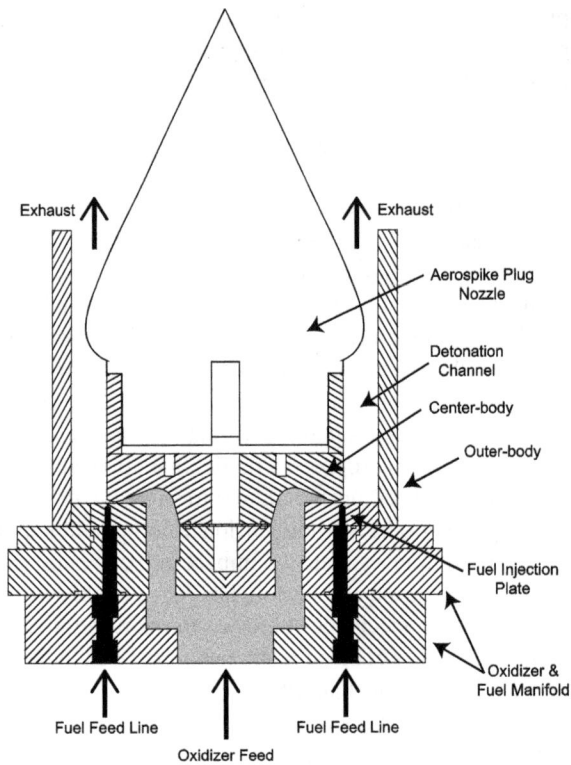

Fig. 1.1 Schematic of the six-inch diameter modular research rotating detonation engine device coupled to an aerospike plug-nozzle with propellant feed paths shown (Fotia et al. 2016)

of oxidizer and the modular nature of both the outer-body and center-body components, allowing for many physical arrangements to be investigated based on the same global radially fed device. The fuel and air mixing scheme at the base of the detonation channel is similar to that used by Shank et al. (2012) and is shown in Fig. 1.2.

Thrust measurements were made on a horizontal thrust stand, using a hydraulic load cell. The calibration of the installed test-section showed good linear behavior with hysteresis limited to a maximum of 12 N (3 lbf). The gas flow rates to the test-section were metered upstream of the feed manifolds through the use of sonic nozzles.

Testing was conducted on five different nozzle configurations shown schematically in Fig. 1.3. These included two variations on a bluff body exhaust, three variations on an aerospike nozzle and one classical converging-diverging nozzle configuration. A comparison between the bluff body exhaust and the plug-nozzle configurations was made by Fotia et al. (2016). The focus here will be kept on the behavior of the plug-nozzle as these results are more relevant to actual direct thrust production applications of rotating detonation engines. The length of the outer-body of the rotating detonation engine was kept constant at 11.4 cm (4.5 inch) for all configurations tested, unless otherwise noted. The converging-diverging nozzle configuration is considered for comparison reasons to ensure that the discussion will be on the behavior of the rotating detonation engine and not on any peculiarities of the nozzling geometry implemented.

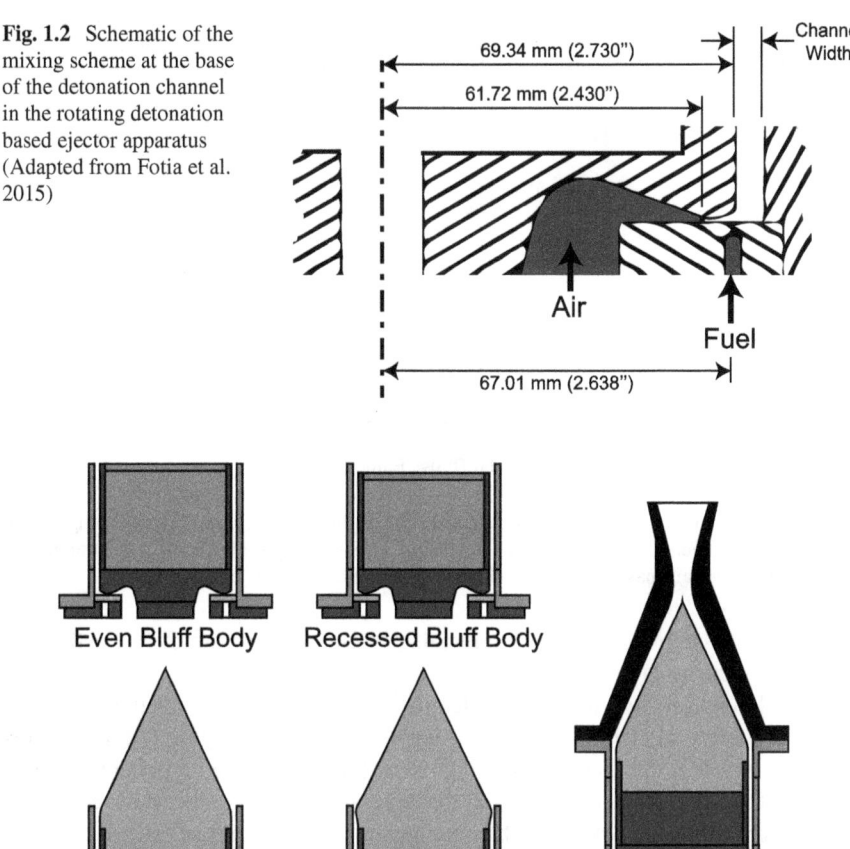

Fig. 1.2 Schematic of the mixing scheme at the base of the detonation channel in the rotating detonation based ejector apparatus (Adapted from Fotia et al. 2015)

Fig. 1.3 Schematic of the various experimental nozzle configurations (Adapted from Fotia et al. 2016)

3 Thrust Production and Scaling

The fundamental performance metrics used to describe fuel efficiency and effectiveness of working fluid use in a thrust-producing device are the specific impulse, I_{sp}, and specific thrust, $F_g/\dot{m}_{Air}$, respectively. The specific impulse will be defined consistent with the "airbreathing" definition, where

$$I_{sp} = \frac{F_g}{\dot{m}_{Fuel} g_o} \tag{1.1}$$

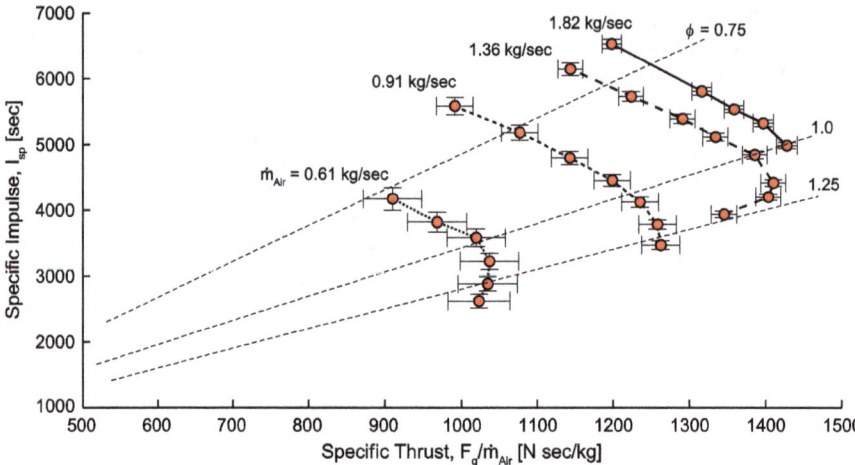

Fig. 1.4 Performance of a six-inch rotating detonation engine coupled to an aerospike plug-nozzle operated on hydrogen/air, shown in terms of specific impulse, I_{sp}, as a function of specific thrust, $F_g/\dot{m}_{Air}$, for various air mass flow rates and equivalence ratios (Fotia et al. 2017)

with F_g being the gross measured thrust, $\dot{m}_{Fuel}$ the mass flow rate of fuel and g_o the acceleration due to gravity.

Considering a set internal rotating detonation engine geometry, an increase in air mass flow rate is matched with increasing specific thrust. Figure 1.4 shows this behavior, which is due to the increasing initial pressure of the propellants prior to being detonated and the increasing backpressurization of the device due to combustion-induced blockage in the finite geometry of the engine. The data shown is for combustion of gaseous hydrogen and air, with an air injection area ratio, A_{Air}/A_{RDE}, of 0.435 and a nozzle area convergence ratio, A_{Noz}/A_{RDE}, of 0.6.

To provide a framework in which to discuss the production of thrust by various geometric configurations of the rotating detonation engine, a simplified thrust equation can be written by considering the rotating detonation engine as a simple thrust producing device. This can be expressed in terms of specific impulse as

$$I_{sp} = \frac{F_g}{\dot{m}_{Fuel} g_o}$$

$$= \frac{1}{g_o}\left(1 + \frac{1}{f_s \phi}\right)\sqrt{\gamma R T_e} M_e \qquad (1.2)$$

$$= \frac{1}{g_o}\left(1 + \frac{1}{f_s \phi}\right)\sqrt{\gamma R T_{oNoz}} \left(\frac{2}{\gamma - 1}\right)^{\frac{1}{2}} \left[1 - \left(\frac{P_{oNoz}}{P_e}\right)^{\frac{\gamma-1}{\gamma}}\right]^{\frac{1}{2}}$$

where f_s is the stoiciometric fuel-air mass ratio, γ is the ratio of specific heats of the exhaust products, R is the gas constant of the exhaust products, ϕ is the equivalence ratio, M is the Mach number, with T representing temperature and P, pressure. A subscript e has been used to denote the state of the gas post-nozzle expansion, o denotes stagnation conditions, and Noz denotes fluid conditions at the entrance to the nozzle.

Care was taken to intentionally label the nozzle entrance stagnation quantities as in the particular case of stagnation pressure; these quantities will be different from those entering the test-section due to the pressure-gain nature of the detonative combustion process. This allows for the driving physical behaviors to be separated out without fully defining the nature of the rotating detonation combustion process.

To maintain the focus on the operation of the rotating detonation combustor, it has been assumed that the aerospike plug-nozzles implemented at the exhaust plane of the annular detonation channel provide near-perfect expansion of the exhaust products to ambient pressure conditions. This assumption is justified as this class of nozzle shows this property at lower nozzle pressure ratios.

If the detonation channel width of the engine is changed, it follows that to provide the same performance the stagnation-to-ambient pressure ratio, $P_{o\,Noz}/P_e$, across the nozzle must be held constant for operation at a given equivalence ratio, or stagnation temperature. To do this the state of the flow entering and exhausting from the combustor must be examined for a set detonation channel mass flux.

When considered in a fundamental one-dimensional sense, the area expansion ratio between the air injector and the detonation channel, A_{Air}/A_{RDE}, must be held constant between the different channel width geometries to hold the propellant pre-detonation pressure losses constant. For the nozzle pressure ratio to be held constant between geometries, setting the Mach number of the exhaust products post-expansion, the mass flux through the nozzle must be held constant to make appropriate comparisons between different channel widths. This occurs if the rotating detonation engine channel-to-nozzle throat area contraction ratio is set, since if

$$\frac{\dot{m}}{A_{RDE}} = \text{Constant} \tag{1.3}$$

so must

$$\frac{\dot{m}}{A_{Noz}} \frac{A_{Noz}}{A_{RDE}} = \text{Constant} \tag{1.4}$$

As can be seen in Fig. 1.5, the experimental performance for differing channel width configurations of the rotating detonation engine show consistent trends when the appropriate one-dimensional analysis based parameters are fixed. The figure shows data for an air injection area ratio, A_{Air}/A_{RDE}, of 0.435, a detonation channel mass flux, $\dot{m}/A_{RDE}$, of 173.0 kg/s/m^2 and a nozzle area contraction ratio, A_{Noz}/A_{RDE}, of 0.6.

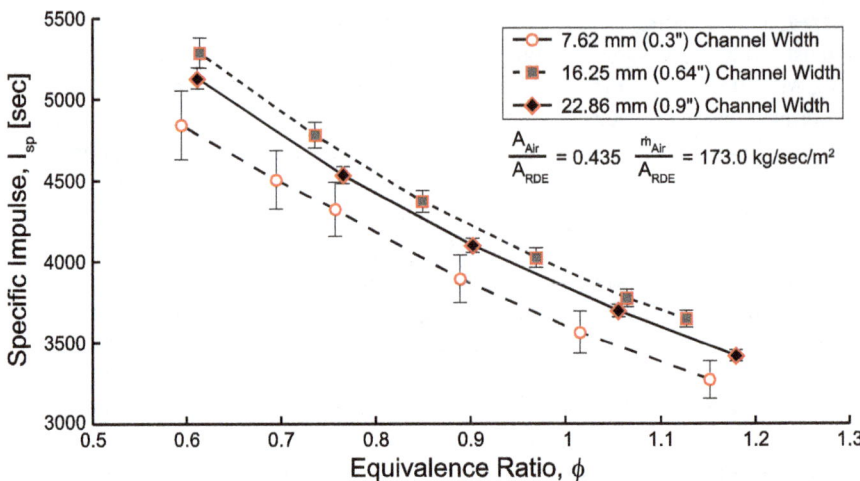

Fig. 1.5 Performance of a six-inch rotating detonation engine operated on hydrogen/air, given in term of specific impulse, I_{sp}, as a function of equivalence ratio, ϕ, for various detonation channel widths (Fotia et al. 2017)

While there is some slight vertical shifting of the specific impulse curves between detonation channel widths, the slope of these curves remain the same relative to each other. Some deviation is present due to three-dimensional effects, such as differences in propellant mixing and flow recirculation losses at the base of the detonation channel. These are the most likely sources of loss as in the true three-dimensional arrangements it is the radial length scales across the channel that are being altered.

Contributing to the production of thrust, these differing internal flow structures and levels of fuel mixing have the potential to provide alternative internal wave arrangements to the propagating detonation and the expansion waves present in the device.

Specific impulse gives a measure of fuel usage in the creation of thrust. The normalized corrected thrust, $F_g P_{o\,Ref}/P_{o\,Air}$, describes thrust production from the stagnation pressure supplied to the engine through the air feed manifold. The difference in the use of pressure between channel configurations can be directly seen in Fig. 1.6a where the higher corrected thrust is found for the 22.86 mm (0.9 inch) channel width, while the 16.25 mm (0.64 inch) channel width configuration shows more effective pressure usage than the 7.62 mm (0.3 inch) channel.

The corrected thrust obtained from a given channel configuration can be manipulated by changing the three parameters held constant to this point in the comparison. Those parameters being the air injection area ratio, the nozzle area contraction ratio and the mass flux through the detonation channel.

If a lower detonation channel mass flux is examined, as in Fig. 1.6b, the relative relationship between the different configurations stays the same, while simply decreasing the levels of corrected thrust created by the device. The 22.86 mm

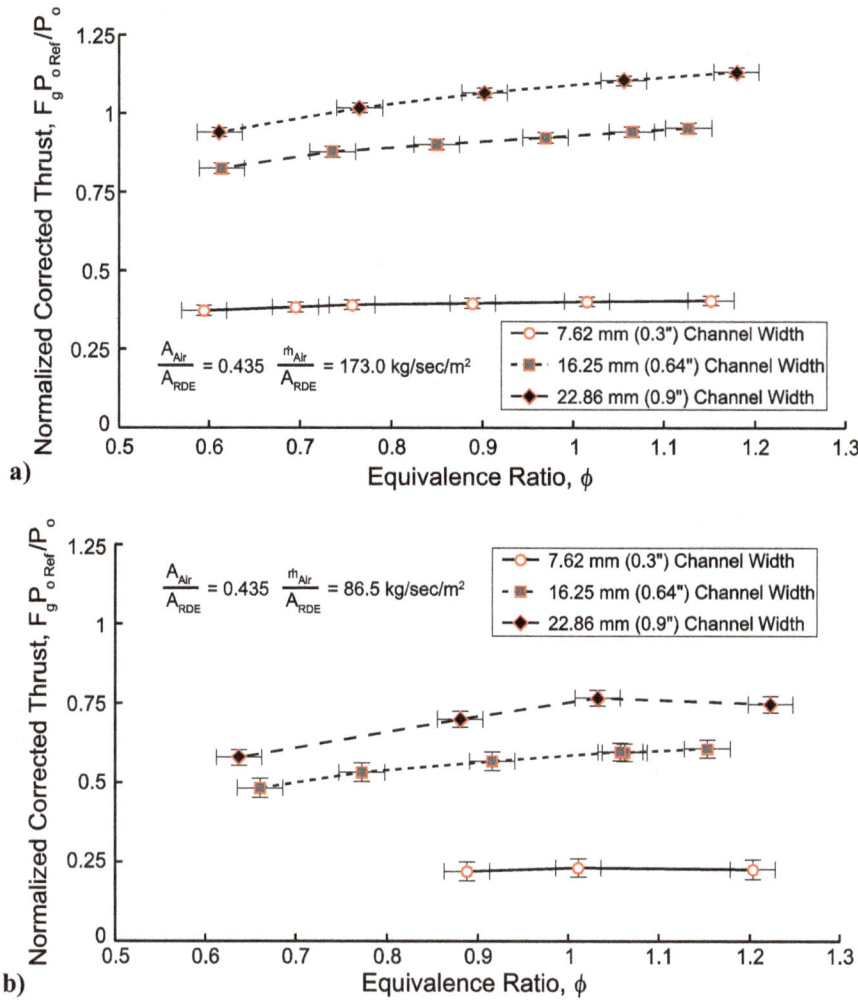

Fig. 1.6 Influence of various operating parameters on thrust production in a six-inch rotating detonation engine operated on hydrogen/air, shown for a particular air injection area ratio, A_{Air}/A_{RDE} and two detonation channel mass fluxes, $\dot{m}/A_{RDE}$ (Fotia et al. 2017)

channel width can be seen to still perform the most effectively, while the 7.62 mm channel configuration produces the least thrust per unit of stagnation pressure fed to the engine. The lower mass flux through the detonation channel leads to reduced local pre-detonation pressures, and then reduced post-detonation pressures.

Examining the impact of the nozzle area contraction ratio, A_{Noz}/A_{RDE}, requires a shift in the variables examined through the inclusion of the nozzle throat area into both the air injection area ratio and detonation channel mass flux parameters. Previously, this was not required as the nozzle area contraction ratio for all the cases being considered was 0.6. The appropriate air injection area ratio becomes

$$\frac{A_{Air}}{A_{Noz}} = \frac{A_{Air}}{A_{RDE}} \frac{A_{RDE}}{A_{Noz}} \tag{1.5}$$

as this ratio will impact the backpressuring of the annular detonation chamber and through it the mechanics of the mixing process and formation of recirculation zones at the base of the detonation channel.

To hold the pressure ratio across the nozzle constant the nozzle throat mass flux, $\dot{m}/A_{Noz}$, must be held constant. It should be noted that these variables reduce to the A_{Air}/A_{RDE} and $\dot{m}/A_{RDE}$ parameters for instances of comparison between configurations of constant A_{RDE}/A_{Noz}, which has been the case for the discussion presented thus far.

Figure 1.7a shows a comparison of the normalized corrected thrust obtained from a rotating detonation engine with a 0.6 area constriction, A_{Noz}/A_{RDE}, and an unconstricted nozzle throat, $A_{Noz}/A_{RDE} = 1.0$. Interestingly, the non-constricted nozzle shows improved pressure conversion to thrust over the constricted configuration, with the trend holding for both of the detonation channel widths shown.

The reverse of this behavior is found when the fuel consumption is factored in, as in Fig. 1.7b. As is to be expected from the earlier discussion, the 16.25 mm channel configuration displays a higher efficiency of thrust production from the combustion of fuel, which is now factored into the normalized corrected specific impulse, $F_g P_{o\,Ref}/\dot{m}_{Fuel} g_o P_{o\,Air}$. However, in the case of this pressure-gain combustion device this leads to higher levels of flow blockage and a reduction in the effective use of feed pressure. The Mach number of the exhaust products are still set by the stagnation-to-ambient pressure ratio across the nozzle.

The interplay between the effective usage of fuel and pressure creates a parameter space in which the performance of the engine can be tailored to a specific flight regime, or varied to address the changing dynamic properties along a flight profile.

To add another dimension to the analysis of rotating detonation engine performance, Paxson (2016) showed that excessive nozzle area contraction can lead to unstable rotating detonation engine operation, if not offset by appropriate pressure losses upstream in the air injection scheme. This unstable operation was correlated in the simulations to a reduction in specific impulse of up to 10%.

Along a similar avenue of investigation, combustion chamber length can be used to adjust the strength of the interaction of the downstream propagating oblique shockwave produced by the detonation and the wave reflected from the nozzle throat that propagates back upstream towards the air injector. To show the impact of this behavior the measured specific thrust from three different physical combustion chamber/nozzle arrangements is given in Fig. 1.8a, where there is no appreciable difference apparent between the conventional converging-diverging nozzle with a 21 cm (8.25") length outer-body, an aerospike plug-nozzle with the same 21 cm (8.25") length outer-body and a plug-nozzle with a shorter 11.4 cm (4.5") length outer-body.

The impact of the physical alterations to the combustion chamber become evident when the normalized corrected thrust, $F_g P_{o\,Ref}/P_{o\,Air}$, is examined as in Fig. 1.8b,

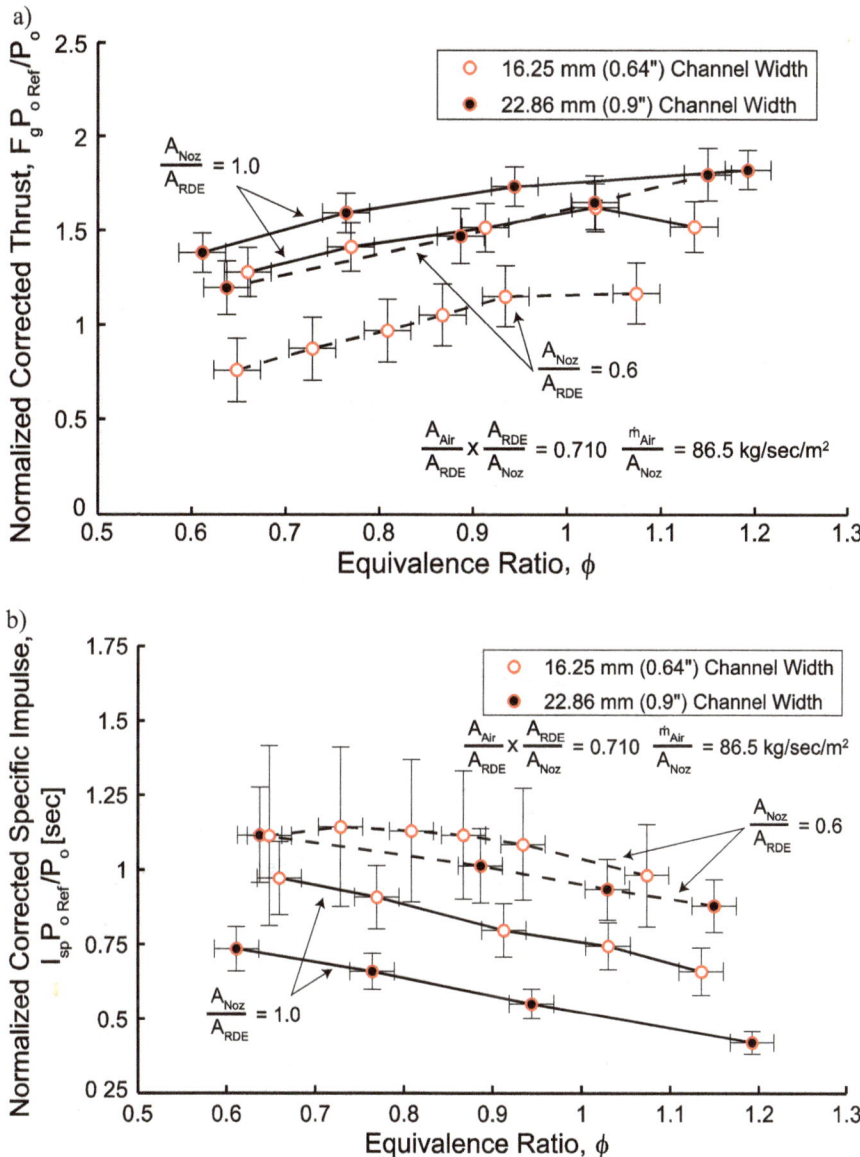

Fig. 1.7 Influence of nozzle area ratio on (**a**) normalized corrected thrust, $F_g\, P_{o\,Ref}/P_{o\,Air}$, and (**b**) normalized corrected specific impulse, $I_{sp}\, P_{o\,Ref}/P_{o\,Air}$, in a six-inch rotating detonation engine operated on hydrogen/air, shown for two different nozzle area ratios, A_{Noz}/A_{RDE}, set nozzle mass flux, $\dot{m}/A_{Noz}$, and air injection area ratio, A_{Air}/A_{Noz} (Fotia et al. 2017)

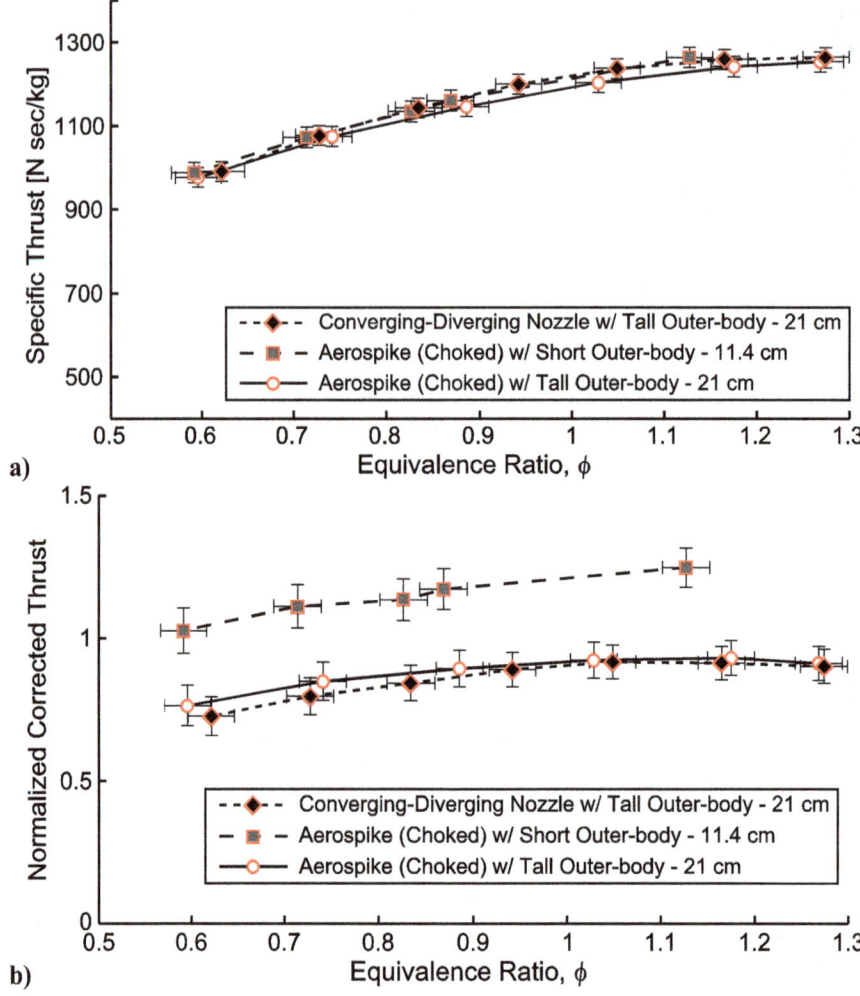

Fig. 1.8 Comparison of experimental performance shown in terms of (**a**) specific thrust, $F_g/\dot{m}_{Air}$ and (**b**) normalized corrected thrust, $F_g P_{o\,Ref}/P_{o\,Air}$, as functions of equivalence raito, ϕ, for three different implementations of nozzles with an area ratio, A_{Noz}/A_{RDE}, of 0.6

where the same data for the same physical configurations as Fig. 1.8a is plotted. The plug-nozzle configuration with the shorter outer-body clearly makes better use of the stagnation pressure supplied to the combustor to produce thrust than either of the longer outer-body configurations. The exact nature of this behavior, and the key physical mechanisms that drive it, are open areas of research. It is clear, however, that while it is tempting to visualize the rotating detonation engine combustor as a one-dimensional device, to do so would ignore the role that the internal non-steady state wave mechanics play in the production of thrust and the utilization of the stagnation pressure and fuel supplied to the device.

4 Stagnation Pressure-Gain

The true promise that rotating detonation engines hold for enabling advances in the performance of aerospace propulsion systems is the pressure-gain nature of the detonative combustion process. This potential advantage is due to the near-constant volume combustion a propagating detonation wave produces. However, to actually directly measure and analyze this form of pressure-gain combustion is extremely difficult due to the complex unsteady flow fields present in a rotating detonation engine. To begin to examine this, the internal flow field must be considered first.

A notional schematic of the basic wave structure that follows the detonation front is shown in Fig. 1.9. There are five basic flow regions indicated in the schematic. Region A consists of the unburnt fill gas that is not yet correctly mixed to allow the transit of a detonation wave front.

Region B is the detonation products, or burnt gas while Region C is the post-expansion/shock interaction exhausting burnt gas. The gases in Region C will be subjected, in the next wave cycle, to a shock wave due to the Riemann problem created between Regions B and C, to create Region D.

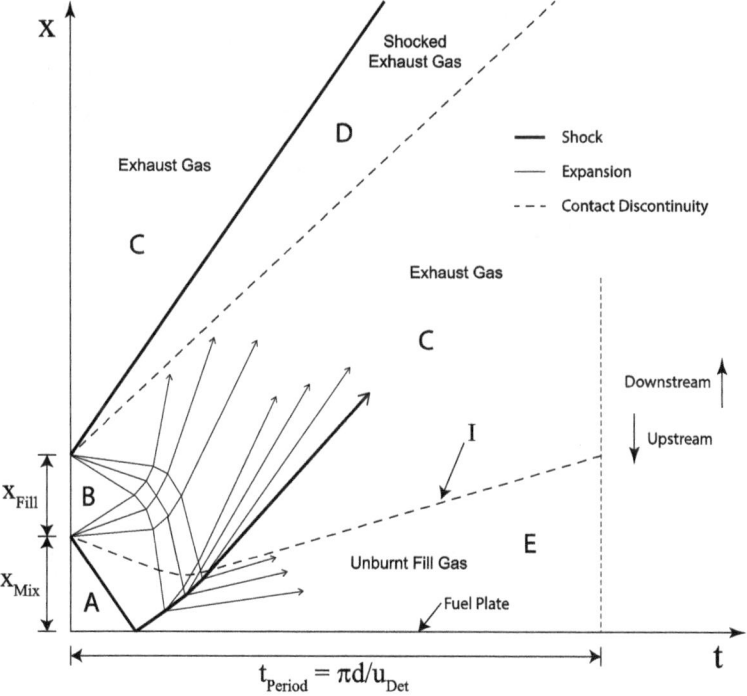

Fig. 1.9 Notional Schematic of the wave structure that follows the detonation front, shown in terms of the axial, x, and temporal/circumferential coordinates (Fotia et al. 2016)

Finally, Region E is the same gases as in Region A but subjected to shock/expansion conditioning prior to the subsequent transit of the next cycle of detonation and refill. When considering how to measure the stagnation pressure creation, or pressure-gain, provided by a rotating detonation device the fact that the time-resolved axial pressure and temperature distributions along the channel are unknown becomes a difficulty. However, the stagnation conditions upstream of the device in the air-feed manifold are known. Considering the schematic in Fig. 1.9, note that for the device to successfully refill the upstream propagating streamline between the burnt and unburnt gases, marked as I in Fig. 1.9, must at some point begin to flow back downstream to exit the combustor.

Making the assumption of axial flow only in the combustion chamber, it could be reasoned that the work-potential created through stagnation pressure gain in the detonation event and present in the expansion of the hot burnt gases upstream towards the propellant feed manifolds must be stagnated, and turned back downstream, by the incoming cold pre-combustion gases. This hypothesis requires that the air-feed manifold stagnation pressure be equal to, or greater, than any stagnation pressure gained downstream. However, as will be shown later in this section, this is not the case for correctly operating rotating detonation engine combustors.

By design, the Mach number of the air flow within the air-feed manifold is kept low, ≈0.09 at mass flow rate of 0.76 kg/s. The total air mass flow through the device is metered upstream by a sonic nozzle arrangement allowing the static pressure measured in the manifold to approach that of the stagnation pressure of the flow. Following the hypothesis that this incoming flow must have the same peak potential to do work as that created within the detonation wave, it is also a direct measurement of the stagnation pressure of that same detonation process. Note that unlike some pulsed detonation devices there is no pressure wave arresting mechanism present in the propellant injectors to absorb the overpressure of the detonation.

This seems like an over simplified suggestion, however, if a flow is considered in which there exists an upstream metering point and a downstream blockage, such as a nozzle contraction. It is accepted that as the blockage increases the pressure present in the segment of tube upstream of that point will increase to allow the metered mass flow to continue to transit the blockage.

Similarly, in the case of the rotating detonation engine, the stagnation pressure measured upstream of the device will adjust to meet the demands of the downstream obstruction, which in this case requires that the flow produced by the detonation be stagnated and turned back downstream.

To assess this assertion, a typical parameter to examine is the mass flow function, mff, which is simply a re-arrangement of the usual mass flow per unit area equations.

$$\mathrm{mff} = \frac{\dot{m}\sqrt{T_o}}{P_o A} = \sqrt{\frac{\gamma}{R}} M \left(1 + \frac{\gamma-1}{2} M^2\right)^{-\frac{\gamma+1}{2(\gamma-1)}} \qquad (1.6)$$

where $\dot{m}$ is the mass flow rate, T_o is the stagnation temperature, P_o is the stagnation pressure, A is the minimum cross-sectional area in the configuration and M the local

1 Performance of Rotating Detonation Engines for Air Breathing Applications 15

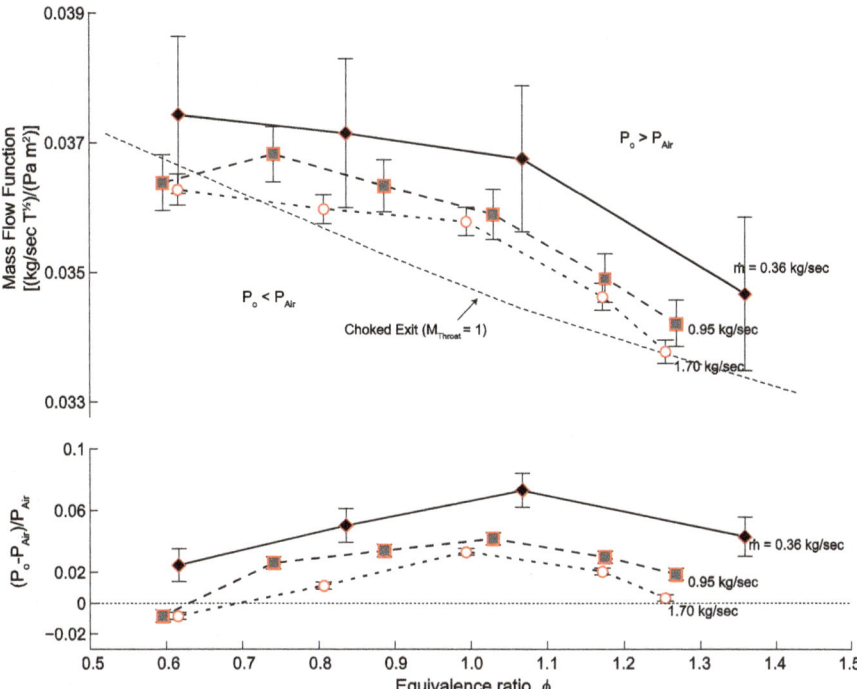

Fig. 1.10 Measurements for the 60% area contraction aerospike configuration at air mass flow rates of 1.7, 0.95 and 0.36 kg/s, showing above the mass flow function, mff, in comparison to an ideal choked nozzle throat and below the stagnation pressure-gain required to meet the observed flow conditions (Fotia et al. 2016)

Mach number of the flow. The ratio of specific heats, γ, and gas constant, R, are also parameters. The mass flow function is a non-unique function of Mach number, reflecting the subsonic and supersonic flow regimes. The stagnation pressure, measured in the air-feed manifold, created in the device can now be used with the stagnation temperature of the flow known from the equivalence ratio, given by the upstream fuel and air flow metering, and the known geometry to provide a mass flow function value for the exit choke on the nozzle coupled to the device. The upper part of Fig. 1.10 shows these experimental measured mass flow function values compared to the expected theoretical value that would be present if an ideal choke were located at the nozzle exit.

The measured values are higher than those expected for an ideal choke, over a large range of the observed conditions. This result conflicts with the defined behavior of the mass flow function which is at a maximum for a Mach number of unity. If the experimental conditions were to be located below the ideal choke line it could be explained by the presence of a "partial" choke at the nozzle throat, a condition where the bulk Mach number across the throat is to some degree subsonic. This effect would also be similar to those observed for nozzle discharge coefficients of less than unity. Alternatively, a point located above the curve indicates that there is a potential issue with the mass flow function calculation.

These high values of mass flow function can only be a result of one of the input parameters being inconsistent with the assumptions of the calculation. Since mass flow rate is metered external to the experiment and the geometry is fixed and known, only the stagnation quantities remain. The stagnation temperature was calculated from the known gas composition and it is possible that there is combustion efficiency to the process. However, to correct the mass flow function such that the mass flow function matches the ideal, or equivalently the local Mach number is a real value, it would require an extremely low combustion efficiency in the device, approximately 85%, to account for differences. This condition is unrealistic as this would subsequently certainly combust unconfined in the laboratory.

This leaves the most likely explanation for the inconsistency being the assumption that the air-feed manifold static pressure is the same as the stagnation pressure potential created by the detonation in the channel. Points located on the upper portion of Fig. 1.10 must have a higher observed apparent stagnation pressure than that present in the air-feed manifold.

Our earlier hypothesis that the air-manifold pressure will be equal to the stagnation pressure created in the detonation process has now been shown to be incorrect. Instead, there is an equivalence-ratio-dependent measured stagnation pressure-gain occurring. The lower portion of Fig. 1.10 shows the difference between the measured air-feed manifold pressure and the stagnation pressure required to ideally choke the nozzle throat. The peak pressure-gain occurs around an equivalence ratio of unity, and ranges from approximately 3% to 7%, dependent on feed conditions, as compared to an ideal choke.

5 Gaseous Hydrocarbon Fuels

To this point, only rotating detonation engines operated on gaseous hydrogen fuel have been considered. A change in either fuel or oxidizer will directly influence both the operability and performance of any combustion driven device. This effect applies doubly for a detonating system as not only the lower heating value of the propellants is now different, but also the detonation cell size which sets the geometric scalings within the device. Such a change will also have an impact on the performance attainable by detonating a particular fuel/oxidizer combination through a set geometry, as compared to any other propellant pairing for that particular geometry.

Figure 1.11 shows the resulting specific impulse if hydrogen and air propellants are used. It can be seen that the rotating detonation engine test data provides a fair match to the performance levels previously measured by Schauer et al. (2001a) for hydrogen/air operation of a pulsed detonation engine, as well as the theoretical performance of a pulsed detonation engine calculated by Wintenberger et al. (2003) for the same fuel combination.

1 Performance of Rotating Detonation Engines for Air Breathing Applications

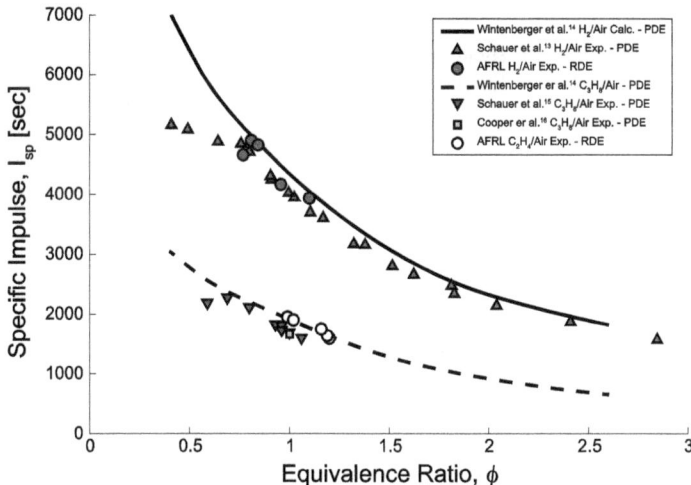

Fig. 1.11 Comparison of experimental rotating detonation engine performance to experimental pulsed detonation engine performance measurements by Schauer et al. (2001a, b) and Cooper et al. (2001) on both hydrogen and gaseous hydrocarbon fuel, as well as theoretical performance calculations by Wintenberger et al. (2003) (Fotia et al. 2017)

The results of testing conducted using gaseous ethylene and air are shown in the figure as well. These measured values again match closely the ethylene/air pulsed detonation engine performance reported by Schauer et al. (2001b) and Cooper et al. (2001) and the calculated performance levels of Wintenberger et al. This is an encouraging result, in that the rotating detonation engine is capable of producing thrust with similar levels of fuel efficiency as a pulsed detonation system when operated with gaseous hydrocarbon fuels, but now in a continuous manner.

The measured thrust data shows the expected drop in specific impulse, when considering a change from hydrogen to ethylene fuel. This same data also shows, in Fig. 1.12, that the usage of air in the production of thrust in a hydrocarbon fueled rotating detonation engine is comparable to that of the hydrogen fueled device, with the ethylene/air data actually showing slightly higher levels of specific thrust for a given equivalence ratio. When converting to hydrocarbon fuels from hydrogen this is the expected relationship and is a good indicator that the rotating detonation combustor is operating as a detonating device consistent with the hydrogen fueled cases.

The question of whether an ethylene fueled rotating detonation engine exhibits similar levels of efficiency in the conversion of the supplied stagnation pressure to thrust, as found in the hydrogen fueled devices, is valid due to the differing physical properties of the two fuels. As can be seen in Fig. 1.13, the corrected thrust measured for the hydrocarbon fueled cases matches that of the hydrogen fueled cases. This indicates that in the instance of this particular physical geometric arrangement the considerable difference in volumetric and energy density properties between

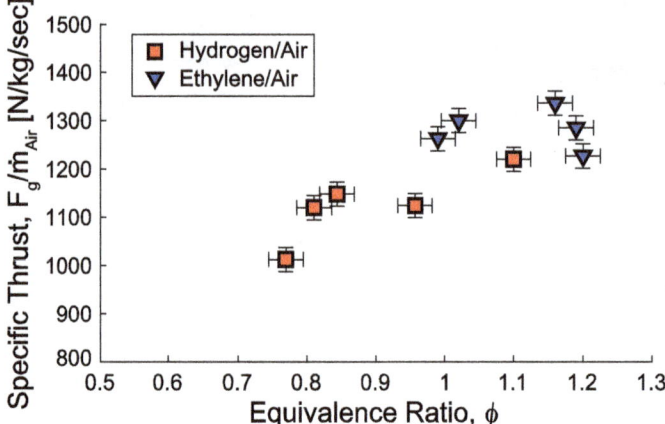

Fig. 1.12 Experimental measured specific thrust, $F_g/\dot{m}_{Air}$, shown as a function of equivalence ratio, ϕ, for both hydrogen and gaseous ethylene fuel, from a similarly configured six-inch diameter rotation detonation engine (Fotia et al. 2017)

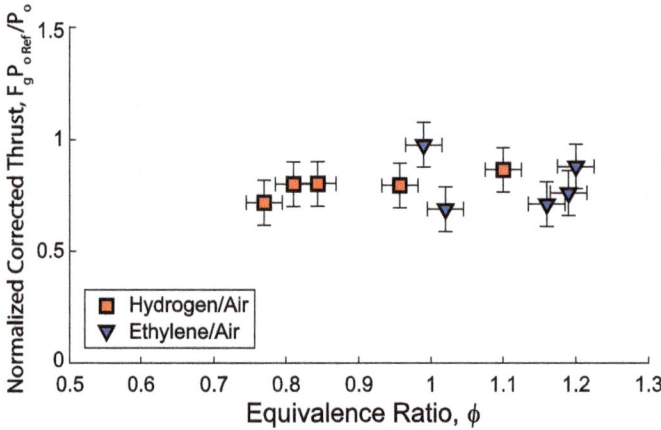

Fig. 1.13 Experimental normalized corrected thrust, $F_g\, P_{o\,Ref}/P_{o\,Air}$, shown as a function of equivalence ratio, ϕ, for both hydrogen and gaseous ethylene fuel (Fotia et al. 2017)

gaseous ethylene and gaseous hydrogen does not adversely impact the basic operating mechanics of the combustor.

Looking forward to future applications and the desired operation of rotating detonation engines on heavier hydrocarbon fuels, it is important to know that for a particular device geometry it is possible to obtain the same pressure utilization as seen for the canonical hydrogen fueled test cases. Hydrocarbon fuels have been experimentally shown to provide acceptable levels of specific thrust when combusted in a rotating detonation engine, with measured specific impulses matching the expected values from previous experiments and calculations for pulse detonation engine systems.

6 Conclusions

In an effort to better characterize the potential performance of air breathing rotating detonation engines, a large body of experimental test results has been collected, across a range of geometric and flow conditions. These test results have been used to identify the driving system variables that adequately describe the device. This allows comparisons to be made between different geometric configurations of the engine. The basic parameters include the air injection area expansion ratio, the mass flux through the detonation channel and the area contraction into the nozzle throat.

The first-order mechanisms that drive the production of thrust, either through the combustion of fuel or expansion of detonation magnified pressure, in these differing physical arrangements are examined with this approach. The current data shows that a trade-off space exists in which the fuel efficiency of rotating detonation engines can be exchanged for efficient stagnation pressure utilization through alterations in the engine geometry.

A mass flow function analysis is used to investigate the stagnation state conditions required to ideally choke the nozzle of these systems. From the experimental data, it is determined that for the nozzle to be choked, or even under a "partial" choke condition, stagnation pressure creation must be present in the combustor. This pressure-gain effect was found to be a function of equivalence ratio, with maximum stagnation pressure increases of between 3% and 7% dependent on the air mass flow rate though the engine.

The operation of a six-inch nominal diameter rotating detonation engine on gaseous hydrocarbon fuel has been demonstrated. The measured experimental performance of rotating detonation engines match well with previous experimental data and theoretical predictions for pulsed detonation engine devices. Hydrocarbon/air operation is found to compare appropriately with the performance measured for hydrogen/air propellants. Even though there are a number of technical challenges still to be met if heavier hydrocarbon fuels are to be implemented in rotating detonation engine systems, from the perspective of achievable performance, operation on gaseous hydrocarbon/air propellant provides the expected levels of performance which is encouraging if this technology is to be applied to future aerospace propulsion systems.

References

Bykovskii, F. A., Zhdan, S. A., & Vedernikov, E. F. (2010). Continuous detonation in the regime of self-oscillatory ejection of the oxidizer. 1. Oxygen as a oxidizer. *Combustion, Explosion, and Shock Waves, 46*(3), 344–351.

Bykovskii, F. A., Zhdan, S. A., & Vedernikov, E. F. (2011). Continuous detonation in the regime of self-oscillatory ejection of the oxidizer. 1. Air as an oxidizer. *Combustion, Explosion, and Shock Waves, 47*(2), 217–225.

Cooper, M., Jackson, S., Austin, J., Wintenberger, E., and Shepherd, J. E. (2001). *Direct experimental impulse measurements for detonations and deflagrations.* Paper presented at the 37th AIAA Joint Propulsion Conference, AIAA Paper No. 2001–3812, Salt Lake City, 8–11 July 2001.

Davidenko, D. M., Gokalp, I., and Kudryavtsev, A. N. (2008). *Numerical study of the continuous detonation wave rocket engine.* Paper presented at the 15th International Space Planes and Hypersonic Systems and Technologies Conference, AIAA Paper No. 2008–2680, Dayton, 28 April – 1 May 2008.

Dyer, R., Naples, A., Kaemming, T., Hoke, J., and Schauer, F. (2012). *Parametric testing of a unique rotating detonation engine design.* Paper presented at the 50th AIAA Aerospace Sciences Meeting, AIAA Paper No. 2012–0121, Nashville, 9–12 January 2012.

Fotia, M. L., Hoke, J. L., and Schauer, F. (2014). *Propellant plenum dynamics in a two-dimensional rotating detonation experiment.* Paper presented at the 52nd AIAA Aerospace Sciences Meeting, AIAA Paper No. 2014–1013, National Harbor, 13–17 January 2014.

Fotia, M. L., Schauer, F., and Hoke, J. L. (2015). *Experimental ignition characteristics of a rotating detonation engine under backpressured conditions.* Paper presented at the 53rd AIAA Aerospace Sciences Meeting, AIAA Paper No. 2015–0632, Orlando, 5–9 January 2015.

Fotia, M. L., Schauer, F., Kaemming, T., & Hoke, J. (2016). Experimental study of the performance of a rotating detonation engine with nozzle. *Journal of Propulsion and Power, 32*(3), 674–681.

Fotia, M. L., Hoke, J., & Schauer, F. (2017). Experimental performance scaling of rotating detonation engines operated on gaseous fuels. *Journal of Propulsion and Power, 33*(5), 1187–1196.

Frolov, S. M., Aksenov, V. S., Ivanov, V. S., & Shamshin, I. O. (2015). Large-scale hydrogen-air continuous detonation combustor. *International Journal of Hydrogen Energy, 40*, 1616–1623.

Hishida, M., Fujiwara, T., & Wolanski, P. (2009). Fundamentals of rotating detonations. *Shock Waves, 19*, 1–10.

Kaemming, T., Fotia, M. L., Hoke, L., & Schauer, F. (2017). Thermodynamic modeling of a rotating detonation engine through a reduced order approach. *Journal of Propulsion and Power, 33*(5), 1170–1178.

Kindracki, J., Wolanski, P., & Gut, Z. (2011a). Experimental research on the rotationg detonation in gaseous fuels-oxygen mixtures. *Shock Waves, 21*, 75–84.

Kindracki, J., Kobiera, A., Wolanski, P., Gut, Z., Folusiak, M., & Swiderski, K. (2011b). Experimental and numerical research on the rotating detonation engine in hydrogen-air mixtures. Advances in Propulsion Physics. Edited by S. Frolov. *Torus Press, 2011*, 35–62.

Lin, W., Zhou, J., Liu, S., Lin, Z., & Zhuang, F. (2015). Experimental study on propagation mode of H2/Air continuously rotating detonation wave. *International Journal of Hydrogen Energy, 40*, 1980–1993.

Naples, A., Hoke, J., Karnesky, J., and Schauer, F. (2013). *Flowfield characterization of a rotating detonation engine.* Paper presented at the 51st AIAA Aerospace Sciences Meeting, AIAA Paper No. 2013–0278, Grapevine, 7–10 January 2013.

Paxson, D. (2014). *Numerical analysis of a rotating detonation engine in the relative reference frame.* Paper presented at the 52nd AIAA Aerospace Sciences Meeting, AIAA Paper No. 2014–0284, National Harbor, 13–17 January 2014.

Paxson, D. E. (2016). *Impact of an exhaust throat on semi-idealized rotating detonation engine performance.* Paper presented at the 54th Aerospace Sciences Meeting, AIAA Paper No. 2016–1647, San Diego, 4–8 January 2016.

Rankin, B., Hoke, J., and Schauer, F. (2014). *Periodic exhaust flow through a converging-diverging nozzle downstream of a rotating detonation engine.* Paper presented at the 52nd AIAA Aerospace Sciences Meeting, AIAA Paper No. 2014–1015, National Harbor, 13–17 January 2014.

Russo, R., King, P. I., Schauer, F., and Thomas, L. M. (2011). *Characterization of pressure rise across a continuous detonation engine.* Paper presented at the 47th AIAA/ASME/SAE/ASEE Joint Propulsion Conference, AIAA Paper No. 2011–6046, San Diego, 31 July – 3 August 2011.

Schauer, F., Stutrud, J., & Bradley, R. (2001a). *Detonation initiation studies and performance results for pulsed detonation engine applications*. Paper presented at the 39th AIAA Aerospace Sciences Meeting, AIAA Paper No. 2001-1129, Reno, 8–11 January 2001.

Schauer, F., Stutrud, J., Bradley, R., Katta, V., Hoke, J. (2001b). Detonation initiation and performance in complex hydrocarbon fuel pulsed detonation engines. JANNAF, July 2001, Paper I-05.

Schwer, D. A. and Kailasanath, K. (2010). *Numerical investigation of rotating detonation engines*. Paper presented at the 46th Joint Propulsion Conference and Exhibit, AIAA Paper No. 2010-6880, Nashville, 25–28 July 2010.

Schwer, D. A. and Kailasanath, K. (2011). *Numerical study of the effects of engine size on rotating detonation engines*. Paper presented at the 49th AIAA Aerospace Sciences Meeting, AIAA Paper No. 2011-581, Orlando, 4–7 January 2011.

Shank, J. C., King, P. I., Karnesky, J., Schauer, F., and Hoke, J. L. (2012). *Development and testing of a modular rotating detonation engine*. Paper presented at the 50th AIAA Aerospace Sciences Meeting, AIAA Paper No. 2012–0120, Nashville, 9–12 January 2012.

Wintenberger, E., Austin, J. M., Cooper, M., Jackson, S., & Shepherd, J. E. (2003). Analytical model for the impulse of single-cycle pulse detonation tube. *Journal of Propulsion and Power, 19*(1), 22–38.

Chapter 2
Development of Gasturbine with Detonation Chamber

Piotr Wolański, Piotr Kalina, Włodzimierz Balicki, Artur Rowiński, Witold Perkowski, Michał Kawalec, and Borys Łukasik

Abstract Extensive and complex studies of the application of continuously rotating detonation (CRD) to gasturbine are presented. Special installation of high pressure preheated air supply system was constructed which allows to supply air at rate of a few kg/s, preheated to more than 100 °C and at initial pressure up to 2.5 bar. Supply system for Jet-A fuel which could be preheated to 170 °C was also constructed. Additionally gaseous hydrogen supply system was added to the installation. Measuring system for control air flow and measurements of detonation parameters were installed and data acquisition and control system implemented. Extensive research of conditions in which CRD could be established and supported in open flow detonation chambers, throttled chambers and finally in detonation chambers attached to the GTD-350 gasturbine engine where conducted. Conditions for which stable detonation was achieved are presented. It was found that for conditions when the GTD-350 engine was supplied by gaseous hydrogen or by dual-fuel, Jet-A and gaseous hydrogen, thermal efficiency of the engine could be improved even by 5–7% as compared to the efficiency of the base engine.

1 Introduction

It is well known that unlike deflagration which results in pressure drops, during detonation pressure is significantly increased, so implementation of detonative combustion into engine may result in improvement of engine efficiency. Such idea was first proposed by Zeldovich (1940), but first possible application of detonation to the propulsion system comes from the University of Michigan where Nicholls et al. built and tested the first pulsed detonation engine (Nicholls et al. 1957). Detonative combustion can be implemented into propulsion systems in many different ways. In engines, detonation can be stationary or nonstationary (pulsating) as

P. Wolański (✉) • P. Kalina • W. Balicki • A. Rowiński • W. Perkowski
M. Kawalec • B. Łukasik
Institute of Aviation, Warsaw, Poland
e-mail: wolanski@itc.pw.edu.pl

related to engine frame, stationary or quasi-stationary as related to moving coordinate system. Application of detonation combustion to engine offers not only higher thermodynamic efficiency, but also higher energy release rate and more compact heat release chamber as compared with conventional engines using deflagration. Such engines can be applied to propel subsonic and supersonic airplanes as well as to spaceplanes or rockets. Detailed description of possible application of detonation to propulsion can be found in (Wolański 2011, 2013). During last ten years more research on detonative propulsion was focused on application of quasi-stationary detonation propulsion (stationary as related to the rotating frame of reference). Such engines are the "Rotating Detonation Engine" (RDE), which is alternately called Continuous Wave Detonation Engine (CWDE).

2 Continuously Rotating Detonation

In the early sixties of the last century, at Novosibirsk Institute of Hydromechanics, stabilized spinning detonation was successively achieved by Voitsekhovskii et al. (1960, 1963). The first attempt of practical applications of stabilized spinning detonation was then undertaken at the University of Michigan on development of continuous detonation propulsion system, but unfortunately no successful operation of such system was achieved. In conclusion of the report authors clearly stated: "Nothing has been found that makes the concept not possible but important questions … remain to be answered. It is believed that much is to be gained from further studies of rotating detonation wave in annular chamber. While successful operation has not been achieved herein, nothing fundamental stands in the way of this accomplishment" (Nicholls et al. 1962). After this, research on application of continuous detonation to propulsion system was interrupted for several years, but at the beginning of this century it was more successfully revitalized in different laboratories (Bykovskii and Vedernikov 2003; Bykovskii et al. 2006; Kindracki et al. 2006; Falempin et al. 2006; Davidenko et al. 2007; Kindracki 2008; Zitoun and Desbordes 2011). Recently experimental and numerical research of CRD are carried out in Russia, Poland, France, USA, Japan, China, Korea and Singapore.

The continuously rotating detonation (CRD) or sometimes called continuous spin detonation, is a basic process for all RDE. It is most often initiated and run in cylindrical chambers, but sometimes also in a disk like chamber or other configurations. Typical configuration of cylindrical detonation chamber is shown in Fig. 2.1 (Wolanski 2010, 2013) and stable CRD pressure record in Fig. 2.2 (Wolanski 2010, 2011, 2013).

The first calculation of rotating detonation structure was performed by Adamson and Olsson (1967), Shen and Adamson (1972). They were examining possibility of using CRD in rocket motor, but even by using very low power computers they were able to recognize basic structure of detonation. The first detailed calculation of the structure in relatively large 2-D cylindrical chamber was performed by Hishida et al. (2009). This was possible for the case when annular cylindrical chamber thickness of channel to radius of the chamber is small. In Fig. 2.3 flow field of the

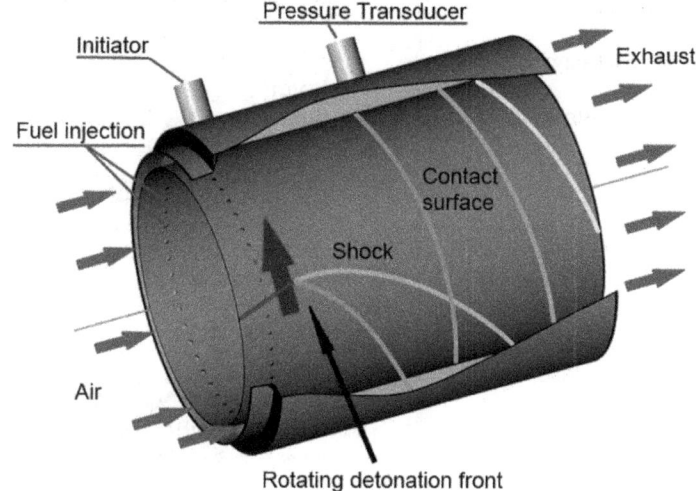

Fig. 2.1 Schematic diagram of detonation chamber for basic research of CRD

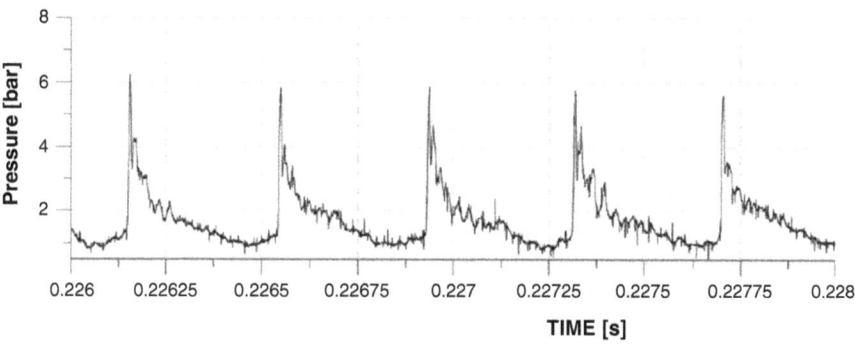

Fig. 2.2 Typical pressure variation for stable CRD in research chamber

CRD is shown in laboratory coordinate system. From this calculation a very important conclusion could be drawn, that even detonation wave rotate along circumference of the detonation chamber, flow of the products from chamber is basically axial, so in RDE there will be very little losses of energy for rotational component of the flow. A detailed description of 2-D structure of CRD can be found in (Hishida et al. 2009). When a chamber depth is larger, the 3-D calculation of CRD performed is necessary. Detailed calculations of 3-D CRD structure are now being performed in many research centers in Poland, France, USA, Russia, Japan, Korea, Singapore and other countries (Kindracki et al. 2011; Folusiak et al. 2011; Schwer and Kailasanath 2010; Shao and Jian-Ping 2010; Kailasanath et al. 2011; Davidenko et al. 2011; Yamada et al. 2010).

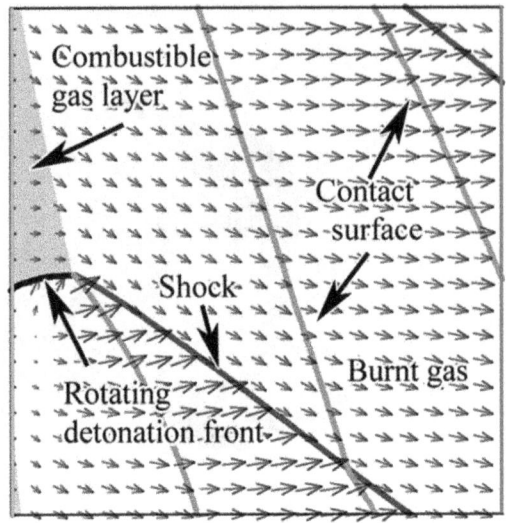

Fig. 2.3 Computational results of 2-D CRD structure showing velocity vector in laboratory coordinate system (Hishida et al. 2009)

3 Experimental facility

Research of application of CRD to gasturbine were carried out at the Institute of Aviation for nearly 5 years (2010–2015). For the research GTD-350 engine, which is used to power Mi-2 helicopter (NATO reporting name "Hoplite") was chosen. The reason for this choice was the location of the combustion chamber, which basically is located outside the rotating part of the engine, compressor and turbines, and made engine combustion chamber modification relatively easy.

RD-350 Gasturbine Engine

GTD-350 engine consist of the seven-stage axial flow compressor plus single-stage centrifugal compressor. Maximum compression ratio is between 4.5–6 (depending on version) and maximum air flow rate about 2 kg/s. The engine has a single reverse flow combustion chamber and two turbines: single-stage compressor turbine and two-stage power turbine. A schematic diagram of the GTD-350 is presented in the Fig. 2.4.

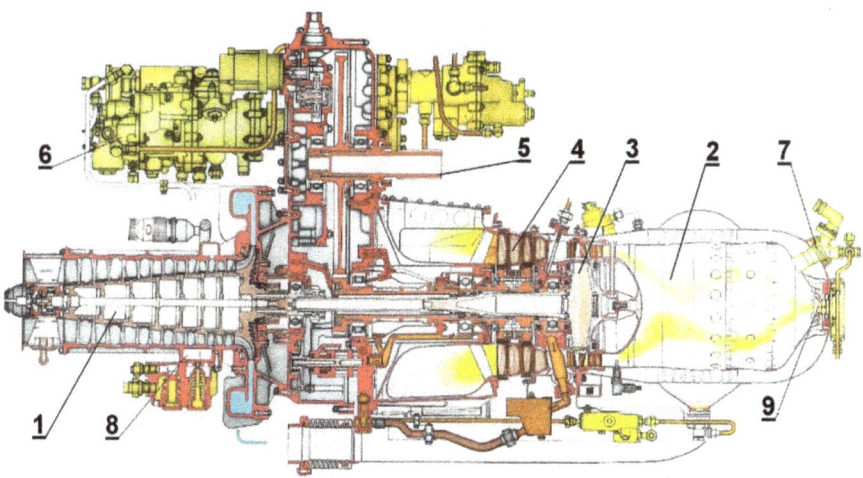

Fig. 2.4 Schematic diagram of the GTD-350 gas turbine engine. *1* compressor, *2* combustion chamber, *3* compressor turbine, *4* power turbine, *5* power shaft, *6* pump-regulator, *7* ignition device, *8* air bleeding valve, *9* main injector

Detonation Chambers

The initial part of the research was to evaluate conditions at which CDW could be established in a chamber of the diameter suitable for connecting the turbine to the engine. To find the optimum configuration of the chamber many different detonation chambers were tested. Since it was found that for pure Jet-A fuel continuous detonation is difficult to establish, and is basically not very stable, two other configurations were tested. Beside the chamber which used only Jet-A fuel, a chamber that was supplied only by gaseous hydrogen as a fuel and dual fuel detonation chambers in which Jet-A and gaseous hydrogen are used simultaneously were tested. In the case of dual fuel supply, the gaseous hydrogen consists usually of less than 25%, as related to energy contribution during combustion, but some time hydrogen concentration can even exceed 50%. Schematic diagrams of such chambers are presented in Fig. 2.5. In all cases chambers were supplied by high pressure preheated air. The initial pressure of air used in experiments was ranging from 1 bar to 2.5 bar and the initial air temperature was in the range of 80–130 °C, at the inlet to the chamber. The flow rate of air could also be changed from 0.4 kg/s up to 2.5 kg/s. All dual fuel chambers were tested in three different modes: supplied only by Jet-A fuel, supplied only by gaseous hydrogen and supplied by both fuels simultaneously. In the case of Jet-A fuel, also two different modes were tested: fuel supplied at ambient temperature and fuel preheated to 170 °C. A schematic diagram of the fuel supply system is shown in Fig. 2.6.

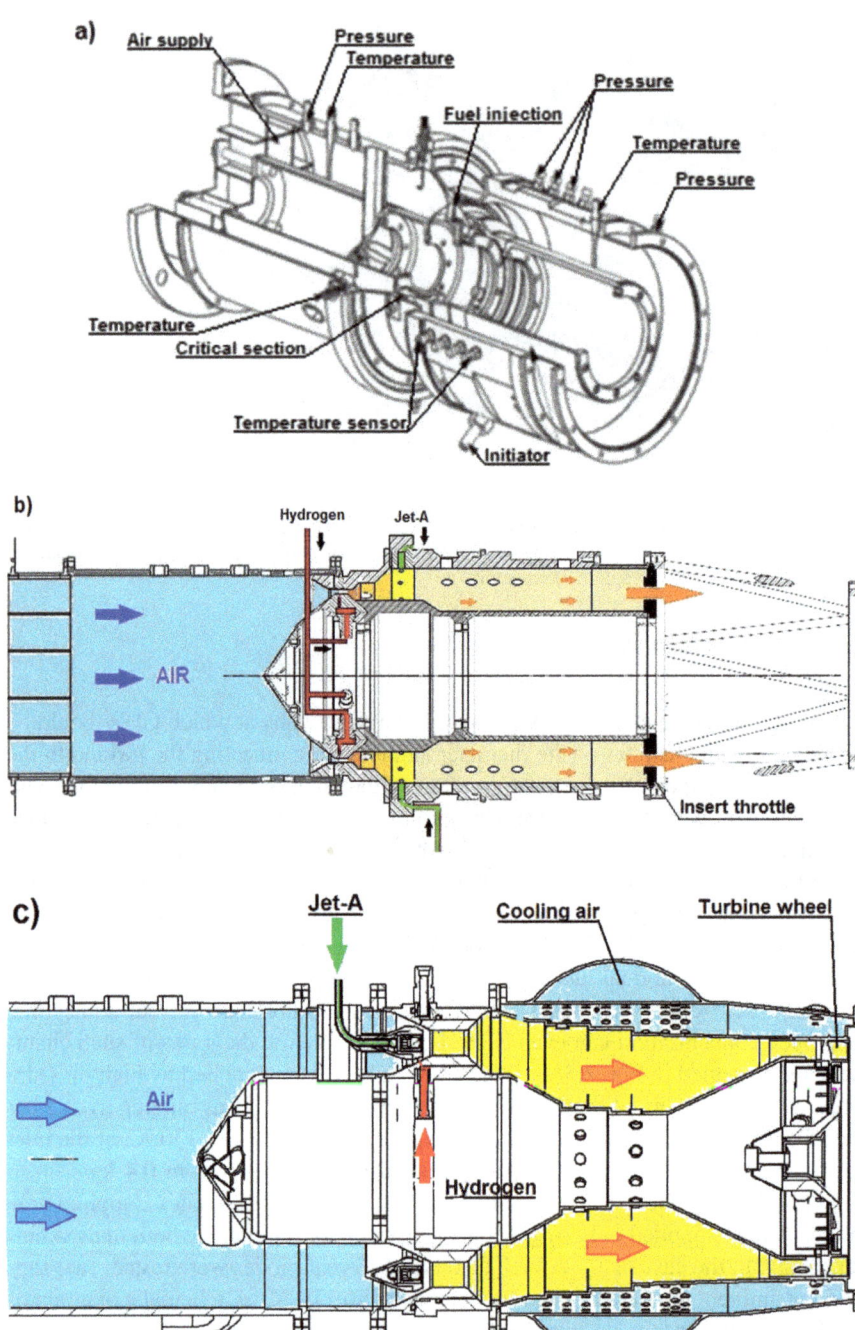

Fig. 2.5 Different configurations of tested chambers: (**a**) schematic diagram of cross-cutting view of detonation chamber with indication of typically used measuring points, (**b**) diagram of dual fuel detonation chamber, (**c**) diagram of dual fuel detonation chamber with supply of cooling air

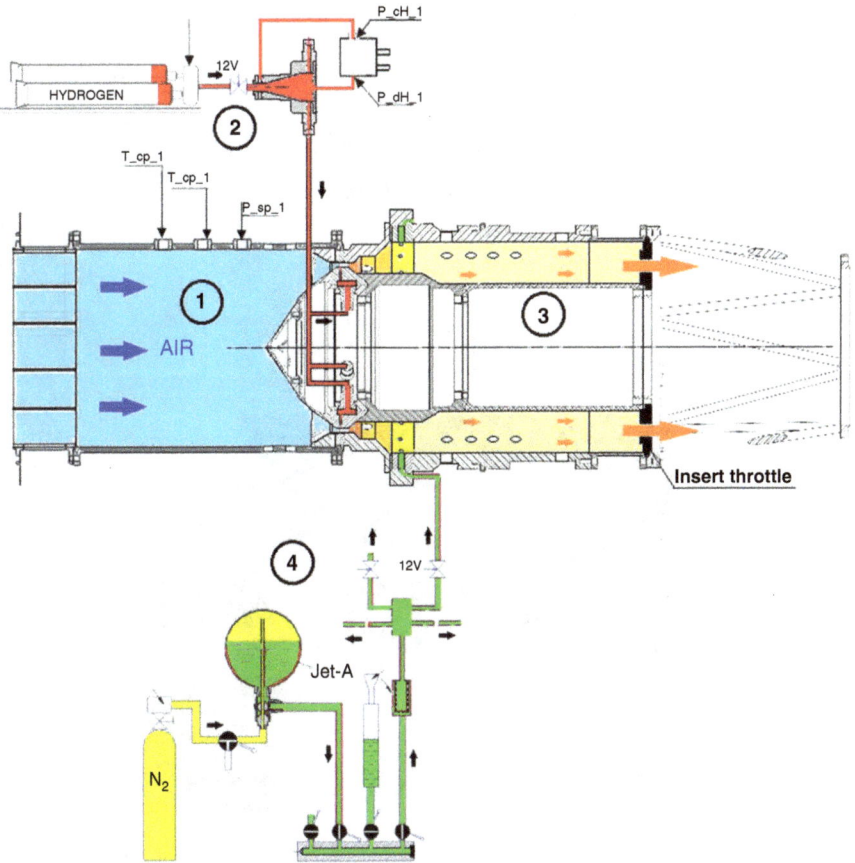

Fig. 2.6 Schematic diagram of the research chamber with dual fuel supply system. *1* air flow, *2* hydrogen supply system, *3* chamber "hot part", *4* kerosene Jet-A supply system

GDT-350 Gasturbine with Detonation Chamber

Final research were carried out at the test stand in which detonation chamber was directly connected to the GTD-350 gasturbine engine. Schematic diagram of the test stand is shown in Fig. 2.7. GTD-350 gasturbine engine is working in this case in the open cycle.

In this case high pressure preheated air from compressor and preheater is supplied to the detonation chamber, which is directly connected to the inlets of the turbines of the GTD-350 engine. Detonation products enter vanes of the turbine stage, which is directly connected to compressor. Then products flow through free turbine which is connected to the brake, which measure the torque, so the power can be calculated. Compressed air in multistage compressor is throttled to the desire pressure, so the power delivered from the turbine to compressor is calculated and

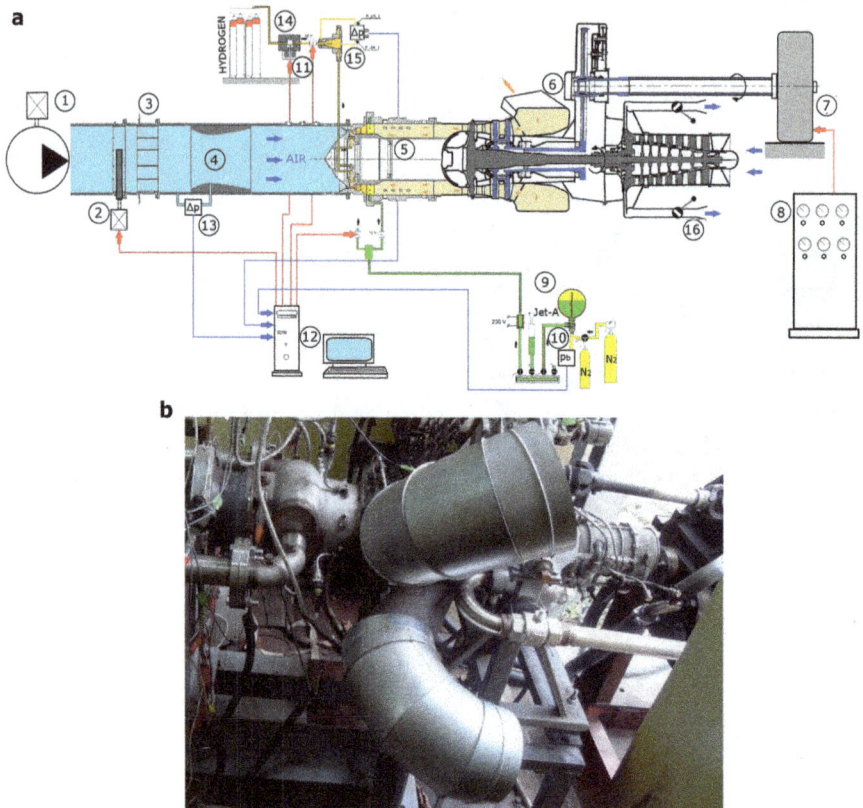

Fig. 2.7 Gas turbine research facility. (**a**) control system of the turboshaft engine with detonation combustion chamber: *1* air compressors unit, *2* air flow controller, *3* inlet flow equalizer grid, *4* venturi air mass flow measurement, *5* detonation combustion chamber, *6* turboshaft engine GTD-350, *7* brake, *8* brake controller, *9* Jet-A injection system (pressurized by nitrogen), *10* Jet-A flow meter, *11* electro hydraulic valve, *12* management computer, *13* air flow differential pressure, *14* hydrogen supply system, *15* hydrogen flow differential sensor, *16* air exhaust with suppression valve. (**b**) picture of test stand

performance of the engine can be obtained. Supply of both fuels (hydrogen and Jet-A) is controlled by computer and the data acquisition system collects and records parameters of the engine test.

4 Experimental Research

Extensive research were carried out at the Institute of Aviation for nearly 5 years. Initial works were directed into preparation of special installation of compressed and preheated air supply system, fuel atomization and mixture formation, selection of the detonation initiator system, design and construction of control and date recording system. Over ten different detonation chambers were tested, initially with an open end and then with a throttle placed at the end of the chamber to simulate turbine inlet as well as pressure drop at the turbines. In the open chamber tests it was possible to evaluate most favorable conditions which support continuously rotating detonation (CRD) in the chamber. Besides pressure measurement which was made for each run, also visual observation of flame, direct photography as well as noise level and frequency, were usually used to identified mode of combustion in the chamber. If CRD was initiated the very short flame was observed accompanied by high frequency noise. For deflagrative combustion long flame emerging from the chamber was easily visible with a low frequency noise. Typical pictures of flame emerging from the chamber for non-detonation (deflagration) and detonation case are shown in Fig. 2.8.

Even for open chambers detonation was much easier to initiate if the Jet-A fuel was preheated to about 170 °C (better atomization and mixture formation). A typical pressure record for detonative combustion of Jet-A fuel mixed with air is presented in Fig. 2.9.

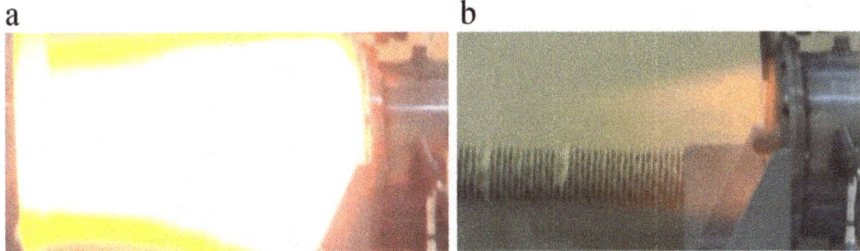

Fig. 2.8 Direct picture of flame emerging from the combustion chamber for (**a**) deflagrative combustion, (**b**) for detonation

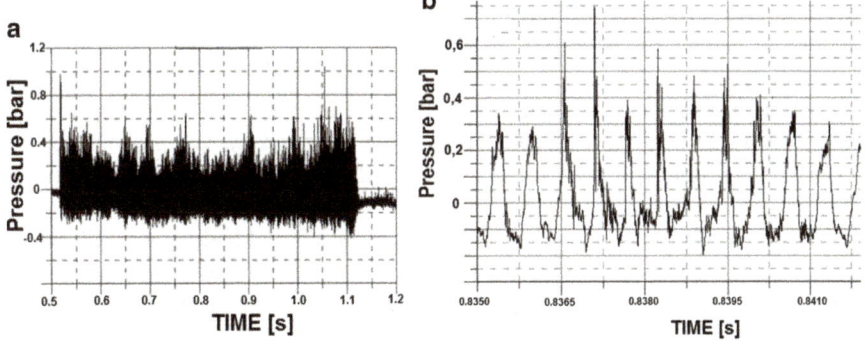

Fig. 2.9 Typical pressure records of rotating detonation in cylindrical chamber for liquid kerosene, air supply rate – 0.7 kg/s (100 °C), Jet-A – 0.032 kg/s (170 °C), (**a**) the whole test record, (**b**) the enlargement of selected interval

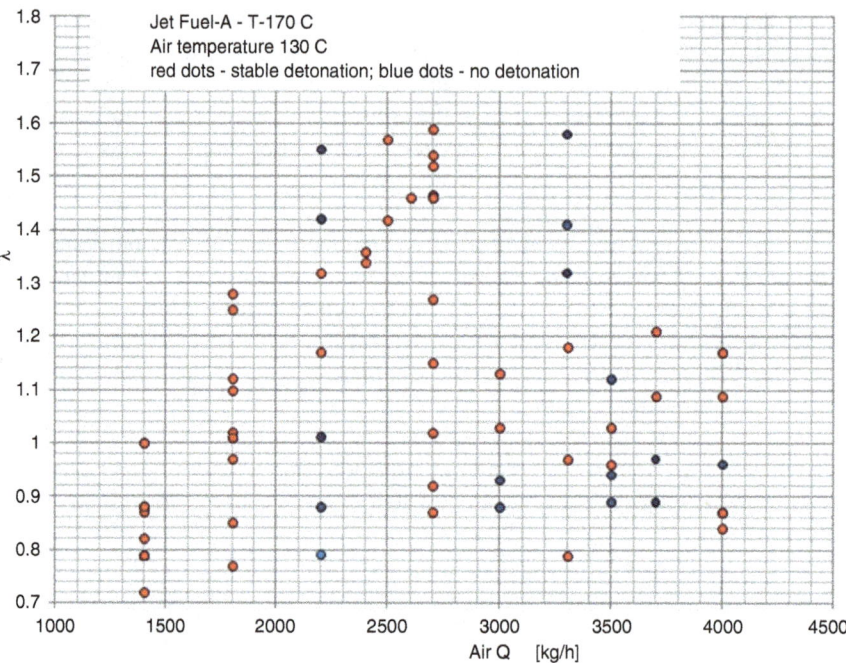

Fig. 2.10 The results of stable detonation and no detonation for Jet-A fuel –air mixture. Fuel was preheated to 170 °C and air to 130 °C

However, stability and repeatability of detonation for those tests were rather poor. It is clearly visible in Fig. 2.10, which present collection of many test results for such a mixture, at different air supply rate as well as different fuel supplies λ (λ – ratio of air supply rate to the theoretical stoichiometric air supply rate). For low rates of air supply, which results in lower velocity at the inlet and probably better mixture formation was obtained and detonation was stable, but for higher rate of air flow appearance of stable detonation is not certain and rather stochastic. To improve stability of detonation of Jet-A fuel, detonation chambers were modified to allow dual-fuel supply.

Gaseous hydrogen was added as the sensitizer for the jet fuel. This drastically improved stability of initiated detonation in the chamber. Typical pressure record for dual-fuel supply system is presented in Fig. 2.11. It is clearly seen that addition of hydrogen to the mixture improved stability and intensity of the detonation process. It is also clearly seen that when jet fuel is cut off and hydrogen is still being supplied, detonation combustion is continued, but intensity of such a process is much weaker. The best results are, however, obtained for hydrogen-air mixtures. As it is seen in Fig. 2.12, pressure picks are very even and very stable. 3-D numerical calculations were also carried out to verify the structure and stability of such detonation. Contours of pressure, temperature and mass fraction of hydrogen as well as particles streamlines are depictured in Fig. 2.13.

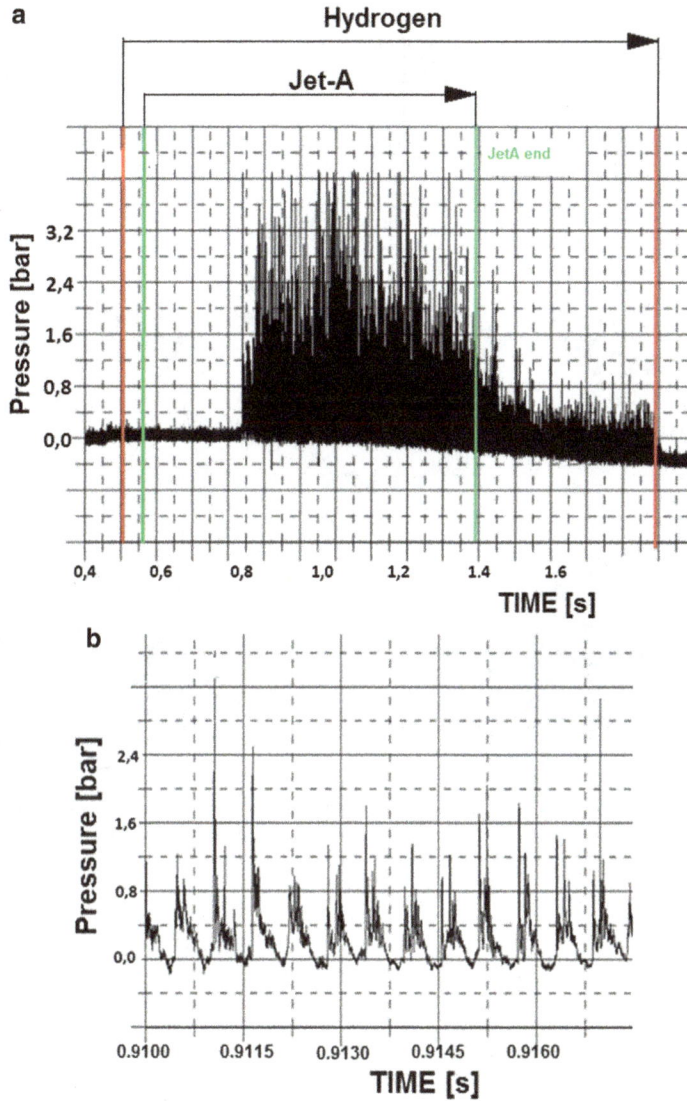

Fig. 2.11 Typical pressure record for dual fuel (hydrogen and kerosene), air supply rate – 2 kg/s, H_2–0.022 kg/s, Jet-A – 0.021 kg/s.(**a**) the whole test record, (**b**) the enlargement of selected interval

Such calculations are very helpful in understanding the details of flow field in chamber at given condition, and as a result, help to modify geometry of the detonation chamber. In tested chambers detonation in hydrogen-air mixture was usually very stable as compared to dual-fuel mixtures.

It is very well seen in Fig. 2.14, where variation of rotating detonation velocity with time is clearly visible.

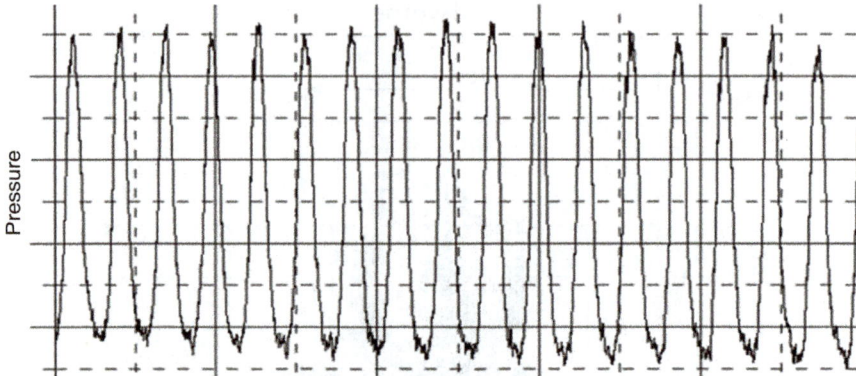

Fig. 2.12 Typical pressure records for hydrogen-air mixture, the enlargement of selected interval, air supply rate – 1 kg/s, H_2–0.017 kg/s

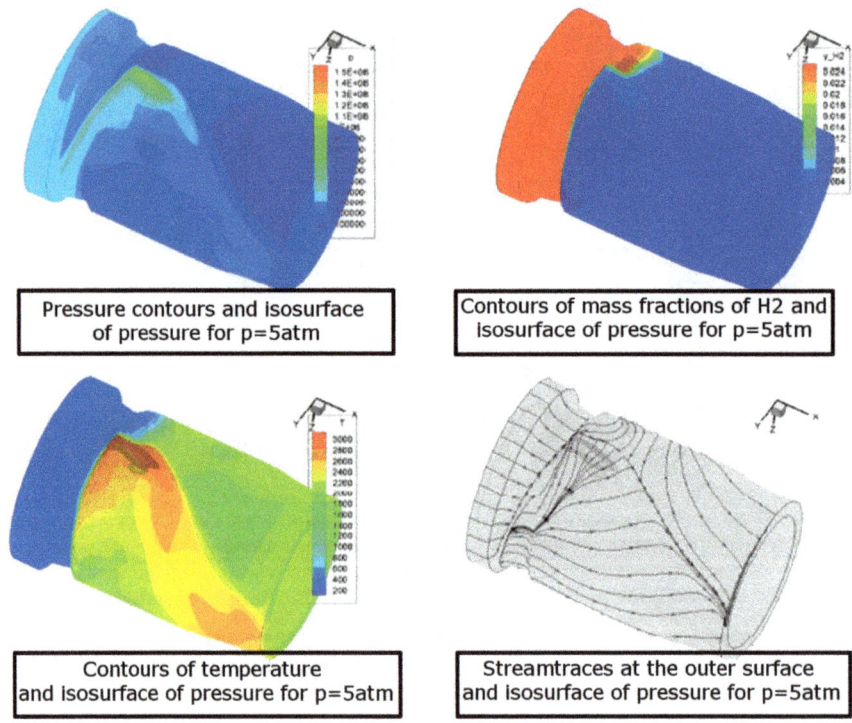

Fig. 2.13 Numerical calculations of CRD structure in 3-D complex geometry for hydrogen-air mixture (Folusiak et al. 2011)

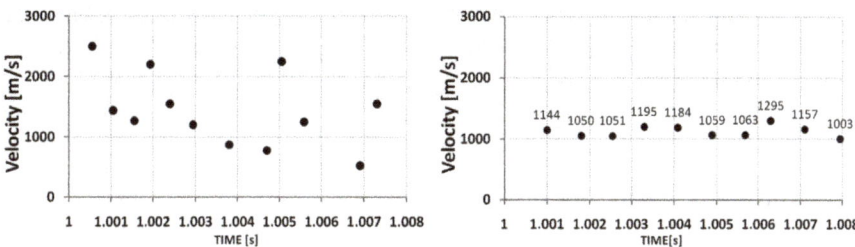

Fig. 2.14 Stability of the CRD, (**a**) Hydrogen and Jet Fuel and Air mixture, (**b**) Hydrogen and Air mixture

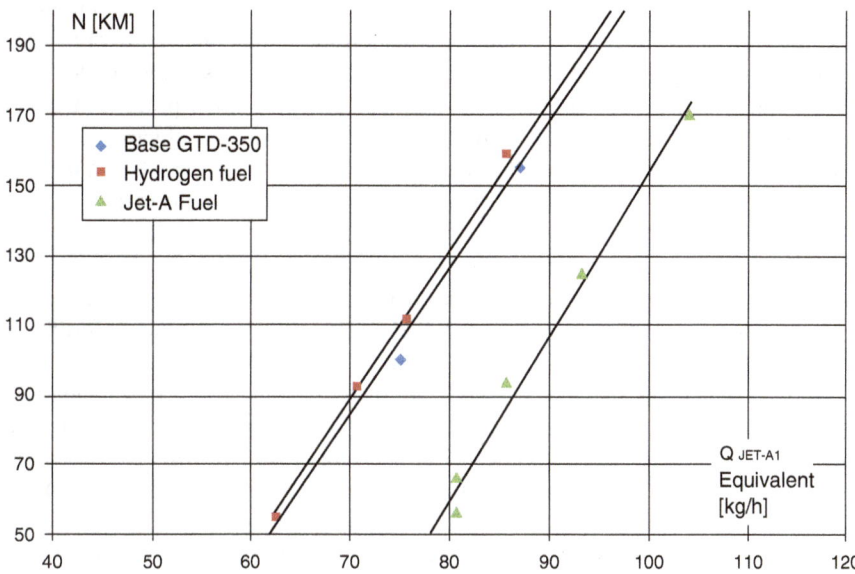

Fig. 2.15 Engine shaft power as a function of fuel consumption (for Hydrogen and Jet-A in equivalent of total heat)

Mixtures of hydrogen with jet fuel exhibit rather big instability, while for hydrogen-air mixture continuously rotating detonation is very stable. Finally, many experiments were conducted with GTD-350 engine attached to detonation chamber. At those tests not only stability of detonation was monitored but also performance of the engine calculated. Experiments were conducted for lean mixture conditions at which maximum temperature of the products, at chamber exit, were below temperature allowed for the first stage of high pressure turbine. It was found that on Jet-A fuel supply system engine can operate, but the performance of such engine is even below base GTD-350 characteristic. Best performance of the engine was obtained for hydrogen-air mixture. Due to limitation of hydrogen supply test were carried out for the duration less than one minute, but this was sufficient for the engine to operate at the steady state conditions. At such case engine operates steadily and shows higher efficiency by 5–7%. Slightly less performance was recorded for dual fuel supply system (gaseous hydrogen and liquid preheated Jet-A fuel). Results from the experiments are presented in Fig. 2.15.

5 Summary and Conclusions

Complex research on application of continuously rotating detonation to gasturbine engine were performed at the Institute of Aviation in Warsaw. Special experimental facility was built and equipped with control and measurement systems. Initially research was focused on the mixture formations well as on the process of initiation of detonation in chamber. The conducted research shows that preheating of the liquid fuel is very helpful for better mixture formation, but it does not guarantee successful initiation of detonation as well as stability of detonation process. A much better performance was obtained for dual fuel supply. An addition of gaseous hydrogen to the liquid fuel improves the initiation and stability of the CRD. The best results, however, were obtained for the mixtures of air with gaseous hydrogen.

It was found that performance of GDT-350 with detonation chamber can improve efficiency of the engine even by 5–7%, when engine is supplied with gaseous hydrogen, but also some improvements were recorded for dual fuel mixtures. Unfortunately, when engine was supplied only by Jet-A fuel improvement of engine performance was not possible. However, paraphrasing the statement of Michigan group from sixties of the last century, that nothing has been found that makes the concept not possible but higher pressure and larger diameter of the chamber will be required to stabilize rotating detonation in Jet-A fuel-air mixture. As it is known that detonation stability is related to the cell size, and cell size is decreasing with increase of the pressure, so one can expect that for conditions under which modern jet engines are operating – stable continuously rotating can be achieved for jet fuel air mixtures. However, more research in higher initial pressure and in larger dimensions of the detonation chamber are necessary to implement continuously rotating detonation to gasturbines and jet combustors, which will result in higher engine efficiency and higher engine performance.

Acknowledgements Research conducted during 2010-2015 under the project UDA-POIG.01.03.01-14-071 "Gasturbine with detonative combustion chamber" supported by EU and Ministry of Regional Development, Poland.

References

Adamson, T. C., Jr., & Olsson, G. R. (1967). *Performance analysis of a rotating detonation wave rocket engine. Acta Astronautica, 13*, 405–415.
Bykovskii, F. A., & Vedernikov, E. F. (2003). Continuous detonation of a subsonic flow of a propellant. *Combustion, Explosion, and Shock Waves, 39*(3), 323–334.

Bykovskii, F. A., Zhdan, S. A., & Vedernikov, E. F. (2006). Continuous spin detonation of fuel-air mixtures. *Combustion, Explosion, and Shock Waves, 42*(4), 451–460.
Davidenko, D., Jouot, F., Kudryavtsev, A., Dupré, G., Gökalp, I., Daniau, E., & Falempin, F. (2007). *Continuous detonation wave engine studies for space application*. In: 2nd European Conference for Aero-Space Sciences (EUCASS 2007), Brussels, 1–6 Jul. 2007.
Davidenko, D., Eude, Y., Gökalp, I., & Falempin, F. (2011). *Theoretical and numerical studies on continuous detonation wave engines*. In: Detonation Wave Propulsion Workshop, Bourges, 2011, July 11–13.
Falempin, F., Daniau, E., Getin, N., Bykovskii, F. A., & Zhdan, S. (2006). *Toward a continuous detonation wave rocket engine demonstrator*. In: 14th AIAA/AHI Space Planes and Hypersonic Systems and Technologies Conference, AIAA2006–7956, Canberra, 6–9 Nov.
Folusiak, M., Swiderski, K., Kobiera, A., & Wolanski, P. (2011). *Graphics Processors as a tool for rotating detonation simulations*. In: 23rd ICDERS, Irvine, USA.
Hishida, M., Fujiwara, T., & Wolanski, P. (2009). *Fundamentals of rotating detonation. Shock Waves, 19*, 1–10.
Kailasanath, K., Cheatham, C. Li, S., Patnaik, G., & Schwer, D. (2011). *Detonation wave engine research at NRL*. In: Detonation Wave Propulsion Workshop, Bourges, 11–13 July 2011.
Kindracki, J. (2008). *Badania eksperymentalne i symulacje numeryczne procesu inicjacji wirującej detonacji gazowej*. Praca doktorska, Politechnika Warszawska. PhD thesis, (in Polish).
Kindracki, J., Fujiwara, T., & Wolanski, P. (2006). *An experimental study of small rotating detonation engine*. In: Pulsed and continuous detonation propulsion. Edited by G. Roy and S. M. Frolov. Torus Press. pp. 332–338.
Kindracki, J., Kobiera, A., Wolański, P., Gut, Z., Folusiak, M., & Świderski, K. (2011). Experimental and numerical research on the rotating detonation engine in hydrogen-air mixture. In S. Frolov (Ed.), *Advances in propulsion physics*. Moscow: Torus Press.
Nicholls, J. A., Wilkinson, H. R., & Morrison, R. B. (1957). Intermittent detonation as a thrust-producing mechanism. *Journal of Jet Propulsion, 27*(5), 534–541.
Nicholls, J. A., et al. (1962). *The feasibility of a rotating detonation wave rocket motor*. In: The Univ. of Mich., ORA Report 05179-1-P.
Schwer, D. A., Kailasanath, K. (2010). *Numerical investigation of rotating detonation engines*. In: 46th AIAA/ASME/SAE/ASEE Joint Propulsion Conference & Exhibit, 25–28, July 2010, Nashville, AIAA Paper 2010–6880.
Shao, Y.-T., & Jian-Ping, W. (2010). *Continuous detonation engine and the effects of different types of nozzles on its propulsion performance. Chinese Journal of Aeronautics, 23.*, 2010, 647–652.
Shen, P. I., & Adamson, T. C., Jr. (1972). Theoretical analysis of a rotating two-phase detonation in liquid rocket motors. *Acta Astronautica, 17*, 715–728.
Voitsekhovskii, B. V. (1960). Stationary spin detonation. *Journal of Applied Mechanics and Technical Physics, 3*, 157–164. (in Russian).
Voitsekhovskii, B. V., Mitrofanov, V. V., & Topchiyan, M. E. (1963). *Structure of the detonation front in gases*. In: .Izdatielstvo SO AN SSSR, Novosibirsk (in Russian).
Wolanski, P. (2010). *Development of the continuous rotating detonation engines*. In G. D. Roy & S. M. Frolov (Eds.), *Progress in pulsed and continuous detonations* (pp. 395–406). Moscow: Torus Press.
Wolański, P. (2011). Detonation engines. *Journal of KONES, 18*(3), 515–521.
Wolański, P. (2013). Detonative propulsion. *Proceedings of the Combustion Institute, 34*, 125–158. Elsevier.
Yamada, T., Hayashi, A. K., Yamada, E., Tsuboi, N., Tangirala, V. E. (2010). *Numerical analysis of threshold of limit detonation in rotating detonation engine*. In: 48th AIAA Aerospace Sciences Meeting Including the New Horizons Forum and Aerospace, 4–7 January 2010, Orlando, AIAA 2010–153.
Zeldovich, Y. B. (1940). On the use of detonative combustion in power engineering. *Journal of Technical. Physics., 10*, 1453–1461. (in Russian).
Zitoun, R., & Desbordes, D. (2011). *PDE and RDE studies at PPRIME*. In: Detonation Wave Propulsion Workshop, Bourges, 11–13 July 2011.

Chapter 3
Flow Structure in Rotating Detonation Engine with Separate Supply of Fuel and Oxidizer: Experiment and CFD

Sergey M. Frolov, Viktor S. Aksenov, Vladislav S. Ivanov,
Sergey N. Medvedev, and Igor O. Shamshin

Abstract The experimental and computational investigations of detonation liquid rocket engine (DLRE) operating on natural gas (NG) – oxygen mixture have been performed to examine the impact of the DLRE configuration and fuel supply parameters on the operation process and thrust performance. In experiments, the absolute pressures of NG and oxygen supply were up to 30 and 15 atm, respectively; the mass flow rate of the reactive mixture was varied from 0.05 to 0.7 kg/s; the overall mixture composition was varied from fuel lean (with equivalence ratio 0.5) to fuel rich (with equivalence ratio 2.0). The maximum thrust and the maximum specific impulse obtained in this experimental series was 75 kgf and 160 s, respectively, at the maximum average pressure in the combustor of about 10 atm. It is shown that the increase of static pressure in the combustor results in the increase of both engine thrust and specific impulse. With the growth of the specific mass flow rate of reactive mixture, the operation process, on the one hand, becomes more stable, and on the other hand, the number of detonation waves simultaneously rotating in one direction in the combustor annulus increases. The results of DLRE fire tests were used to explore the predictive capabilities of the Semenov Institute of Chemical Physics (ICP) computational technology designed for full-scale simulation of the operation process in continuous-detonation combustors. Comparison of the predicted results with measurements proved that the calculations accurately predict the number of detonation waves circulating in the tangential direction of the annular DLRE combustor and the chaotic near-limiting operation mode resembling the mode with longitudinally pulsating detonation in the DLRE with CD nozzle extension. Calculations predict with reasonable accuracy both the detonation propagation velocity and detonation rotation frequency. In addition, calculations correctly

S.M. Frolov (✉) • V.S. Aksenov • V.S. Ivanov • S.N. Medvedev • I.O. Shamshin
Center for Pulsed Detonation Combustion, Semenov Institute of Chemical Physics (ICP), National Research Nuclear University MEPhI, Moscow, Russia
e-mail: smfrol@chph.ras.ru

predict the trends in the variation of DLRE operation parameters in an engine of a particular design. As in the experiments, the use of nozzle extension increases thrust. As for the thrust values, the calculations were shown to systematically overestimate them by at least 27% compared with measurements.

1 Introduction

At present, space propulsion engineering addresses a number of promising areas of development. One of them is the use of liquefied natural gas (LNG) as a propellant, and the other is the use of continuous-detonation combustion of the reactive mixture in a liquid rocket engine (LRE). The expediency of the transition to LNG – oxygen fuel couple is mainly due to (a) an increased specific impulse compared to kerosene – oxygen LRE; (b) the availability and low cost of LNG; (c) significantly less soot formation during combustion; and (d) high environmental characteristics compared with kerosene (Belov et al. 2000). The expediency of transition to a continuous-detonation combustion is mainly due to higher efficiency of the thermodynamic cycle with detonative combustion as compared with the conventional cycle using relatively slow combustion at constant pressure as shown by Zel'dovich (1940). Other advantages of a detonation LRE (DLRE) are: (a) a compact combustion chamber with an increase in the total pressure; (b) short nozzle; (c) high combustion efficiency; and (d) low concentrations of harmful substances in the exhaust gas. In theory, the replacement of kerosene by LNG in a traditional LRE promises the gain in the specific impulse of 3–4%, and the transition from a traditional LRE to the engine with detonative combustion promises the gain of 13–15% (Chvanov et al. 2012). The energy efficiency of the detonative LRE was proved experimentally by Frolov et al. (2014, 2015a, b) for the hydrogen–oxygen fuel couple. The first experiments with continuous-detonation combustion of methane–oxygen mixture in the annular combustor were made by Bykovskii and Zhdan (2013). Similar experiments were carried out by Kindracki et al. (2011), Frolov et al. (2015c) and Ivanov et al. (2015). Described in Frolov et al. (2015c) are the experiments with thrust measurements of the DLRE operating on natural gas (NG) – oxygen mixture at relatively low pressure in the annular combustor. The maximum value of specific impulse obtained by Frolov et al. (2015c) was 107 s. Natural gas used by Frolov et al. (2015c) and Ivanov et al. (2015) contained 92.8% methane, 3.9% ethane; 1.1% propane; 0.4% butane; 0.1% pentane; 1.6% nitrogen; and 0.1% carbon dioxide.

In this paper, we continue the study started by Frolov et al. (2015c) and Ivanov et al. (2015). The purpose of the research is to investigate both experimentally and computationally the impact of the DLRE con and fuel supply parameters on the operation process and thrust performance.

2 Experiments

The test stand consists of a receiver for methane or NG (0.16 m^3) and oxygen (0.32 m^3), high-speed valve system, fuel manifolds of large cross section, a thrust table and precision measurement system of thrust, pressures of fuel components supply, and ionization currents in the combustor (Frolov et al. 2015c; Ivanov et al. 2015). The maximum possible mass flow rate of the reactive mixture at this test stand is about 1.5 kg/s. The test stand is equipped with a remote control system.

Figure 3.1 shows a photo of the DLRE (Fig. 3.1a) and its scheme (Fig. 3.1b). The DLRE is an annular combustor with the injector head on the one side and a jet nozzle with a conical central body on the other side. The annular combustor is formed by two coaxial cylinders of 100 mm height: 90 mm inner diameter cylinder nested within a hollow outer cylinder of 100 mm in diameter, so that the gap between the cylindrical surfaces is 5 mm. The injector head consists of a replaceable thin disc with a sharpened edge attached to the inner cylinder of the combustor so that it forms an annular slit of width δ with the outer combustor wall. Oxidant (gaseous oxygen) is supplied to the combustor through this annular slit in the axial direction. Fuel (NG) is fed through the equally distributed radial holes 0.8 mm in diameter drilled in the outer wall of the combustor in the cross section located at a distance of 0.5 mm downstream from the disc. The number of radial holes is 144. During firing, the DLRE is water cooled.

In one test series, a profiled washer was installed between the inner cylinder of the combustor and a conical central body. This washer was aimed to increase the operation pressure by blocking the cross-section area of the combustor outlet by 50%. In the other test series, a converging-diverging (CD) conical nozzle extension was attached to the edge of the external cylinder of the combustor (Fig. 3.1c). In this DLRE configuration the minimum cross-sectional area was also 50% of the cross-sectional area of the annular combustor channel. In some tests, the position of the replaceable thin disc relative to the fuel injector holes was adjusted using rigid pads of different thickness.

All tests were performed under normal atmospheric conditions.

The fire test began with sending a digital signal to the opening of the oxygen supply valve, then (after 100 ms) to the opening of the NG supply valve, then (after 100 ms) to the ignition, and was normally continued for 1 s, and thereafter followed by successive cut-off of oxygen and NG supply.

The registration system included three consecutive ionization probes and the low-frequency sensor of absolute static pressure, all located in one combustor cross section 10 mm downstream from the injector head with a relative angular position of 90°. This registration system allowed the identification of the operation mode (continuous detonation or continuous combustion), to measure the rotational speed of detonation waves in the annular gap of the combustor when operating in continuous-detonation mode, determine the direction of detonation rotation, propagation velocity and the number of detonation waves while circulating over the injector head, and to measure the mean static pressure in the vicinity of the bottom of the

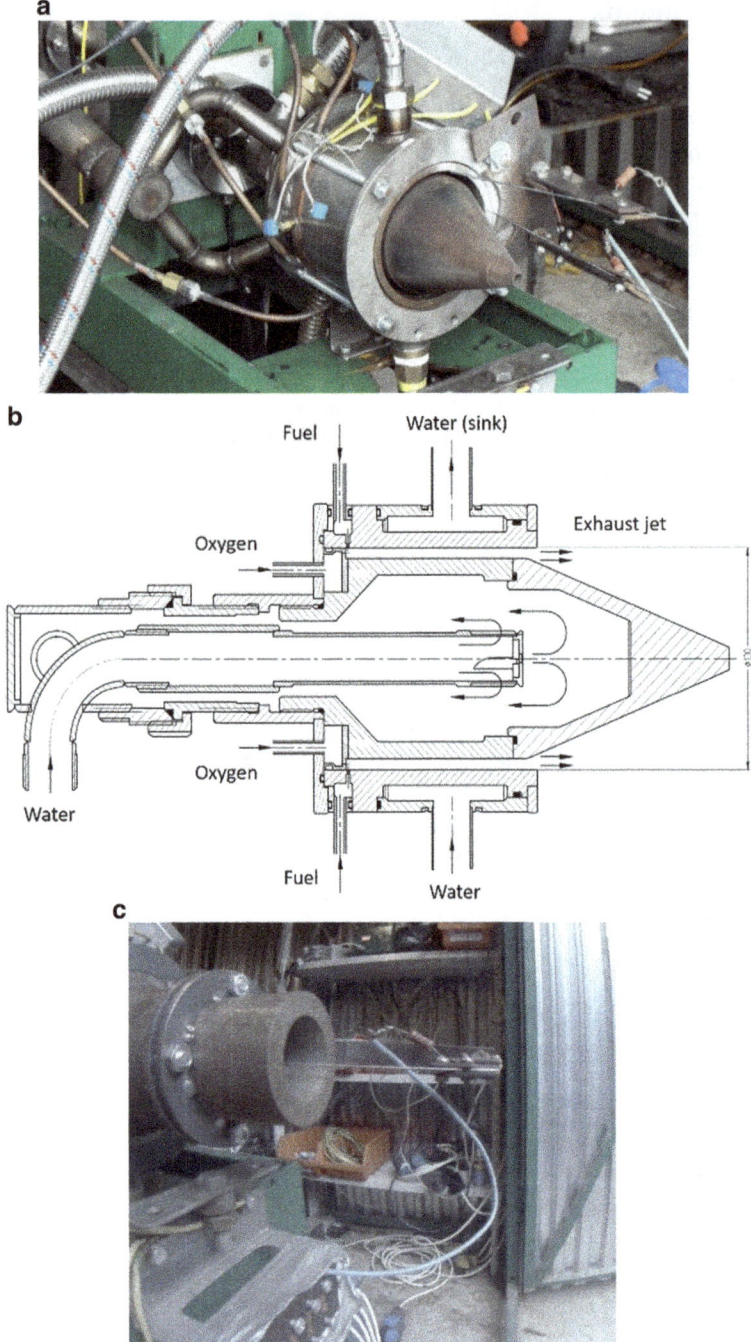

Fig. 3.1 Photograph of the DLRE without nozzle extension (**a**), its schematic (**b**) and a photograph of the DLRE with nozzle extension (**c**)

combustor. Besides measurements of ionization currents, thrust (using a calibrated load cell) and static pressure (using calibrated low-frequency pressure transducers) in the collectors supplying oxygen and NG directly into the DLRE annular gap were measured, as well as high-speed video recording using multiple high-speed digital cameras was conducted.

Mixture composition was determined based on the mass flow rates of fuel components entering the combustor. The latter were determined by the pressure drop in the receivers supplying gaseous oxygen and NG. Figure 3.2 shows an example of the experimental dependences of the pressure in the oxygen and NG receivers when feeding oxygen through the annular slit of $\delta = 1$ mm width. It is seen that after a short transient of ~100 ms duration the pressure in the receivers decreases linearly in time. The sharp drop in pressure during the transient is due to the filling of manifolds after the opening of the shut-off valve. Pressure drop during the experiment was always less than 10% of the initial pressure in the receiver, i.e., the supply pressures of fuel components into the injector head of the combustor was almost constant. The fact that the supply pressure of the fuel components during the test remained approximately constant distinguishes our experiments from experiments of Bykovskii and Zhdan (2013), in which the supply pressure decreased during a test significantly.

To initiate detonation in the DLRE a powerful spark discharge at the outlet section of the combustor was used. Spark discharge was created in the gap between the tungsten electrode and the grounded casing of the combustor. The electrode was mounted directly on the engine body and was connected to an AC generator with a voltage of 10 kV and a frequency of 100 Hz.

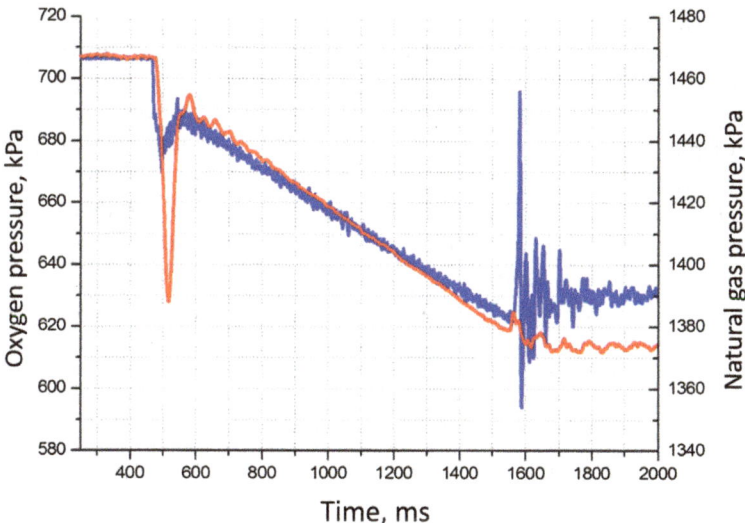

Fig. 3.2 Experimental pressure histories in oxygen and NG receivers. The *blue curve* is the pressure of gaseous oxygen, the *red curve* is the pressure of NG

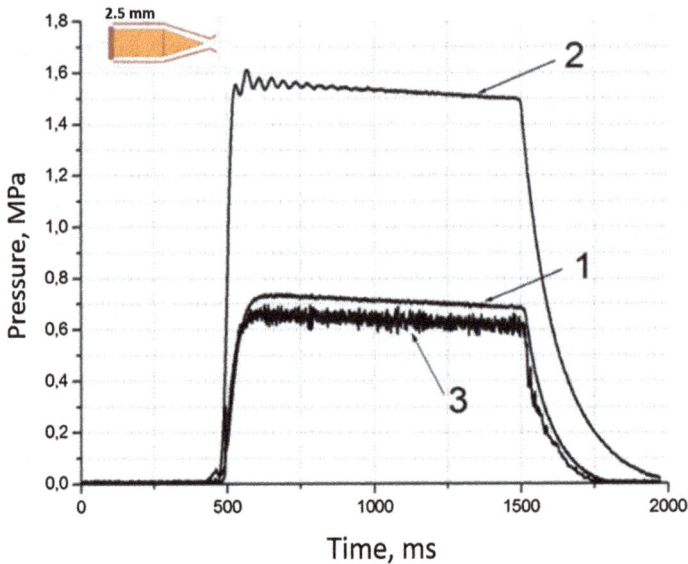

Fig. 3.3 Example of overpressure histories in oxygen (*curve 1*) and NG (*curve 2*) collectors, and in the combustor (*curve 3*) during firing a DLRE of the configuration shown in the insert

Figure 3.3 shows an example of overpressure records in the oxygen and NG collectors, as well as in the combustor in the fire test of DLRE with the annular slit of width $\delta = 2.5$ mm and with a CD nozzle extension attached. Hereinafter, the inserts in figures show schematically the corresponding DLRE configuration and the annular slit width δ. It can be seen that during the test the pressure in the collectors and in the combustor is almost constant. On the one hand, the average chamber pressure ($P_c \approx 0.6$ MPa) is less than the feed pressure of oxygen ($P_{O2} \approx 0.7$ MPa) by 0.1 MPa, i.e. $P_{O2}/P_c \approx 1.17$, which indicates a relatively small pressure drop when supplying oxygen through the annular slit with $\delta = 2.5$ mm (in the experiments of Bykovskii and Zhdan (2013) $P_{O2}/P_c = 2.3$ when $P_c \approx 0.3$ MPa). On the other hand, the ratio of NG feed pressure ($P_{CH4} \approx 1.5$ MPa) to the average pressure in the combustor is significantly higher ($P_{CH4}/P_c \approx 2.5$), i.e. the supply of NG through the radial holes of 0.8 mm in diameter is accompanied by a large pressure loss.

Figure 3.4 shows an example of signal records of three ionization probes during DLRE operation in a continuous-detonation mode. The probe signals are registered in the form of a voltage on the load resistance of 2 Ohm. All three signals have strong repetitive pulses of large amplitude. The pulse frequency (~ 7.6 kHz) and a time shift between signals of different probes show that in this case a single detonation wave circulates in the combustor at a velocity of $D \approx 2200$ m/s.

For express-analysis of probe records a methodology of fast Fourier transform was applied. As a result, a graph of the characteristic frequency f of the operation process vs. time was obtained. Typically, the process frequency is a multiple to the characteristic frequency of detonation wave rotation in the combustor. Based on the

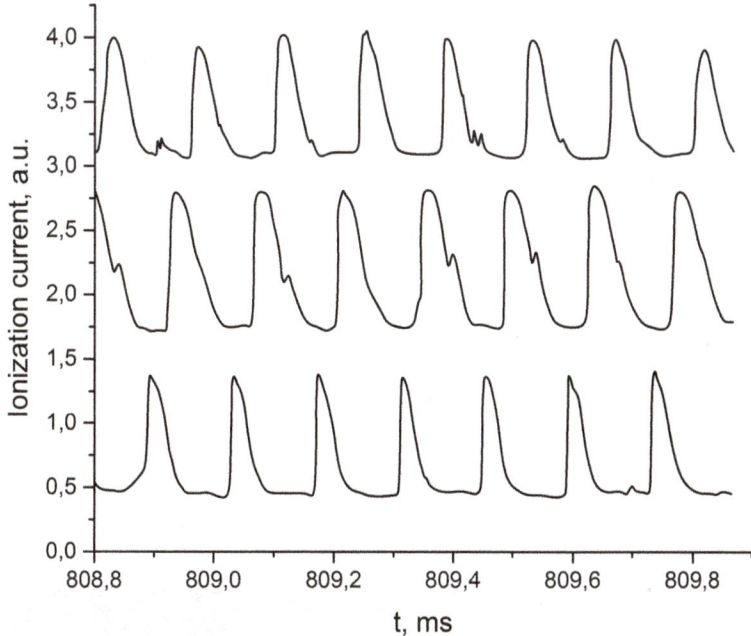

Fig. 3.4 Example of a record of continuous-detonation operation process with one detonation wave in NG – oxygen mixture

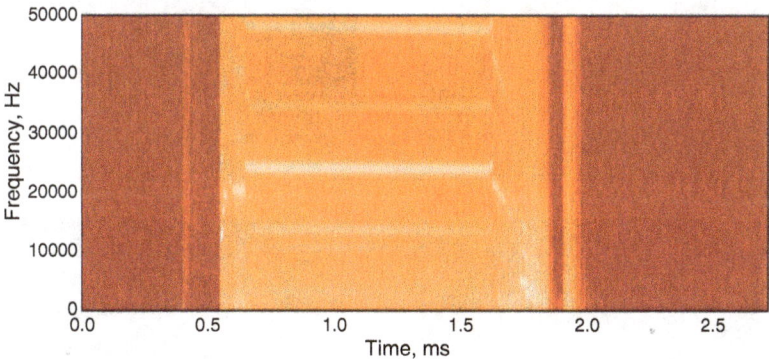

Fig. 3.5 Fast Fourier transform of the data obtained with the ionization probes for DLRE operation with five detonation waves

measured frequency one can readily determine the number n and the velocity D of the detonation waves rotating in the annulus. Figure 3.5 shows an example of an experiment with five detonation waves simultaneously rotating in the annulus of the combustor in one direction. Clearly visible is the base frequency at the level of $f \sim 23$ kHz. At a more detailed examination of Fig. 3.5 one can see transients at the initial and final stages of the DLRE operation when the base frequency and thus the number of detonation waves vary.

To confirm the findings obtained from the analysis of ionization probe records, another method of detonation registration in the DLRE combustor was additionally used, namely high-speed video-recording of the operation process at 200,000 frames/s from the side of the exhaust jet. This registration confirmed the presence of detonation in the annular combustor, the amount and rotation speed of detonation waves, and other information obtained on the basis of ionization probe methodology. Figure 3.6 shows the examples with video-recording of two, three, four, and five detonation waves in the DLRE.

Figure 3.7 shows examples of the exhaust jet images for the DLRE without (a) and with (b) CD nozzle extension.

Finally, Fig. 3.8 shows an example of the measured time history of DLRE thrust in one of the firing tests. The average thrust in this test was $T \approx 73$ kgf.

With the DLRE of different configurations (with different widths of the annular slit δ, with and without washer, and with and without CD nozzle extension), a series of fire tests (about 200 tests all in all) with NG – oxygen fuel was performed at

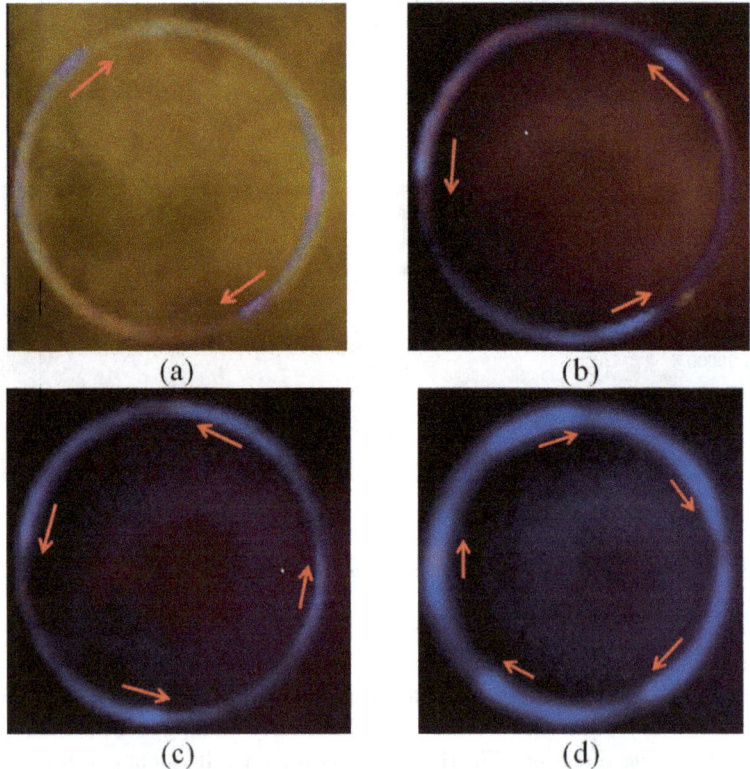

Fig. 3.6 High-speed video frames (200,000 frames/s) in the DLRE operating with two (**a**), three (**b**), four (**c**), and five (**d**) detonation waves

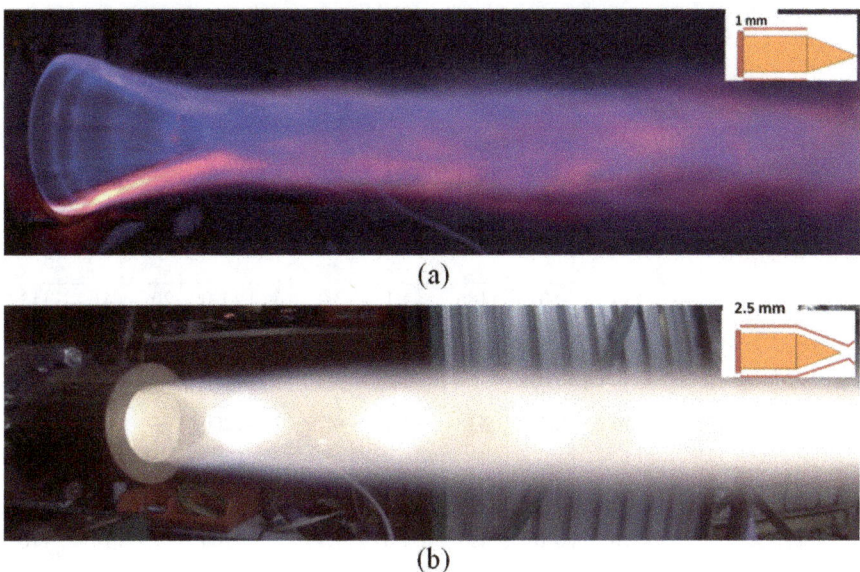

Fig. 3.7 Photos of exhaust jets in firing tests of DLRE without (**a**) and with (**b**) CD nozzle extension

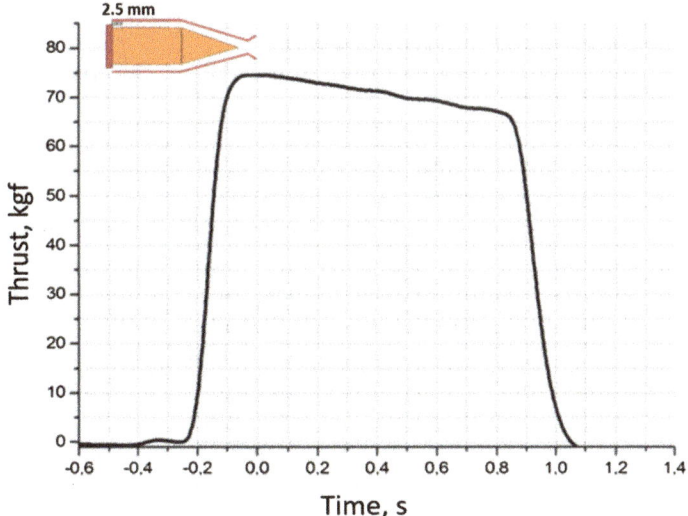

Fig. 3.8 Example of the measured time history of DLRE thrust

reactive mixture mass flow rates ranging from 0.05 to 0.7 kg/s. Varied in the tests were the absolute pressures of NG (up to 30 atm) and oxygen (up to 15 atm) supply. The composition of the mixture was changed from fuel lean with fuel-to-oxygen equivalence ratio $\Phi = 0.5$ to fuel rich with $\Phi = 2.0$. The experimental results were

Table 3.1 Results of ten selected tests

N	Config.	P_{O2} atm	P_{CH4} atm	G kg/s	G_{sp} kg/s/m²	Φ	P_c atm	n	D m/s	f kHz	T kgf	I_{sp} s
1[a]	1 mm	8	21	0.70	2243	0.87	1.0	–	–	–	16.5	24
2	1 mm	8	19	0.46	1474	1.1	4.5	4	1880	24	56	122
3	1 mm	6	14	0.35	913	1.15	3.3	4	1650	21	41	116
4	1 mm	6	14	0.37	1186	1.1	3.2	3	2090	20	41	111
5[b]	1 mm	6	13	0.30	962	1.12	3.2	1	2320	7	33	110
6	2.5 mm	7	16	0.35	418	1.1	6.7	1	2200	7	46	130
7	2.5 mm	8	15	0.36	430	1.1	7.6	1	1500	5	52	145
8[c]	2.5 mm	3	10	0.16	191	1.0	1.9	1	2200	7	18	110
9	2.5 mm	10	21	0.46	550	1.2	9.3	1	1500	5	73	157
10[c]	2.5 mm	7	15	0.36	430	1.15	6.6	1	–	5	51	141

[a]Test without ignition
[b]Replaceable thin disc was displaced 2 mm downstream
[c]Near-limiting mode with a detonation wave pulsating longitudinally rather than circulating in the tangential direction

well reproducible, except for some tests with the initial conditions at the edge of existence of different operation modes.

Table 3.1 shows the conditions of 10 sample tests with the indication of DLRE configuration, the width of the annular slit δ, oxygen and NG absolute pressures P_{O2} and P_{CH4} in collectors, reactive mixture mass flow rate G, specific (per unit area of injector holes for reactive mixture supply) mass flow rate G_{sp}, fuel-to-oxygen equivalence ratio Φ, as well as the results of the tests in terms of the average absolute pressure in the combustor P_c near the injector head, the number of detonation waves n, detonation velocity D, the frequency f of detonation rotation, thrust T, and specific impulse I_{sp} calculated as thrust T divided by gG, where g is the acceleration of gravity. Note that the Chapman – Jouguet detonation velocity for the homogeneous stoichiometric NG – oxygen mixture is approximately 2350 m/s, however this value can be treated only as a reference since reactive mixture composition in the combustor is not known due to separate supply of fuel and oxidizer.

The values of thrust T shown in Table 3.1 were obtained by averaging the thrust curve (see, e.g., Fig. 3.8 corresponding to Test 9 in Table 3.1) over the time interval from 0 to 0.4 s, i.e. after the establishment of a quasi-stationary mode of DLRE operation.

In Tests 1 to 5 the DLRE of basic design with the annular slit width of δ = 1 mm without washer and CD nozzle extension was used. In Test 5 the position of the replaceable thin disc relative to the fuel injector holes was displaced 2 mm downstream, so that NG was injected radially 1.5 mm upstream from the annular slit.

In Tests 6 to 10 the annular slit was expanded to δ = 2.5 mm. In addition, in Test 6 a washer blocking 50% of the combustor outlet cross section was installed. In Test 7 the washer was replaced by a CD nozzle extension (see Fig. 3.1c), in which the minimum cross-sectional area was also 50% of the cross-sectional area of the annular combustor channel. In Test 8 neither washer, nor CD nozzle extension was used. Finally, Tests 9 and 10 were made with CD nozzle extension at different supply pressures of NG and oxygen.

In Test 1 the reactive mixture components were purged through the DLRE without ignition to measure the baseline thrust created by the cold flow. In Tests 2 to 4 stable operation modes with four and three equidistant detonation waves simultaneously circulating in the same tangential direction were detected. Despite Tests 3 and 4 were made at similar initial conditions in terms of DLRE configuration and absolute pressures P_{O2} and P_{CH4} in collectors, the experimental results were somewhat different: in Test 3 four detonation waves were detected whereas in Test 4 the number of detonation waves was three. This difference is attributed to the fact that the initial conditions for Tests 3 and 4 are at the edge of existence of the operation mode with four detonation waves. In Test 5 with the DLRE of basic configuration a steady operation mode with a single detonation wave circulating in a tangential direction was observed.

In tests 3, 6 and 7 the mass flow rate of reactive mixture was maintained approximately constant (0.35–0.36 kg/s). With an increase in pressure in the combustor both thrust and specific impulse increased monotonically: in Test 3 with the combustor pressure of 3.3 atm the thrust and the specific impulse were 41 kgf and 116 s, respectively, and in Test 7 these values increased to 52 kgf and 145 s respectively at an average pressure in the combustor of 7.6 atm. With the growth of the specific mass flow rate of reactive mixture (from 191 kg/s/m² in Test 8 to 418, 430 and 913 kg/s/m² in Tests 6, 7 and 3) the operation process in the combustor, on the one hand, was becoming more stable and, on the other hand, the number of detonation waves simultaneously rotating in the combustor in one direction increased. Thus, in Test 8 a near-limiting operation mode with rotating-to-pulse detonation transition was observed: in this test, the detonation wave was running in the longitudinal rather than in tangential direction, periodically recovering near the DLRE outlet. At the specific mass flow rate on the level of 420–430 kg/s/m² (Tests 6 and 7) the operation mode with one detonation wave stably circulating in the tangential direction was detected. Finally, in Test 3 four equidistant detonation waves were stably circulating in the combustor. Replacement of the flow washer in Test 6 by the CD nozzle extension in Test 7 under otherwise similar conditions led to increased thrust performance, but the detonation velocity was greatly reduced (down to 1500 m/s instead of 2200 m/s) approaching limiting values. Interestingly, the increase of the mass flow rate of reactive mixture from 0.36 kg/s in Test 7 to 0.46 kg/s in Test 9 in the

DLRE of the same configuration with CD nozzle extension did not lead to a change in the operation mode. In both tests the DLRE operated with one detonation wave rotating in the tangential direction at a velocity of 1500 m/s, although the thrust and the specific impulse in Test 9 increased to 73 kg/s and 157 s, respectively. Decreasing of supply pressures of fuel components as compared to Test 9 other conditions being the same led to the establishment of the near-limiting operation mode with the detonation wave pulsating longitudinally. It is worth noting that at the mass flow rate of reactive mixture close to that in Test 4, the thrust in Test 10 appeared to be higher than in Test 4: 51.0 instead of 40.5 kgf.

The reason for the reduction of the detonation velocity in the DLRE with CD nozzle extension will be the subject of further research.

3 Calculations

The objective of calculations described below was to explore the predictive capability of the computational technology developed at the Semenov Institute of Chemical Physics (ICP) in terms of reproducing all DLRE operation modes detected in the experiments.

Figure 3.9 shows a longitudinal section (Fig. 3.9a) and one-quarter perspective view (Fig. 3.9b) of the DLRE of several different configurations with oxygen and NG collectors and a piece of oxygen feeding manifold. In addition to these DLRE configurations, other configurations used in the experiments were also considered.

Three-dimensional numerical simulation of physical and chemical processes in the DLRE of Fig. 3.9 was carried out according to the procedure described in detail by Frolov et al. (2012). The flow of viscous compressible gas is described by the Reynolds-averaged unsteady Navier-Stokes equations supplemented by energy equation and continuity equations of all chemical species of a multicomponent mixture. Turbulent fluxes of mass, momentum, and energy are modeled by a standard k-ε-turbulence model for compressible flow. Because all the physical and chemical processes in the combustor under consideration take place within a very short time, the contribution of frontal (controlled by transport processes) combustion in chemical sources in the energy and species continuity equations was neglected. The contribution of volumetric reactions in these chemical sources were determined by the lagrangian Monte-Carlo based Particle Method (PM) in which the average rates of chemical reactions in a turbulent flow are calculated with due regard for the effect of turbulent fluctuations of temperature and species concentrations (the effect of "turbulence – chemistry interaction"). The governing equations are closed with the caloric and thermal equations of state of an ideal gas mixture with variable specific heats, as well as with initial and boundary conditions. All other thermal parameters of the gas were also considered variable. In contrast to Frolov et al. (2012), where the calculations were performed for continuous-detonation combustor with separate

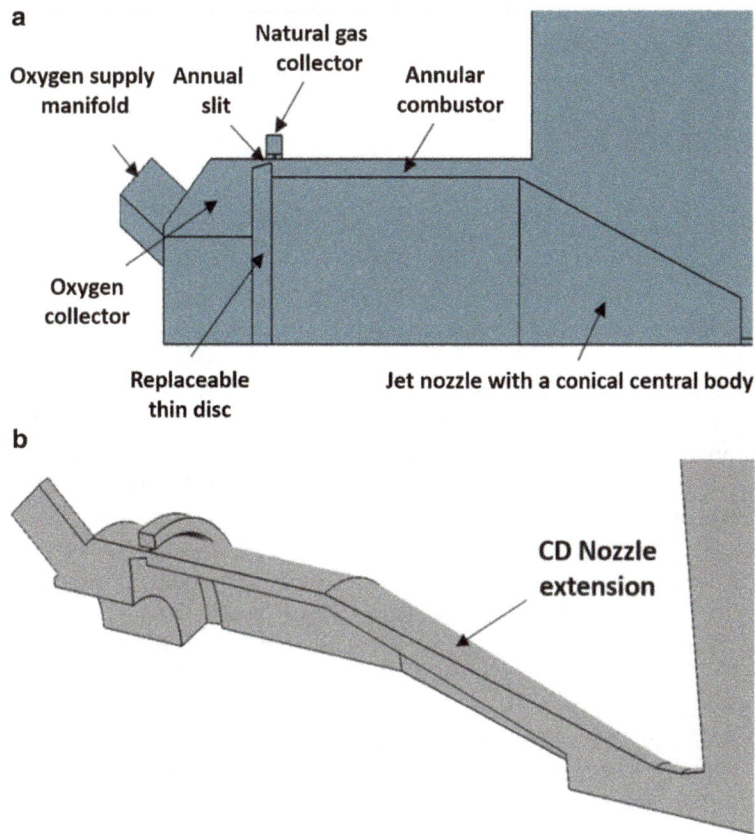

Fig. 3.9 Longitudinal section (**a**) and a perspective view of a quarter (**b**) of DLRE of two different configurations with oxygen and NG collectors and pieces of oxygen feeding manifolds

supply of hydrogen and air, the calculations described herein are made for separate supply of NG and oxygen.

For simulating volumetric chemical transformations in the NG – oxygen mixture we used the technique proposed by Levin and Korobeinikov (1969) and Medvedev et al. (2015). The chemical conversion of the reactive mixture was assumed to occur in two stages: the first stage corresponds to the self-ignition delay, and the second to the establishment of thermodynamic equilibrium in the combustion products. Upon the completion of the first stage, the initial reactive mixture was assumed to immediately convert into the thermodynamically equilibrium reaction products.

(HP-problem is solved for the entire computational cell; here H is the enthalpy, P is the pressure).

To calculate the self-ignition delay of the NG – oxygen mixture we used the overall reaction mechanism containing five reactions, similar to the mechanism proposed by Basevich and Frolov (2006):

$CH_4 + 1.5\, O_2 \to CO + 2\, H_2O$	(I)
$CO + H_2O \to CO_2 + H_2$	(II)
$CO_2 + H_2 \to CO + H_2O$	(III)
$H_2 + H_2 + O_2 \to H_2O + H_2O$	(IV)
$CO + CO + O_2 \to CO_2 + CO_2$	(V)

with tabulated Arrhenius parameters, preexponential factor A and activation energy E, in the expression for the rate of reaction (I):

$$W = -A[O_2][CH_4]\exp(-E/RT)$$

i.e., reaction (I) is considered as a bimolecular reaction. Arrhenius parameters for other reactions (II) – (V) are the same as given by Basevich and Frolov (2006). A similar approach was used by Frolov et al. (2012) and provided good results for all measured characteristics of the operation process in a hydrogen–air continuous-detonation combustor.

Table 3.2 shows the values of Arrhenius parameters A and E for NG – oxygen mixture in the range of initial temperature from 1100 to 1400 K, initial pressure from 1 to 80 atm, and fuel-to-oxygen equivalence ratio Φ from 0.5 to 4.0. These values are derived from the best fit of self-ignition delay predicted by mechanism (I)–(V) with the self-ignition delay predicted by a thoroughly validated detailed kinetic mechanism (DKM) of Basevich et al. (2013). In the calculations with the DKM, NG had the following composition: $[CH_4] = 92.8\%$, $[C_2H_6] = 3.9\%$, $[C_3H_8] = 1.1\%$, $[C_4H_{10}] = 0.4\%$, $[C_5H_{12}] = 0.1\%$, $[N_2] = 1.6\%$, and $[CO_2] = 0.1\%$. Figure 3.10 shows the examples of calculated dependences of the self-ignition delay on the initial temperature at different initial pressures and reactive mixture compositions obtained by mechanism (I)–(V) (curves) and DKM (symbols).

Table 3.2 Arrhenius parameters A and E of reaction (I) in the range of initial temperature from 1100 to 1400 K

Φ	P, atm	A, cm³/mol/s	E, kcal/mol
0.5	1	6e + 14	48
1	1	4e + 14	48
2	1	3e + 14	48
4	1	2e + 14	48
0.5	12	1.5e + 14	46
1	12	1e + 14	46
2	12	8e + 13	46
4	12	8e + 13	46
0.5	50	4e + 13	43
1	50	2.7e + 13	43
2	50	1e + 13	40
4	50	1e + 13	40

3 Flow Structure in Rotating Detonation Engine with Separate Supply of Fuel...

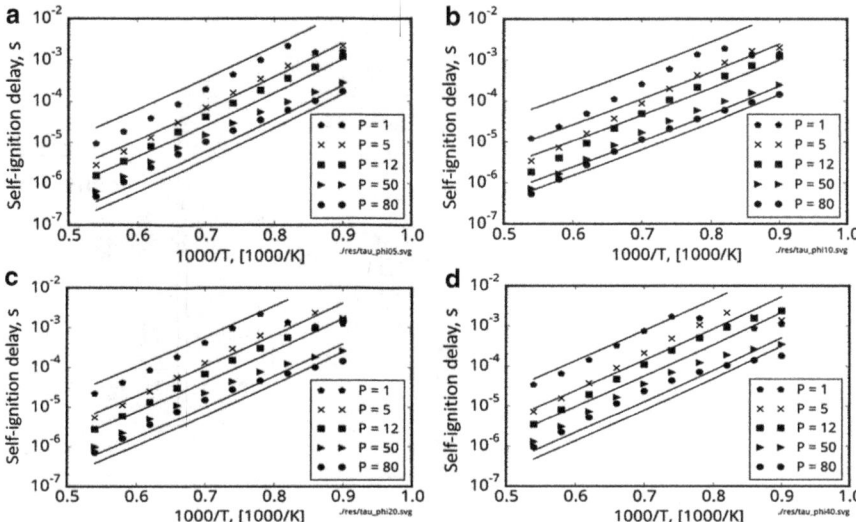

Fig. 3.10 Examples of calculated dependences of self-ignition delay on the initial temperature at different initial pressures and reactive mixture compositions obtained by mechanism (I)–(V) (curves) and DKM (symbols): $\Phi = 0.5$ (**a**), 1.0 (**b**), 2.0 (**c**), and 4.0 (**d**)

For the calculation of equilibrium parameters of the combustion products the approach of Pope (2003) was used.

The above-described model of a two-stage chemical transformation of the reactive mixture was validated on the zero-dimensional problem of gas self-ignition in the adiabatic reactor and on the one-dimensional problem of detonation initiation by a strong shock wave. The results were compared with the results of calculations based on the DKM for NG–oxygen mixture. In both cases, very similar values of self-ignition delays, equilibrium temperature and pressure, and equilibrium concentrations of the main species in the combustion products, as well as the values of the detonation velocity were obtained.

Figure 3.11 shows a part of a structured nonuniform computational grid of the DLRE with separate supply of NG and oxygen. The minimum size of the computational cells in the reaction zone was 0.2 mm. Round radial holes in the outer wall of the combustor (for NG supply) are modeled by a set of rectangular cells with the same hydraulic resistance. The total number of cells, together with the nozzle and the buffer zone attached to the engine outlet was 4.5×10^5–5.5×10^5.

At the entrance of NG collector and oxygen manifold fixed mass flow rates of NG and oxygen were specified. At all solid walls no-slip conditions and a constant temperature of 293 K were adopted. At the outlet boundary of the buffer zone normal to the engine axis and located at a large distance from the nozzle exit, a constant static pressure of 1 atm was adopted, whereas at the side boundaries of the buffer zone the symmetry conditions were specified.

At the beginning of calculation the annular combustor was filled with air under normal conditions, and the NG and oxygen collectors were filled with NG and oxy-

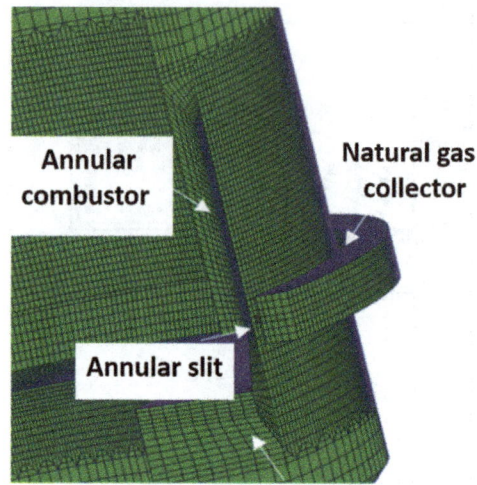

Fig. 3.11 Longitudinal section of the DLRE computational grid

gen at a pressure equal to the corresponding experimentally measured values at the cut-off valves. Initially, all gases were assumed to be quiescent. Each calculation started from purging of the combustor with gases accompanied by turbulence generation, molecular mixing processes, and followed by the procedure of detonation initiation. To initiate detonation, a certain number of ignition sources was placed in the combustor annulus. The ignition sources were the localized finite regions with fast burnout of reactive mixture. After a transition period a quasistationary operation process with one or several detonation waves circulating in the same or in different tangential directions, or a chaotic operation mode resembling a mode with longitudinally pulsating detonation wave was set in the combustor.

Given below is a brief description of the calculations corresponding to selected tests in Table 3.1.

Figure 3.12 shows the predicted quasi-stationary fields of static pressure at the outer wall (a) and in the combustor cross-section 4 mm downstream from the combustor bottom (b), as well as a static temperature at the outer wall (c) under conditions of Test 2 in Table 3.1. As seen, the calculation predicts the operation process with four equidistant detonation waves circulating in the same direction at a velocity of 2210 ± 30 m/s. The predicted frequency of detonation rotation is 28 kHz. According to Table 3.1, the operation process detected in Test 2 also exhibits four detonation waves, however the measured detonation velocity is 14% lower (f = 24 kHz).

The predicted value of thrust developed by the DLRE is 68.6 kgf, whereas the measured thrust was about 56 kgf, i.e. calculation overestimates the thrust by approximately 22%.

Figure 3.13 shows the predicted quasi-stationary fields of static pressure and static temperature under conditions of Test 4 in Table 3.1. The calculation predicts the operation process with three equidistant detonation waves circulating in the

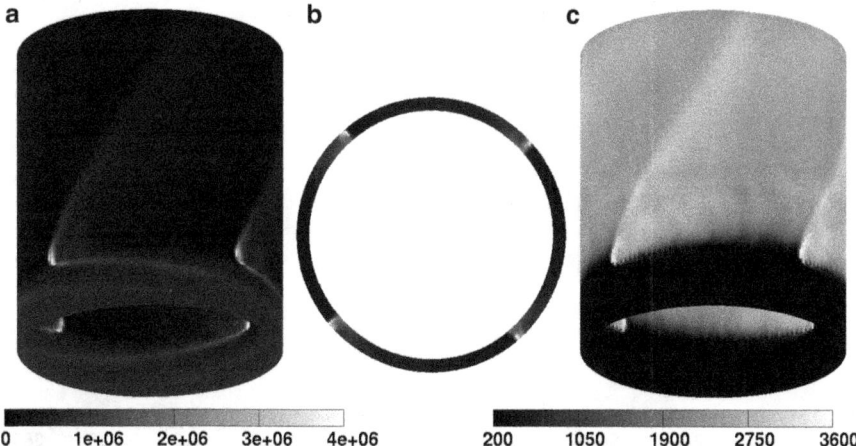

Fig. 3.12 Predicted quasi-stationary fields of static pressure in Pa (**a, b**) and static temperature in K (**c**) under the conditions of experimental Test 2

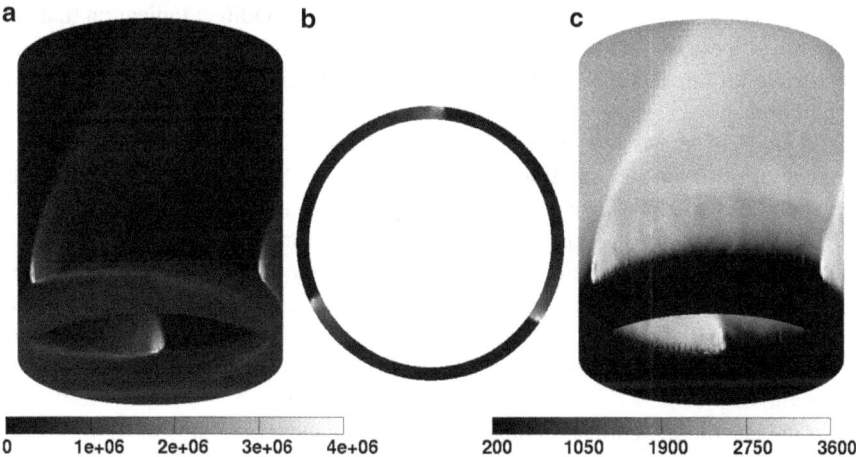

Fig. 3.13 Predicted quasi-stationary fields of static pressure in Pa (**a, b**) and static temperature in K (**c**) under the conditions of experimental Test 4

same direction at a velocity of 2270 ± 20 m/s. The predicted frequency of detonation rotation is 21.8 kHz. According to Table 3.1, the operation process detected in Test 4 also exhibits three detonation waves, however the measured detonation velocity is 8% lower (f = 20 kHz). The predicted value of thrust developed by the DLRE is 64.2 kgf, whereas the measured thrust was about 41 kgf, i.e. calculation overestimates the thrust by approximately 56%. It is worth noting that Test 4 is "identical" in terms of initial conditions to Test 3, in which four detonation waves were detected in the experiment. The calculation predicts the operation process with three rather

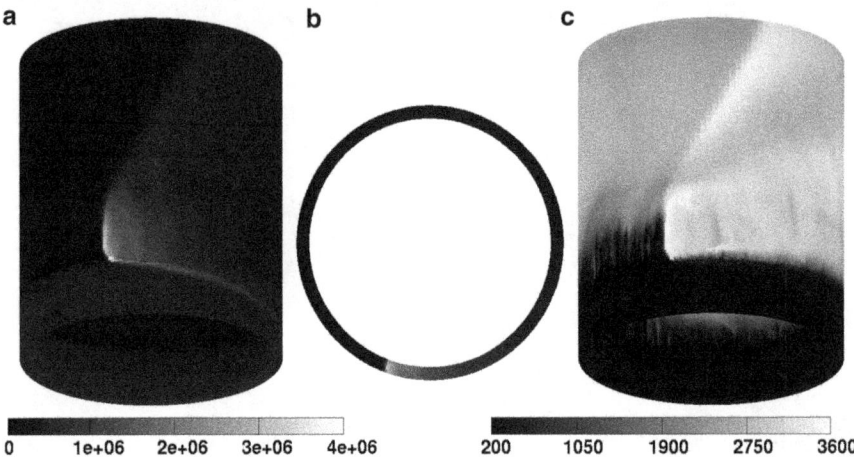

Fig. 3.14 Predicted quasi-stationary fields of static pressure in Pa (**a**, **b**) and static temperature in K (**c**) under the conditions of experimental Test 5

than four detonation waves which could be treated as an indirect indication that the four-wave operation mode is marginal at these conditions and the transition from four-wave to three-wave mode is sensitive to small variations in the initial conditions.

Figure 3.14 shows the predicted quasi-stationary fields of static pressure and static temperature under conditions of Test 5 in Table 3.1. The calculation predicts the operation process with one detonation wave circulating tangentially at a velocity of 2330 ± 20 m/s. The predicted frequency of detonation rotation is 7.4 kHz. According to Table 3.1, the operation process detected in Test 5 also exhibits one detonation wave and the measured detonation velocity is the same as in the calculation. The predicted value of thrust developed by the DLRE is 44.8 kgf, whereas the measured thrust was about 33 kgf, i.e. calculation overestimates the thrust by approximately 36%.

In the calculation corresponding to the experimental conditions of Test 10 with installed CD nozzle extension a chaotic operation process resembling the process with one detonation wave pulsating in the longitudinal direction rather than circulating in tangential direction was obtained. In the calculation, detonation waves propagating in different tangential directions periodically appeared and disappeared in the flow. Detonation reinitiation most often occurred in the outer nozzle due to reflections of decaying shock waves and the resultant detonation waves propagated upstream towards the combustor bottom through the annulus filled with fresh reactive mixture. The mean predicted frequency of detonations in this calculation was 5.1 kHz. According to Table 3.1, the operation process detected in Test 10 also exhibited one detonation wave pulsating longitudinally with the same measured detonation velocity (f = 5 kHz). The predicted value of thrust developed by the DLRE is 64.8 kgf, whereas the measured thrust was about 51 kgf, i.e. calculation overestimates the thrust by approximately 27%.

Thus, comparison of the predicted results with measurements showed that the calculations accurately predict the number of detonation waves circulating in the tangential direction of the annular DLRE combustor of a certain design (four, three, or one detonation wave(s)). Moreover, the calculation predicts a near-limiting chaotic operation mode resembling a mode with detonation pulsations in the longitudinal direction in the DLRE with CD nozzle extension. Also, the calculations predict with reasonable accuracy the detonation velocity and rotation frequency. In addition, the calculations correctly predict the experimental trends in the overall performance of the DLRE of a given design: as in the experiments, the number of detonation waves, the detonation velocity and the engine thrust decrease with decreasing the mass flow rate of reactive mixture. Similar to the experiment, the use of the CD nozzle extension increases thrust. As for the thrust values, they are systematically overestimated by at least 27% as compared with measurements, and a higher value of thrust is obtained even for the cold-flow purging of DLRE.

4 Conclusions

The experimental and computational investigations of DLRE operating on NG – oxygen mixture have been performed to examine the impact of the DLRE configuration and fuel supply parameters on the operation process and thrust performance. In experiments, the absolute pressures of NG and oxygen supply were up to 30 atm and 15 atm, respectively; the mass flow rate of reactive mixture was varied from 0.05 to 0.7 kg/s; the overall mixture composition was varied from fuel lean (with equivalence ratio 0.5) to fuel rich (with equivalence ratio 2.0). The maximum thrust and the maximum specific impulse obtained in this experimental series was 75 kgf and 160 s, respectively, at the maximum average pressure in the combustor of about 10 atm. It is shown that the increase of static pressure in the combustor results in the increase of both engine thrust and specific impulse. With the growth of the specific mass flow rate of reactive mixture, the operation process, on the one hand, becomes more stable, and on the other hand, the number of detonation waves simultaneously rotating in one direction in the combustor annulus increases. The replacement of the washer by the profiled CD nozzle extension with other conditions being similar leads to increased thrust performance, but the detonation velocity is greatly reduced (down to 1500 m/s instead of 2200 m/s), approaching limiting values. The reason for this decrease in the detonation velocity will be the subject of further research.

The results of DLRE fire tests were used to explore the predictive capabilities of the ICP computational technology designed for full-scale simulation of the operation process in continuous-detonation combustors. Comparison of the predicted results with measurements proved that the calculations accurately predict the number of detonation waves circulating in the tangential direction of the annular DLRE combustor (four, three, or one detonation wave(s)) and the chaotic near-limiting operation mode resembling the mode with longitudinally pulsating detonation in the DLRE with CD nozzle extension. Calculations predict with reasonable accuracy both the detonation propagation velocity and detonation rotation frequency. In addition, calculations correctly predict the trends in the variation of DLRE operation parameters with decreasing the mass flow rate of reactive mixture in an engine of a particular design: as in the experiment, the number of detonation waves, detonation velocity and thrust decrease. As in the experiments, the use of nozzle extension increases thrust. As for the thrust values, the calculations were shown to systematically overestimate them by at least 27% compared with measurements. Moreover, a higher value of DLRE thrust is obtained even for the cold-flow purging of the engine. The reasons for this overestimation is currently being investigated.

Acknowledgments The work was supported by the Russian Ministry of Education and Science under the State Contract No. 14.609.21.0002 (Contract ID RFMEFI60914X0002) "Development of technologies for the use of liquefied natural gas (methane, propane, butane) as fuel for rocket and space technology and the creation of a new generation of stand demonstrator rocket engine" under the Federal Target Program "Research and development in priority areas of scientific and technological complex of Russia for 2014–2020."

References

Basevich, V. Y., & Frolov, S. M. (2006). Overall kinetic mechanisms for modeling multistage self-ignition of hydrocarbons in reactive flows. *Russian Journal of Chemical Physics, 25*(6), 54–62.

Basevich, V. Y., Belyaev, A. A., Posvyanskii, V. S., & Frolov, S. M. (2013). Mechanisms of the oxidation and combustion of normal paraffin hydrocarbons: Transition from C1–C10 to C11–C16. *Russian Journal of Physical Chemistry B, 7*(2), 161–169.

Belov, E. A., Bogushev V. Yu., Klepikov I. A., and Smirnov A. M. (2000). *Results of experimental works in NPO Energomash on utilization of methane as a fuel component for LRE*. Herald of NPO Energomash named after Academician V. P. Glushko. Moscow: NPO Energomash named after Academician V. P. Glushko. 18:86–89.

Bykovskii, F. A., & Zhdan, S. A. (2013). *Continuous spin detonation*. Novosibirsk: SB RAS Publ. 423 p.

Chvanov, V. K., Frolov, S. M., & Sternin, L. E. (2012). Liquid-propellant detonation rocket engine. Trans. Of NPO Energomash named after Academician V. P. Glushko. Moscow, NPO Energomash Publ., 29:4–14.

Frolov, S. M., Dubrovskii, A. V., & Ivanov, V. S. (2012). Three-dimensional numerical simulation of the operation of the rotating detonation chamber. *Russian Journal of Physical Chemistry B, 6*(2), 276–288.

Frolov, S. M., Aksenov, V. S., Gusev, P. A., Ivanov, V. S., Medvedev, S. N., & Shamshin, I. O. (2014). Experimental proof of the energy efficiency of the Zel'dovich thermodynamic cycle. *Doklady Physical Chemistry, 459*(2), 207–211.

Frolov, S. M., Aksenov, V. S., & Ivanov, V. S. (2015a). Experimental proof of Zel'dovich cycle efficiency gain over cycle with constant pressure combustion for hydrogen–oxygen fuel mixture. *International Journal of Hydrogen Energy, 40*(21), 6970–6975.

Frolov, S. M., Aksenov, V. S., Dubrovskii, A. V., Ivanov, V. S., & Shamshin, I. O. (2015b). Energy efficiency of a continuous-detonation combustion chamber. *Combustion Explosion, and Shock Waves, 51*(2), 232–245.

Frolov, S. M., Aksenov, V. S., Gusev, P. A., Ivanov, V. S., Medvedev, S. N., & Shamshin, I. O. (2015c). Experimental studies of small samples of bench rocket engines with a continuously-detonation combustors. Goren. Vzryv (Mosk.) —. *Combustion and Explosion, 8*(1), 151–163.

Ivanov, V. S., Aksenov, V. S., Frolov, S. M., & Shamshin, I. O. (2015). Experimental studies of stand sample of rocket engine with continuous-detonation combustion of natural gas – Oxygen mixture. Goren. Vzryv (Mosk.) —. *Combustion and Explosion, 9*(2), 51–64.

Kindracki, J., Wolanski, P., & Gut, Z. (2011). Experimental research on the rotating detonation in gaseous fuels – Oxygen mixtures. *Shock Waves, 21*, 75–84.

Levin, V. A., & Korobeinikov, V. P. (1969). Strong explosion in the combustible gas mixture. *Fluid Mechanics, 6*, 48–48.

Medvedev, S. N., Ivanov, V. S., & Frolov, S. M. (2015). Three-dimensional numerical simulation of operation process and thrust performance of bench rocket engine with continuous detonation combustion of natural gas – Oxygen mixture. *Goren. Vzryv (Mosk.) —Combustion and Explosion, 9*(2), 65–79.

Pope, S. B. (2003). *CEQ: A Fortran library to compute equilibrium compositions using Gibbs function continuation.* Available at: http://eccentric.Mae.cornell.Edu/pope/CEQ. Accessed March 4, 2016.

Zel'dovich, Y. B. (1940). To the question of energy use of detonation combustion. *Journal of Technical Physics, 10*(17), 1455–1461.

Chapter 4
Application of Detonation Waves to Rocket Engine Chamber

Jiro Kasahara, Yuichi Kato, Kazuaki Ishihara, Keisuke Goto, Ken Matsuoka, Akiko Matsuo, Ikkoh Funaki, Hideki Moriai, Daisuke Nakata, Kazuyuki Higashino, and Nobuhiro Tanatsugu

Abstract We present the results of experiments performed with a rotating detonation engine using continuous detonation in an annular combustor to create thrust. Detonation waves propagate in a supersonic and very small region, allowing shortening of the combustor. The combustor of RDE causes high-pressure loss when the propellant is injected, and cooling is necessary due to high heat flux. However, the combustion efficiency of detonation combustion in an annular combustor is the most important, but have not been fully elucidated. In addition, the influence of the injector shape and direct cooling of a rotating detonation combustor require clarification. This paper reports the measurement results of combustor stagnation pressure and thrust, the influence of injector shape on c^* efficiency, and the estimate of heat flux. The c^* efficiency was 88–100% when we used the convergent or convergent-divergent nozzle and the equivalence ratio was less than 1.0. The shape of the injector influenced wave propagation mode, but the mode did not change the c^* efficiency. We estimated time-spatial average heat flux from the terminal temperature, and the heat flux was 8.1 ± 1.8 MW/m^2 in no water injection condition. The rocket RDE sled test was successfully performed. The total mass of the rocket RDE system was 58.3 kg, total time averaged thrust was 201 N, the time averaged mass flow rate was 143 g/s, and the specific impulse was 144 s.

J. Kasahara (✉) · Y. Kato · K. Ishihara · K. Goto · K. Matsuoka
Nagoya University, Nagoya, Aichi, Japan
e-mail: kasahara@nuae.nagoya-u.ac.jp

A. Matsuo
Keio University, Yokohama, Kanagawa, Japan

I. Funaki
Japan Space Exploration Agency, Sagamihara, Kanagawa, Japan

H. Moriai
Mitsubishi Heavy Industries Ltd., Komaki, Aichi, Japan

D. Nakata · K. Higashino · N. Tanatsugu
Muroran Institute of Technology, Muroran, Hokkaido, Japan

Nomenclature

A_{inj} total injector area
A_t throat area
c^* characteristic velocity
ER equivalence ratio
F_t thrust
g gravitational acceleration
I_{sp} specific impulse
M Mach number
$\dot{m}$ mass flow rate
P pressure
P_c combustor stagnation pressure
R gas constant
T temperature
T_c adiabatic flame temperature

Greek Symbols

γ ratio of specific heat
η_{c^*} c^* efficiency

Subscripts

i   ideal
m   measured

1 Introduction

A detonation wave is a combustion wave with pressure gain, which propagates at supersonic speed (2 ~ 3 km/s) into a combustible mixture. There are many fundamental studies of detonation wave engines (Kailasanath 2000, 2003; Wolański 2013; Lu and Braun 2014) and system level research studies (Kasahara et al. 2007; Hoke et al. 2010; Matsuoka et al. 2016; Frolov et al. 2013, 2015a, b; Dubrovskii et al. 2015). The detonation cycle has a higher thermal efficiency than a conventional gas turbine engine (Heiser and Pratt 2002; Wu et al. 2003; Talley and Coy 2002; Frolov et al. 2014; Nordeen 2013). Therefore, it is expected that a high-efficiency propulsion system can be realized using detonation waves.

A rotating detonation engine (RDE) uses continuous detonation in an annular combustor to create thrust. As detonation waves propagate at a supersonic speed in a very small region, the combustor can be shortened. The combustor of RDE causes high pressure loss when a propellant is injected, and cooling system is necessary due to high heat flux from the burned gas to the wall (Theuerkauf et al. 2015).

Some of the earliest RDE research was performed in the 1960s by the Russian scientist Voitsekhovskii (1960), Voitsekhovskii et al. 1967). In the 1990s, Bykovskii et al. (2006) studied air-breathing RDEs for a variety of gaseous and liquid fuels. There have also been many recent studies of RDE. Our research group performed visualization of rotating detonation waves (Nakayama et al. 2012; Nakagami et al. 2017a, b; Fujii et al. 2017; Gawahara et al. 2013), while Kato et al. (2016) and Ishihara et al. (2017) investigated thrust efficiency of the RDE. The CFD analysis (Hishida et al. 2009; Schwer and Kailasanath 2011, 2013), simplified steady state analysis (Fievisohn and Yu 2017), RDE experiments (Kindracki et al. 2011), thermodynamic modeling (Stechmann et al. 2017; Fotia et al. 2016), system level demonstration (Claflin 2012), rotating detonation turbine engine modeling (Paxson and Naples 2017), rotating detonation turbine engine demonstration (Naples et al. 2017) were performed.

The combustion efficiency and specific impulse of detonation combustion in an annular combustor are the most important physical parameters. Although the specific impulse meaning overall thrust performance of the RDE was investigated by many researchers, the combustion efficiency has not been investigated especially in the dependency on the mass flow rate of the propellant and injector geometries. In addition, the influence of the injector shape and direct cooling of a rotating detonation combustor require clarification. This paper reports the measurement results of combustor stagnation pressure and the influence of injector shape on c^* efficiency. We then performed cooling by injecting water directly into the combustor and measured the heat flux on the surface of the outer combustor.

2 Experimental Apparatus

The RDE used in this study is shown in Fig. 4.1. The RDE was made of copper (C1100) in purpose of thermal diffusion. The inner diameter of the annular combustor was 60.5 mm, the width 3.2 mm, and the axial length 48 mm. The RDE had a 30° conical plug, and a detachable convergent nozzle and divergent nozzle. The contraction ratio was 1.25, and the expansion ratio was 2.0 (there is no optimization for this ratios). We used gaseous ethylene and gaseous oxygen as the propellants. Oxidizer (O_2) mass flow rate is much larger than fuel (C_2H_4), then generally oxidizer plenum pressure is higher than fuel. Since oxidizer plenum characteristic length should be smaller, in this experiment the oxidizer (O_2) is injected from the center plenum whereas the fuel (C_2H_4) is injected at a larger outer radius.

Fig. 4.1 Schematic of annular detonation combustor (sectional view)

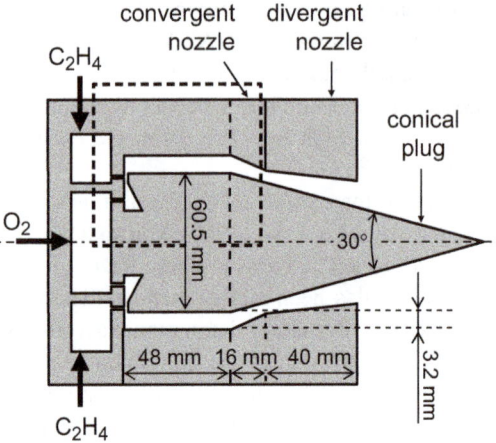

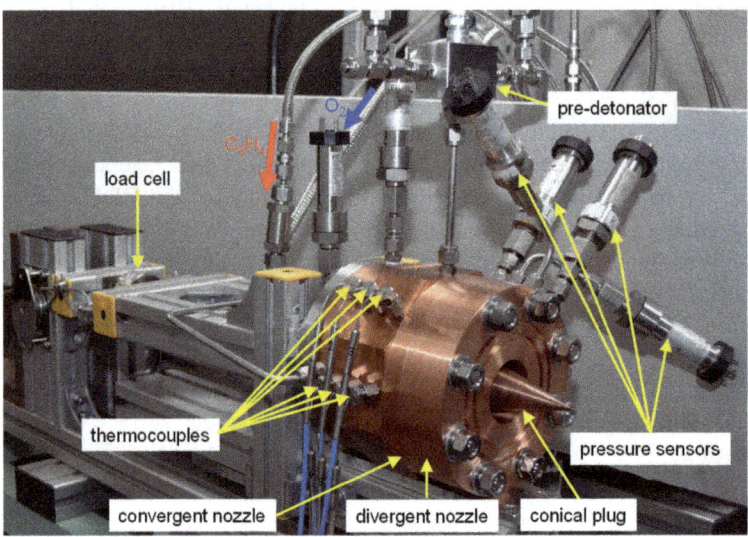

Fig. 4.2 Photograph of experimental apparatus of the RDE

Figure 4.2 shows photograph of the RDE experimental apparatus. We used pre-detonator tube on the side wall of the annular combustor. We also set the load cell on the bottom of the RDE system for measuring the thrust as shown in Fig. 4.2.

Figure. 4.3 is an enlarged view of the RDE. We used two types of injectors (type 1, type 2). Type 1 had 72 fuel injection holes of 0.66-mm diameter and a 1-mm wide oxidizer injection slot. In type 2, the 72 fuel injection holes were 0.50 mm in diameter, and the oxidizer injection slot was 0.3-mm wide. Pressure sensors were inserted in the fuel and oxidizer plenums. Each type had pressure ports (2.0 mm diameter), P_1, P_2, and we measured local static pressure. In type 2, there were pressure ports (0.5 mm diameter), P_{0_1}, P_{0_2}, located at bottom of the combustor to obtain combustor stagnation pressure.

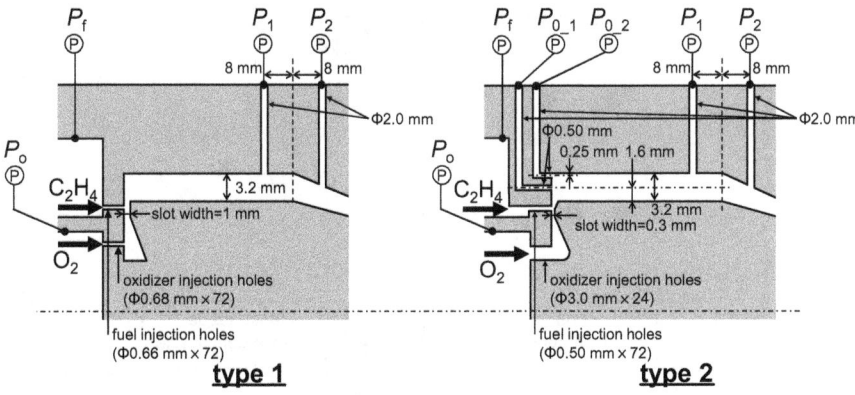

Fig. 4.3 Enlarged view of the RDE combustor: *left*, type 1; *right*, type 2

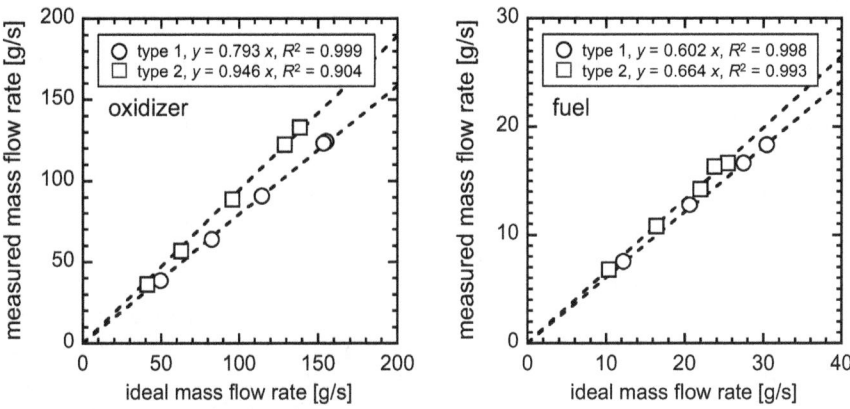

Fig. 4.4 Result of mass flow rate calibration: *left*, oxidizer; *right*, fuel

3 Experimental Results and Discussion for Combustion Chamber

We performed mass flow calibration by blowing down cold gas, fuel and oxidizer separately. The mass flow rate was adjusted by the initial tank pressure or the orifice diameter located in upper flow. Figure. 4.4 shows the results of mass flow rate calibration. Ideal mass flow rate, $\dot{m}_i$, was calculated from plenum pressure in choked condition at the injectors as follows:

$$\dot{m}_i = \frac{PA_{inj}}{\sqrt{RT}}\sqrt{\gamma\left(\frac{2}{\gamma+1}\right)^{\frac{\gamma+1}{\gamma-1}}} \tag{4.1}$$

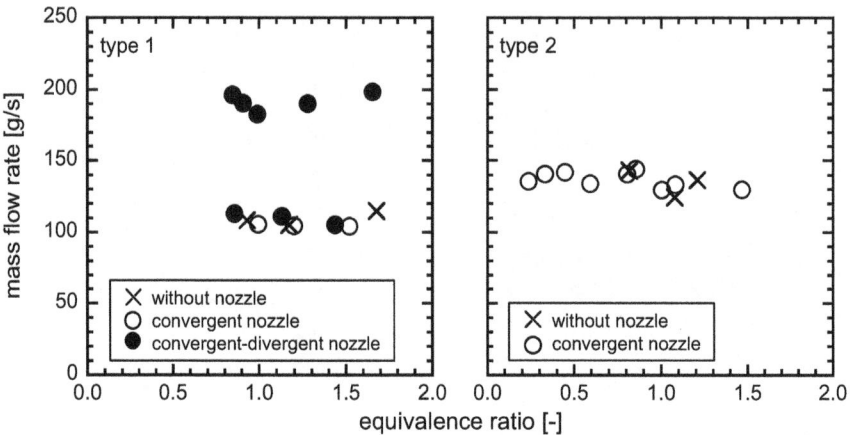

Fig. 4.5 Experimental conditions: *left*, type 1; *right*, type 2

Measured mass flow rate was determined from operation duration and tank mass decrease. The tank mass decrease was determined by an electronic balance or the internal tank pressure decrease. In the combustion test, we determined ideal mass flow rate from plenum pressure, and actual mass flow was obtained using the relationship in Fig. 4.4.

Figure 4.5 shows the experimental conditions. We changed mass flow rate (108 ± 4 g/s, 136 ± 6 g/s, 191 ± 6 g/s), equivalence ratio (0.24–1.68), injector (type 1, type 2), and nozzle (without, convergent, convergent-divergent).

Figure 4.6 presents the visualization result by high-speed camera (SA5, 150,000 fps). Intermittent bright waves were observed in type 1, and thrust was obtained and combustion was kept in the combustor. On the other hand 2 or 3 waves were observed in type 2. Wave propagation speed was 2040 m/s (84% of C-J speed) or 1750 m/s (72% of C-J speed). The lower speed than CJ wave maybe due to insufficient mixing and large wave curvature effect. Thus, injector shape influenced wave propagation mode. There are transient phenomena from two waves to three waves in the type 2 injector case shown in the right of Fig. 4.6. The new wave generates in the two-wave mode. It is due to mass flow rate disturbances at the starting process of the RDE operation. For simplicity, we show only the data points which can be clearly correspond to two or three wave mode.

Figure 4.7 shows the pressure and thrust history in the combustion test using the convergent nozzle. In the type 1 configuration, the injector is hole type only. Otherwise, in the type 2 configuration, the injector of the oxygen is slit type. The slit width is decreasing during the operation due to thermal deformation (expansion) of the inner wall of the RDE. Therefore the pressure in the oxygen plenum increases for the type 2 configuration but remains fairly constant for the type 1 configuration. Thrust was measured by a load cell and average thrust during the operation is drawn as dashed lines. The load cell used in this experiment was AIKOH DUD-200 K, and its maximum frequency response is 500 Hz. The average thrust was 142 N in type

4 Application of Detonation Waves to Rocket Engine Chamber 67

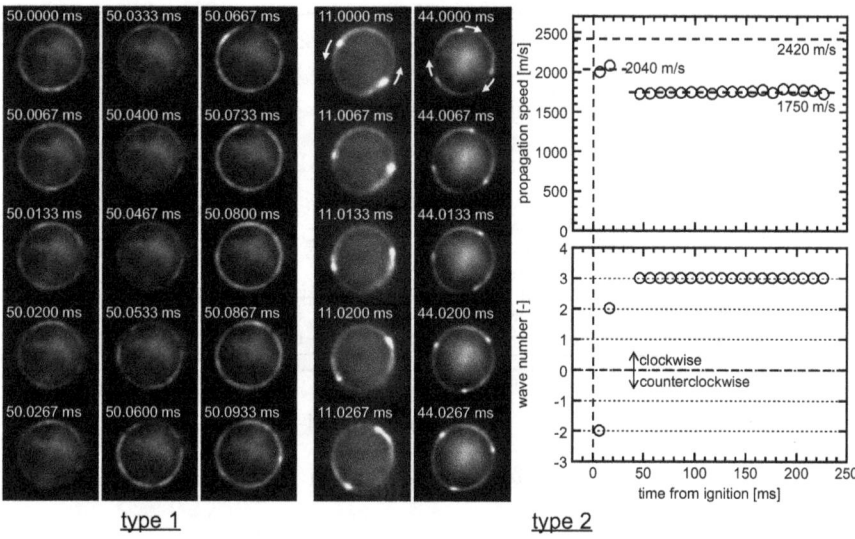

Fig. 4.6 Visualization by high-speed camera: *left*, type 1 ($\dot{m}$ =105 g/s, ER = 1.17, without nozzle); *right*, type 2 ($\dot{m}$ =125 g/s, ER = 1.08, without nozzle)

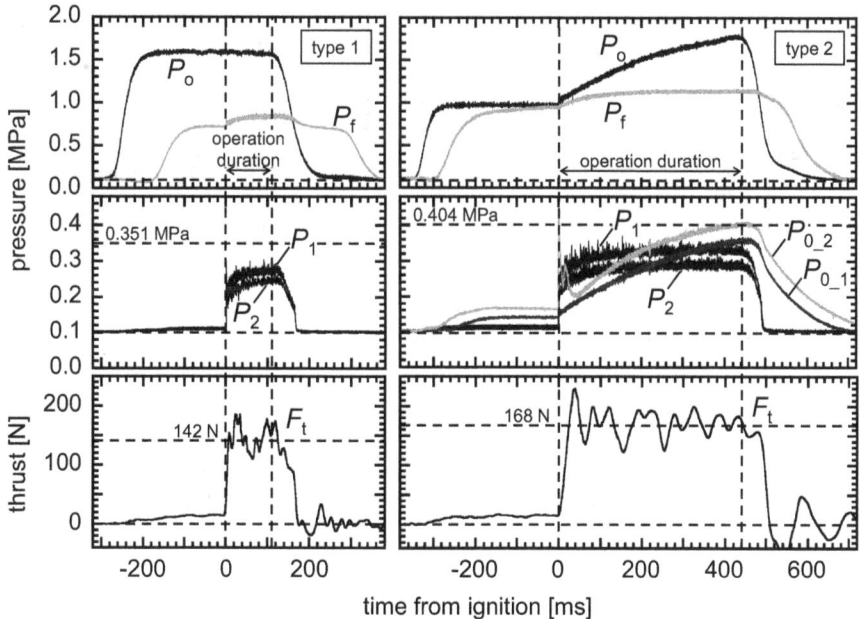

Fig. 4.7 Pressure and thrust history: *left*, type 1 ($\dot{m}$ =105 g/s, ER = 1.00, convergent nozzle); *right*, type 2 ($\dot{m}$ =134 g/s, ER = 0.60, convergent nozzle)

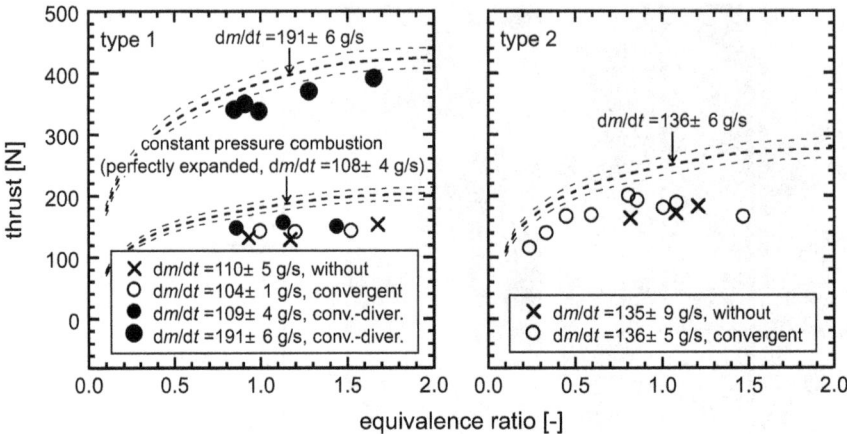

Fig. 4.8 Thrust vs. equivalence ratio: *left*, type 1; *right*, type 2

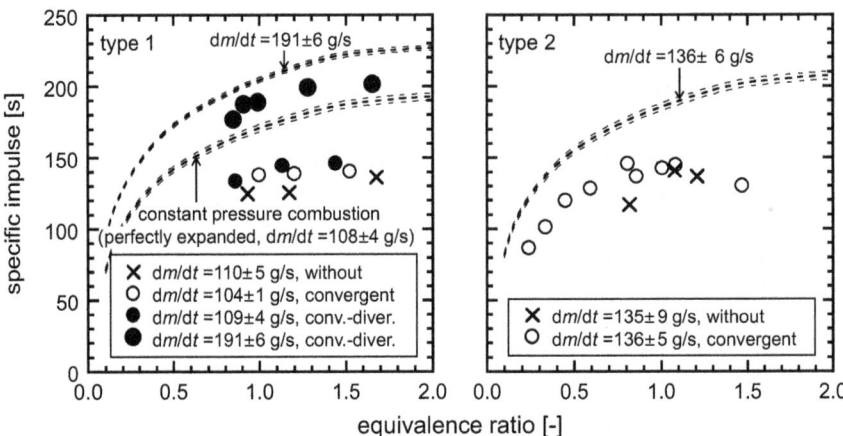

Fig. 4.9 Specific impulse vs. equivalence ratio: *left*, type 1; *right*, type 2

1, and 168 N in type 2. Mass flow rate was determined by using plenum pressure before the ignition.

Figure. 4.8 shows the thrust vs. equivalence ratio. Figure 4.9 shows the specific impulse vs. equivalence ratio. Specific impulse, I_{sp}, was defined as follows:

$$I_{sp} \equiv \frac{F_t}{\dot{m}g} \tag{4.2}$$

Dashed lines indicate the ideal value of the constant pressure combustion rocket engine. In calculating the value, the throat area was equal to the experimental apparatus, and the burned gas was perfectly expanded and frozen in the nozzle. The thrust was 115–391 N, the specific impulse was 87–201 s. By using convergent nozzle thrust increased approximately 10% compared with the combustor without nozzle. Thrust efficiency might increase because of higher combustor pressure. In convergent-divergent nozzle condition, thrust was 10–17% higher than the combustor without nozzle. In case of low mass flow rate (109 ± 4 g/s), overexpansion occurred. When the mass flow rate was 191 ± 6 g/s, we had already achieved 88–94% efficiency as compared to the ideal value so that burned gas was almost perfectly expanded.

In the RDE combustor, the flow is essentially unsteady state and three dimensional. However, in this experiment, experimental input (plenum pressure and mass flow rate) and experimental output (thrust of RDE) were kept almost steady state relative to the RDE unsteady phenomena, as shown in Figs. 4.7 and 4.8. Therefore, for simplicity, the following one or quasi-one steady-state flow in RDE combustor are assumed.

At first, we calculated the ideal combustor stagnation pressure. Gas constant, adiabatic flame temperature, and ratio of specific heat were determined by NASA-CEA. Ideal combustor stagnation pressure, P_{c_i}, was obtained as follows:

$$P_{c\_i} = \frac{\dot{m}\sqrt{RT_c}}{A_t\sqrt{\gamma\left(\frac{2}{\gamma+1}\right)^{\frac{\gamma+1}{\gamma-1}}}} \quad (4.3)$$

Next we obtained combustor stagnation pressure in the experiment. The Mach number at the P_1 port was determined as follows:

$$\frac{A}{A_t} = \frac{1}{M}\left\{\frac{(\gamma-1)M^2+2}{\gamma+1}\right\}^{\frac{\gamma+1}{2(\gamma-1)}} \quad (4.4)$$

The ratio of specific heat was determined by NASA-CEA and assumed to be constant. Measured combustor stagnation pressure, P_{c_m}, was obtained as follows:

$$P_{c\_m} = P_1\left(1+\frac{\gamma-1}{2}M^2\right)^{\frac{\gamma}{\gamma-1}} \quad (4.5)$$

The characteristic velocity, c^*, and characteristic-velocity efficiency, η_{c^*}, were then defined as follows:

$$c^* \equiv \frac{P_c A_t}{\dot{m}} \quad (4.6)$$

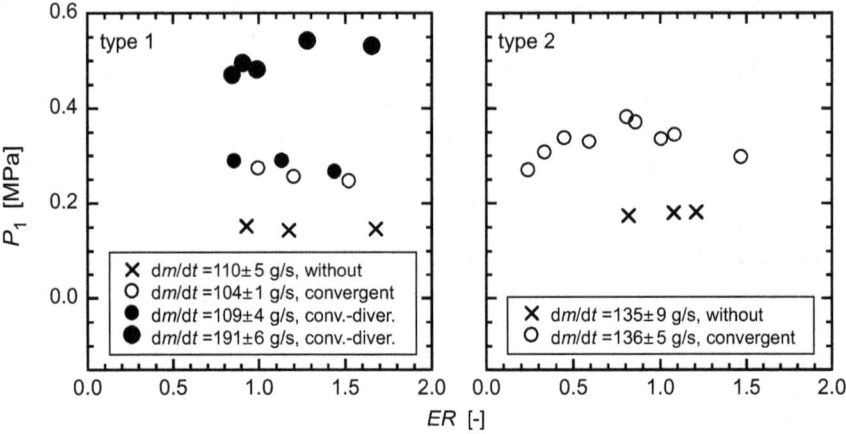

Fig. 4.10 Combustor pressure, P_1 vs. equivalence ratio: *left*, type 1; *right*, type 2

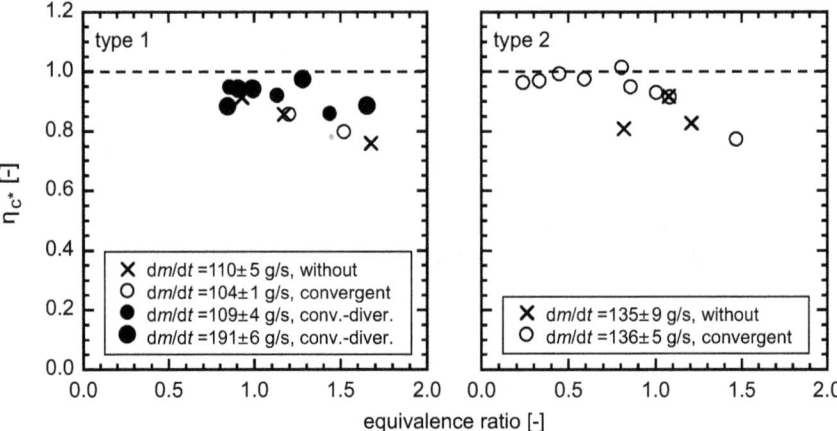

Fig. 4.11 c^* efficiency calculated from P_1 vs. equivalence ratio: *left*, type 1; *right*, type 2

$$\eta_{c^*} \equiv \frac{c^*_m}{c^*_i} = \frac{P_{c\_m}}{P_{c\_i}} \qquad (4.7)$$

Figure 4.10 shows the combustor pressure, P_1 vs. equivalence ratio, and Fig. 4.11 shows the c^* efficiency calculated from P_1 vs. the equivalence ratio. Pressure increase was obtained using either the convergent or convergent-divergent nozzle. We compare type 1 and 2 in Fig. 11; there was little difference in c^* efficiency. The injector shape influenced wave propagation mode, but the mode could not change the c^* efficiency. When we used the convergent or convergent-divergent nozzle and the equivalence ratio was less than 1.0, c^* efficiency was 88–101%. If both combustor pressure and oxidizer mass flow rate were high, there was a tendency for the

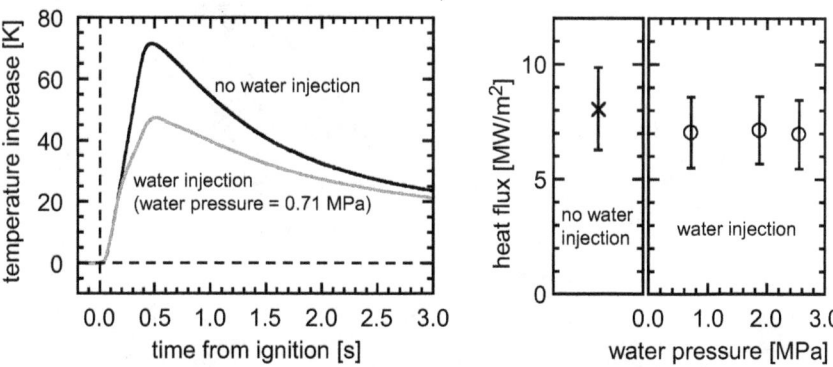

Fig. 4.12 Temperature increase history and heat flux vs. water pressure

efficiency to be high. Then as combustor pressure, P_{0_2}, asymptotically approached the ideal line in Fig. 4.7, we concluded that the combustor stagnation pressure could be directly measured at the bottom corner of the combustor.

Cooling tests by directly injecting water into the combustor through a hole of diameter of 0.5 mm was carried out for development of cooling system of a RDE. Figure 4.12 shows temperature increase history and heat flux versus water pressure. The temperature was measured by a sheath type thermocouple (0.5 mm diameter) which was inserted into a hole in the combustor outer. The distance between the bottom of the hole and the combustor outer surface was 2 mm. Combustion was kept during water injection, and local temperature increase rate was decreased. And we estimated time-spatial average heat flux into the ambient from the terminal temperature, and the heat flux was 8.1 ± 1.8 MW/m^2 with no water injection. There is not much of a difference between no water injection and water injection. The water injector hole is only one, and then the distribution of the water is not homogeneous. The water may be diffused to wider region by the unsteady RDE combustor flow than a conventional relatively steady combustor flow. These effect (small mass flow rate water spreads wider region) causes small heat flux difference between the two cases.

4 Experimental Results and Discussion for RDE System Sled Test

For demonstrating rocket RDE system, the sled test of this system was performed at Muroran Institute of Technology in Japan. Figures 4.13 and 4.14 show the pictures of the rocket RDE system for a sled test. As shown in Fig. 4.13, all the devices of the system were located on the flat aluminum alloy plate (1000 mm length and 30 mm width).

As shown in Fig. 4.14, the rocket RDE system was sledded on 5-inch rails of total length of 100 m. The rocket RDE system was supported by four wheels.

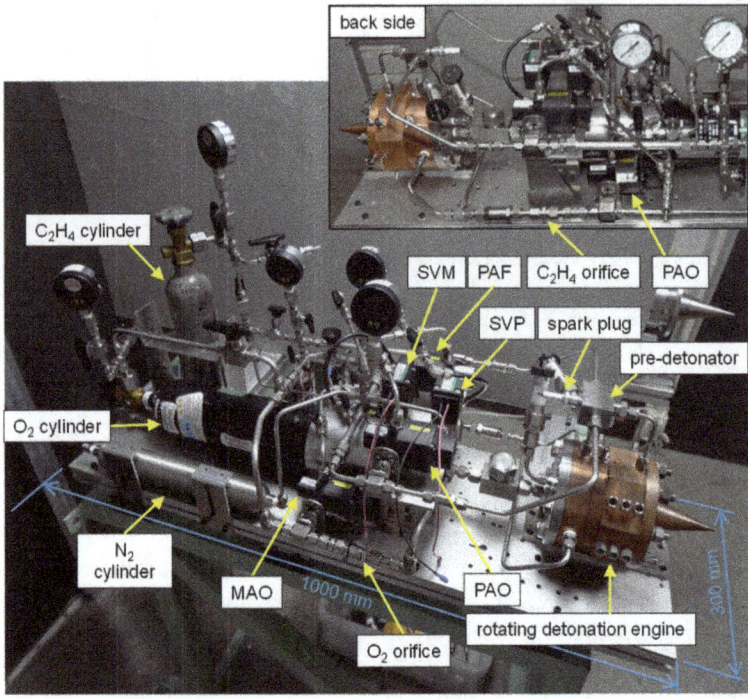

Fig. 4.13 Piping system picture of the rocket RDE system

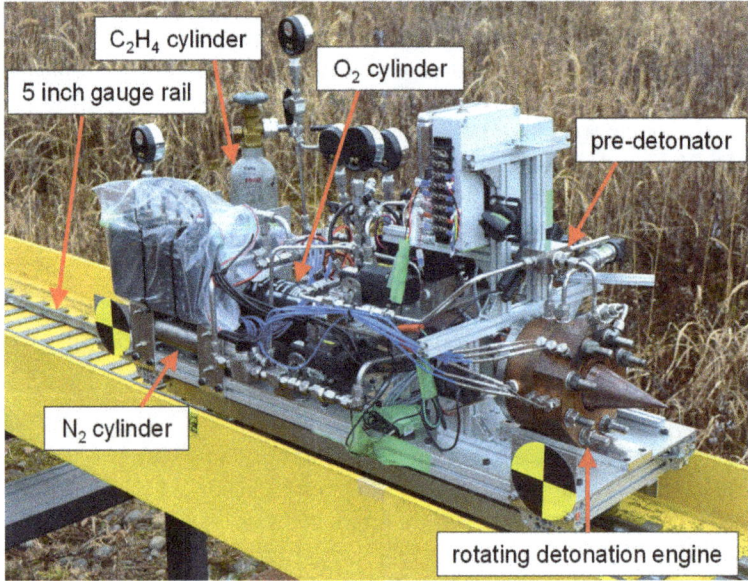

Fig. 4.14 Picture of the rocket RDE system for a sled test

4 Application of Detonation Waves to Rocket Engine Chamber

Fig. 4.15 The RDE sled test was successfully performed

Fig. 4.16 The acceleration and position of the system. The total time averaged acceleration was 2.94 = 2.78 + 0.16 (m/s^2)

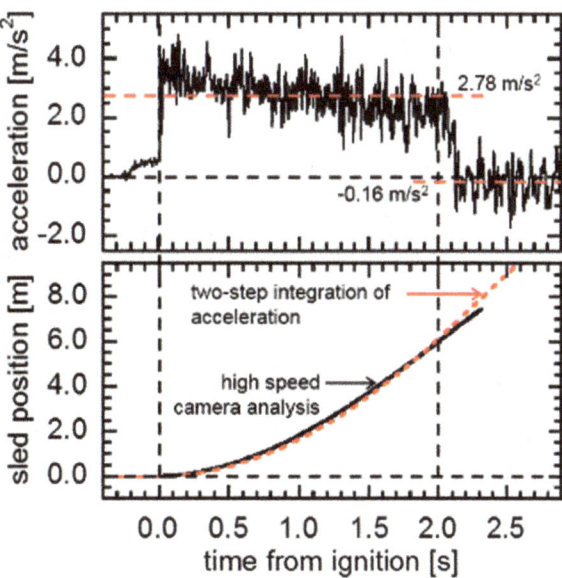

The rocket RDE sled test was successfully performed as shown in Fig. 4.15. The operation time was two seconds and the total sled distance was 70.22 m. As shown in Fig. 4.16, the time averaged acceleration was 2.78 m/s^2. But, time averaged deceleration by friction was estimated as −0.16 m/s^2 using observation of the deceleration after closing the engine thrust. Thus, total thrust can be estimated 2.93 m/s². As the total mass of the system was 58.3 kg, total time averaged thrust was 201 N. The time averaged mass flow rate was 143 g/s. The specific impulse was 144 s.

5 Conclusions

We performed combustion tests of RDE with convergent-divergent nozzle using gaseous ethylene and oxygen as the propellant, measured the combustor pressure, and determined the c^* efficiency. The efficiency was 88–100% when we used the convergent or convergent-divergent nozzle and the equivalence ratio was less than 1.0. The shape of the injector influenced wave propagation mode, but the mode did not change the c^* efficiency. We estimated time-spatial average heat flux from the terminal temperature, and the heat flux was 8.1 ± 1.8 MW/m² in no water injection. The rocket RDE sled test was successfully performed. The total mass of the rocket RDE system was 58.3 kg, total time averaged thrust was 201 N, the time averaged mass flow rate was 143 g/s, and the specific impulse was 144 s.

Acknowledgements The present RDE development was subsidized by a "Study on Innovative Detonation Propulsion Mechanism," Research-and-Development Grant Program (Engineering) from the Institute of Space and Astronautical Science, the Japan Aerospace Exploration Agency. The fundamental device development was subsidized by a Grant-in-Aid for Scientific Research (A), No. 24246137.

References

Bykovskii, F. A., Sergey, Z. A., & Vedernikov, E. F. (2006). Continuous spin detonations. *Journal of Propulsion and Power, 22*(6), 1204–1216.

Claflin, S. (2012). *Recent progress in continuous detonation engine development at Pratt & Whitney Rocketdyne*. Paper presented at International Workshop on Detonation for Propulsion 2012, Tsukuba, 3–5 September 2012.

Dubrovskii, A. V., Ivanov, V. S., & Frolov, S. M. (2015). Three-dimensional numerical simulation of the operation process in a continuous detonation combustor with separate feeding of hydrogen and air. *Russian Journal of Physical Chemistry B, 9*(1), 104–119.

Fievisohn, R. T., & Yu, K. H. (2017). Steady-state analysis of rotating detonation engine flowfields with the method of characteristics. *Journal of Propulsion and Power, 33*(1), 89–99.

Fotia, M. L., Kaemming, T. A., Hoke, J. L., and Schauer, F. (2016). *Thermodynamics modelling and the operation of rotating detonation engines at elevated inlet temperatures*. Paper presented at 2016 International Workshop on Detonation for Propulsion, Temasek Laboratories, National University of Singapore, 12–15 July 2016.

Frolov, S. M., Dubrovskii, A. V., & Ivanov, V. S. (2013). Three-dimensional numerical simulation of the operation of a rotating-detonation chamber with separate supply of fuel and oxidizer. *Russian Journal of Physical Chemistry B, 7*(1), 35–43.

Frolov, S. M., Aksenov, V. S., Gusev, P. A., Ivanov, V. S., Medvedev, S. N., & Shamshin, I. O. (2014). Experimental proof of the energy efficiency of the Zel'dovich thermodynamic cycle. *Doklady Physical Chemistry, 459*(2), 207–211.

Frolov, S. M., Aksenov, V. S., & Ivanov, V. S. (2015a). Experimental proof of Zel'dovich cycle efficiency gain over cycle with constant pressure combustion for hydrogen–oxygen fuel mixture. *International Journal of Hydrogen Energy, 40*(21), 6970–6975.

Frolov, S. M., Aksenov, V. S., Ivanov, V. S., & Shamshin, I. O. (2015b). Large-scale hydrogeneair continuous detonation combustor. *International Journal of Hydrogen Energy, 40*(3), 1616–1623.

Fujii, J., Kumazawa, Y., Matsuo, A., Nakagami, S., & Kasahara, J. (2017). Numerical investigation on velocity deficit of detonation wave in RDE chamber. *Proceedings of the Combustion Institute, 36*(2), 2665–2672.

Gawahara, K., Nakayama, H., Kasahara, J., Tomioka, S., & Hiraiwa, T. (2013). *Detonation engine development for reaction control systems of a spacecraft.* Paper presented at 49th AIAA/ASME/SAE/ASEE Joint Propulsion Conference & Exhibit and 11th International Energy Conversion Engineering Conference, AIAA-2013-3721, San Jose Convention Center, San Jose, July 15–17, 2013.

Heiser, W. H., & Pratt, D. T. (2002). Thermodynamic cycle analysis of pulse detonation engines. *Journal of Propulsion and Power, 18*(1), 68–76.

Hishida, M., Fujiwara, T., & Wolański, P. (2009). Fundamentals of rotating detonations. *Shock Waves, 19*(1), 1–10.

Hoke, J. L., Bradley, R. P., Brown, A. C., Litke, P. J., Stutrud, J. S., and Schauer, F. R. (2010). *Development of a pulse detonation engine for flight.* Paper presented at Symposium on Shock Waves in Japan (pp. 239–246).

Ishihara, K., Nishimura, J., Goto, K., Nakagami, S., Matsuoka, K., Kasahara, J., Matsuo, A., and Funaki, I. (2017). *Study on a long-time operation towards rotating detonation rocket engine flight demonstration.* Paper presented at SciTech 2017, 55th AIAA Aerospace Science Meeting, AIAA 2017–1062, Grapevine, Texas, USA, January 8–12, 2017.

Kailasanath, K. (2000). Review of propulsion applications of detonation waves. *AIAA Journal, 38*(9), 1698–1708.

Kailasanath, K. (2003). Recent developments in the research on pulse detonation engines. *AIAA Journal, 41*(2), 145–159.

Kasahara, J., Hasegawa, A., Nemoto, T., Yamaguchi, H., Yajima, T., & Kojima, T. (2007). Performance validation of a single-tube pulse detonation rocket system. *Journal of Propulsion and Power, 25*(1), 173–180.

Kato, Y., Ishihara, K., Matsuoka, K., Kasahara, J., Matsuo, A., and Funaki, I. (2016). *Study of combustion chamber characteristic length in rotating detonation engine with convergent-divergent nozzle.* Paper presented at 54th AIAA Aerospace Sciences Meeting, AIAA 2016–1406, San Diego, January 4–8, 2016.

Kindracki, J., Wolański, P., & Gut, Z. (2011). Experimental research on the rotating detonation in gaseous fuels–oxygen mixtures. *Shock Waves, 21*(2), 75–84.

Lu, F. K., & Braun, E. M. (2014). Rotating detonation wave propulsion: experimental challenges, modeling, and engine concepts. *Journal of Propulsion and Power, 30*(5), 1125–1142.

Matsuoka, K., Morozumi, T., Takagi, S., Kasahara, J., Matsuo, A., & Funaki, I. (2016). Flight validation of a rotary-valved four-cylinder pulse detonation rocket. *Journal of Propulsion and Power, 32*(2), 383–391.

Nakagami, S., Matsuoka, K., Kasahara, J., Kumazawa, Y., Fujii, J., Matsuo, A., & Funaki, I. (2017a). Experimental visualization of the structure of rotating detonation waves in a disk-shaped combustor. *Journal of Propulsion and Power, 33*(1), 80–88.

Nakagami, S., Matsuoka, K., Kasahara, J., Matsuo, A., & Funaki, I. (2017b). Experimental study of the structure of forward-tilting rotating detonation waves and highly maintained combustion chamber pressure in a disk-shaped combustor. *Proceedings of the Combustion Institute, 36*(2), 2673–2680.

Nakayama, H., Moriya, T., Kasahara, J., Matsuo, A., Sasamoto, Y., & Funaki, I. (2012). Stable detonation wave propagation in rectangular-cross-section curved channels. *Combustion and Flame, 159*(2), 859–869.

Nordeen, C. A. (2013). *Thermodynamics of a rotating detonation engine*, doctoral dissertations, University of Connecticut, 2013.

Naples, A., Hoke, J., Battelle, R., Wagner, M., and Schauer, F. (2017). *Rotating detonation engine implementation into an open-loop T63 gas turbine engine*. Paper presented at 55th AIAA Aerospace Sciences Meeting, AIAA 2017–1747, Grapevine, 9–13 January, 2017.

Paxson, D. E. and Naples, A. (2017). *Numerical and analytical assessment of a coupled rotating detonation engine and turbine experiment*. Paper presented at 55th AIAA Aerospace Sciences Meeting, AIAA 2017–1746, Grapevine, 9–13 January, 2017.

Schwer, D., & Kailasanath, K. (2011). Numerical investigation of the physics of rotating-detonation-engines. *Proceedings of the Combustion Institute, 33*(2), 2195–2202.

Schwer, D., & Kailasanath, K. (2013). Fluid dynamics of rotating detonation engines with hydrogen and hydrocarbon fuels. *Proceedings of the Combustion Institute, 34*(2), 1991–1998.

Stechmann, D., Heister, S. D., and Sardeshmukh, S. (2017). *High-pressure rotating detonation engine testing and flameholding analysis with hydrogen and natural gas*. Paper presented at 55th AIAA Aerospace Sciences Meeting, AIAA 2017–1931, Grapevine, 9–13 January, 2017.

Theuerkauf, S. W., Schauer, F. R., Anthony, R., and Hoke, J. L. (2015). *Experimental characterization of high-frequency heat flux in a rotating detonation engine*. Paper presented at 53rd AIAA Aerospace Science Meeting, AIAA 2015–1603, Kissimmee, 5–9 January, 2015.

Talley, D. G., & Coy, E. B. (2002). Constant volume limit of pulsed propulsion for a constant γ ideal gas. *Journal of Propulsion and Power, 18*(2), 400–406.

Voitsekhovskii, B. V. (1960). Stationary spin detonation. Soviet. *Journal of Applied Mechanics and Technical Physics, 3*, 157–164.

Voitsekhovskii, B. V., Mitrofanov, V. V., & Topchian, M. E. (1967). Investigation of the structure of detonation waves in gases. *Symposium (International) on Combustion, 12*(1), 829–837.

Wolan'ski, P. (2013). Detonative propulsion. *Proceedings of the Combustion Institute, 34*(1), 125–158.

Wu, Y., Ma, F., & Yang, V. (2003). System performance and thermodynamic cycle analysis of air-breathing pulse detonation engines. *Journal of Propulsion and Power, 19*(4), 556–567.

Chapter 5
Numerical Simulation on Rotating Detonation Engine: Effects of Higher-Order Scheme

Nobuyuki Tsuboi, Makoto Asahara, Takayuki Kojima, and A. Koichi Hayashi

Abstract The implementation and simulations of the robust weighted compact nonlinear scheme (RWCNS) for the two-dimensional rotating detonation engine are performed using the detailed chemistry model. The comparison of the MUSCL and the 5th-order RWCNS (WCNS5MN) indicates that the shock front and the contact surface for the WCNS5MN can be improved with the better resolution than those for the MUSCL and that both rotating velocities are approximately 97% of the CJ value. I_{sp} for the WCNS5MN is approximately 5 s larger than I_{sp} for the MUSCL because the mass flow rates for the WCNS5MN are 2–4% smaller than those for the MUSCL.

1 Introduction

Detonation is a strong explosion phenomenon that propagates at supersonic speed (e.g., a hydrogen/air detonation travels at supersonic). Detonation simulation requires to predict (i) a large discontinuous wavefront, such as a shock wave and a contact surface, and (ii) rapid chemical reaction near a combustion front. Because the detonation consists of strong explosion due to triple point collisions, the numerical scheme is required to be robust as well as numerical accuracy.

In the 1990s, the simulation of detonation was performed by the Harten-Yee total variation diminishing (TVD) scheme (Yee 1987) and the Godunov scheme

---

N. Tsuboi (✉)
Department of Mechanical and Control Engineering, Kyushu Institute of Technology, Kitakyushu, Fukuoka, Japan
e-mail: tsuboi@mech.kyutech.ac.jp

M. Asahara
Department of Mechanical Engineering, Gifu University, Yanagido, Gifu, Japan

T. Kojima
Chofu Aerospace Center, Japan Aerospace Exploration Agency, Chofu-shi, Tokyo, Japan

A. Koichi Hayashi
Department of Mechanical Engineering, Aoyama Gakuin University, Sagamihara, Kanagawa, Japan

© Springer International Publishing AG 2018
J.-M. Li et al. (eds.), *Detonation Control for Propulsion*, Shock Wave and High Pressure Phenomena, https://doi.org/10.1007/978-3-319-68906-7_5

(Godunov 1959). In both cases, spatial accuracy reduces first-order near the discontinuities to add large numerical viscosity. In the 2000s, various numerical methods, such as the HLLE scheme (Einfeldt 1988), HLLC scheme (Batten et al. 1997), AUSMDV scheme (Wada et al. 1994), the weighted essentially non-oscillatory (WENO) scheme (Jiang et al. 1996), and the weighted compact nonlinear scheme (WCNS) (Deng and Zhang 2000) were applied to the simulation of detonation with the development of high-performance computers and computational methods.

Wintenberger and Shepherd (2003) simulated a pulse detonation engine (PDE) with the HLLE scheme. When we neglect the fine details of the phenomenon being simulated, a robust numerical method such as the HLLE scheme is effective. On the other hand, Togashi et al. (2009); Asahara et al. (2012), and Kurosaka and Tsuboi (2014) used the AUSMDV scheme for numerical simulation of detonation and showed the 2D and 3D wavefront structure in detail. Hu et al. (2005) and Henrick et al. (2006) showed the detonation front structure in further detail using a fifth-order WENO scheme. Schwer and Kailasanath (2013); Zhou and Wang (2012) simulated a rotating detonation engine (RDE) with the fifth-order WENO scheme.

Although WENO scheme is robust and superior scheme to simulate for strong shock waves with higher-order spatial accuracy, it is difficult to maintain a uniform freestream on an arbitrary grid system (Nonomura et al. 2010). WENO scheme is also probably not suitable for the simulation using a complex and crusterd grid(Nonomura et al. 2010). These problems do not arise in the detonation simulations because most of the detonation simulations with WENO scheme use a orthogonal grid system. Furthermore, large numerical dissipation affects the detonation structure because the WENO scheme uses the Lax-Friedrich scheme for numerical flux evaluation.

The authors simulated the multi-dimensional hydrogen-fueled detonations (Asahara et al. 2012; Tsuboi et al. 2002, 2007, 2009, 2008a, b, 2013a, b, 2017; Niibo et al. 2016; Eto et al. 2005; Tsuboi and Koichi Hayashi 2007; Eto et al. 2016) and hydrocarbon-fueled detonation (Araki et al. 2016) in a micro-scale size, which is the order of a millimeter and centimeter. However, three-dimensional simulations on an experimental scale are difficult because of a huge grid points and computational time. The authors pay their attention to a high-resolution scheme that can provide enough resolution for a simulation with a reduced number of grid points. The present study adopts the weighted compact nonlinear scheme (WCNS) (Deng and Zhang 2000; Nonomura et al. 2010; Zhang et al. 2008; Nonomura and Fujii 2009) in our in-house detonation simulation program, where the WCNS allows the AUSMDV scheme to be used. Some recent simulation results of the detonation using the fifth- and seventh-order WCNS scheme are reported by Iida et al. (2014) and Niibo et al. (2016). Nonomura et al. (2010) reported that the WCNS has three advantages: (i) the WCNS can select a wide variety of flux evaluations, (ii) the WCNS provides high resolution, and (iii) the WCNS can maintain a uniform flow on generalized curvilinear coordinates. Although the WCNS has such the advantages comparing with the WENO, the WCNS is sometimes failed to simulate the strong explosion just after the triple point collision in the detonation. The robust weighted compact nonlinear scheme (RWCNS) developed by Nonomura and Fujii (2013), which improves robustness near a large discontinuity, succeeds to simulate the detonation with a detailed chemical reaction model (Niibo et al. 2016; Iida et al. 2014).

The present research discusses to improve the RWCNS for the reactive compressible fluid codes with multi-component gases and comparison of the two-dimensional numerical results between the MUSCL and the WCNS on the rotating detonation engine.

2 Governing Equations

Simulations of detonations are generally performed using the Euler equations, which include the equations for various chemical species associated with the relevant chemical reactions. As far as species diffusion speed, the effect of gas molecular diffusion on the momentum and energy equations is low because the propagation speed of detonation exceeds $3 < Ma$. Based on a consideration of the small effect of the boundary layer behind the detonation front, we consider the possibility that the effect of viscosity on detonation is small. On the other hand, it is easy to compare between numerical schemes in the inviscid flow because the amount of the numerical viscosity, which depend on the choice of the numerical scheme, affects the resolution in the flowfields. For the above reasons, this study applies the compressible Euler equations to the modeling of shock waves and detonation.

The governing equations (the compressible Euler equations) consist of the mass conservation law of gases, the momentum conservation law, the energy conservation law, and the conservation law of each species. In the case of the two-dimensional Cartesian coordinate system, the compressible Euler equations are

$$\frac{\partial \mathbf{Q}}{\partial t} + \frac{\partial \mathbf{E}}{\partial x} + \frac{\partial \mathbf{F}}{\partial y} = \mathbf{S} \tag{5.1}$$

$$\mathbf{Q} = \begin{pmatrix} \rho \\ \rho u \\ \rho v \\ e \\ \rho_1 \\ \vdots \\ \rho_N \end{pmatrix}, \mathbf{E} = \begin{pmatrix} \rho u \\ \rho u^2 + p \\ \rho uv \\ (e+p)u \\ \rho_1 u \\ \vdots \\ \rho_N u \end{pmatrix}, \mathbf{F} = \begin{pmatrix} \rho v \\ \rho uv \\ \rho v^2 + p \\ (e+p)v \\ \rho_1 v \\ \vdots \\ \rho_N v \end{pmatrix}, \mathbf{S} = \begin{pmatrix} 0 \\ 0 \\ 0 \\ 0 \\ \dot{\omega}_1 \\ \vdots \\ \dot{\omega}_N \end{pmatrix}, \tag{5.2}$$

where ρ is the density in kg/m³, u and v are the velocity in the x and y directions in m/s, e is the total energy per specific volume in J/m³, i is the index of the chemical species ($i = 1, 2, \cdots, N$), N is the total number of species, ρ_i is the density of the ith species in kg/m³, p is the pressure in Pa, and ω_i is the production rate of the ith species by the chemical reaction in kg/(m³ · s), respectively. The total energy per specific volume, e, is defined as

$$e = \sum_{i=1}^{N} \rho_i h_i - p + \frac{1}{2}\rho(u^2 + v^2), \qquad (5.3)$$

where h_i is the specific enthalpy of the ith species in J/kg.

An additional equation is needed to combine with Eq. (5.1) to form a closed system. We assume a thermally perfect gas and the equation of state of which is given by

$$p = \rho \bar{R} T = \sum_{i=1}^{N} \rho_i R_i T = \sum_{i=1}^{N} \rho_i \frac{R}{W_i} T, \qquad (5.4)$$

where $\bar{R}$ is the gas mixture constant in J/(kg · K), R is the universal gas constant in J/(kmol · K), T is the temperature in K, and R_i is the gas constant for the ith species in J/(kg · K), which is given by

$$R_i = \frac{R}{W_i} \qquad (5.5)$$

where W_i is the molecular weight in kg/kmol. The detonation simulation is calculated using the system of equations formed by the compressible Euler equations (Eq. 5.1) and the equation of state (Eq. 5.4).

The specific enthalpy of the ith species in Eq. (5.6) is given by a fifth-order polynomial function of temperature,

$$\frac{h_i}{R_i T} = a_{1i} + a_{2i}T + a_{3i}T^2 + a_{4i}T^3 + a_{5i}T^4 + \frac{a_{6i}}{T} \qquad (5.6)$$

In the same way, the specific heat at constant pressure of the ith species, $C_{p,i}$, is defined by a fourth-order polynomial function of the temperature,

$$\frac{C_{p,i}}{R_i} = a_{1i} + a_{2i}T + a_{3i}T^2 + a_{4i}T^3 + a_{5i}T^4 \qquad (5.7)$$

where a_{1i}, a_{2i}, a_{3i}, a_{4i}, a_{5i}, and a_{6i} are the coefficients calculated from the data in the JANAF tables (Stull and Prophet 1971). The specific heat ratio γ is written as

$$\gamma = \frac{\bar{C}_p}{\bar{C}_v} = \frac{1}{1 - \bar{R}\bar{C}_p}. \qquad (5.8)$$

The gas mixture constant, $\bar{R}$, the gas mixture specific heat at constant pressure, $\bar{C}_p$, and the gas mixture specific enthalpy, $\bar{h}_i$, are defined as

$$\bar{R} = \sum_{i=1}^{N} Y_i R_i, \bar{C}_p = \sum_{i=1}^{N} Y_i C_{p,i}, \bar{h} = \sum_{i=1}^{N} Y_i h_i, \quad (5.9)$$

where

$$Y_i = \frac{\rho_i}{\rho}. \quad (5.10)$$

The speed of sound, c, is calculated from the following frozen speed of sound:

$$c^2 = p_\rho + \sum_{i=1}^{N} Y_i p_{\rho_i} + p_e \left(H - u^2 - v^2\right), \quad (5.11)$$

where

$$p_\rho = \frac{\partial p}{\partial \rho}, p_{\rho_i} = \frac{\partial p}{\partial \rho_i}, p_e = \frac{\partial p}{\partial e}, H = \bar{h} + \frac{1}{2}\left(u^2 + v^2\right). \quad (5.12)$$

H in Eq. (5.12) is the total enthalpy.

3 Chemical Kinetic Model

One- and Two-Step Reaction Model

The most simplified reaction model used for detonation calculation is one-step reaction model. This model describes the combustion with a heat release Q behind a shock wave by using the reaction progress parameter Z which varies continuously from 1 to 0. On the other hand, the two-step model was developed by Korobeinikov et al. (1972) having a structure of two sequences; the induction reaction period and heat release period. This model applies for the conservation equation using an induction reaction progress parameter α and recombination reaction progress parameter β instead of the conservation equation of species, where α and β are unity at the initial state before reaction happens. The recombination reaction starts after the induction reaction is over and α becomes zero. K_j, n_j, l_j, m_i, E_i (i = 1, 2 and j = 1,2,3), and Q are the constants which are obtained empirically. The following Eqs. (5.13) and (5.14) are used instead of the species conservation equation in Eq. (5.1).

$$\frac{\partial \rho \alpha}{\partial t} + \frac{\partial \rho \alpha u}{\partial x} + \frac{\partial \rho \alpha v}{\partial y} = \rho \omega_\alpha \qquad (5.13)$$

$$\frac{\partial \rho \beta}{\partial t} + \frac{\partial \rho \beta u}{\partial x} + \frac{\partial \rho \beta v}{\partial y} = \rho \omega_\beta \qquad (5.14)$$

$$\omega_\alpha \equiv \frac{d\alpha}{dt} = -\frac{1}{\tau_{ind}} = -k_1 p^{n_1} \rho^{l_1} \exp\left(-\frac{E_1}{RT}\right) \qquad (5.15)$$

$$\omega_\beta \equiv \frac{d\beta}{dt} = \begin{cases} -k_2 p^{n_2} \rho^{l_2} \beta^{m_2} \exp\left(-\frac{E_2}{RT}\right) + \\ k_3 p^{n_3} \rho^{l_3} (1-\beta)^{m_3} \exp\left(-\frac{E_2}{RT}\right), \alpha \leq 0 \\ 0, \alpha > 0 \end{cases} \qquad (5.16)$$

It is not necessary, in the case that the mixture is assumed as a simple ideal gas, to add a change in the energy conservation equations if the heat release Q is put in the enthalpy and energy equations as follows:

$$h = c_v T + \beta Q, e = \rho h - p + \frac{1}{2}\rho(u^2 + v^2) \qquad (5.17)$$

where the chemical reaction is not stiff using such two-step mechanism and the equations are integrated explicitly.

One- and Two-Step Reaction Model

In the case using a detailed chemical reaction mechanism for the small mechanism of hydrogen/oxygen reaction, the intermediate species such as H, O, OH, HO_2, and H_2O_2 are necessary for the calculation besides H_2, O_2, and H_2O. In this case, the chemical species are eight and their elementary reactions are about twenty. Two-dimensional calculation provides triple calculation time comparing with a non-reacting flow case because of five governing equations and nine more species equations. Since the equation system is stiff with a detailed reaction model, its production term is integrated implicitly and since heat capacities of species are a function of temperature, the calculation time becomes totally at least more than three times comparing with that without the reaction term.

As far as detonation propagating through a hydrogen/air mixture, it is necessary to analyze it using a reaction mechanism with nitrogen, but mostly nitrogen is considered as a third body because the nitrogen related reactions are rather slow comparing with the oxyhydrogen reactions. As for the hydrogen/air reaction model used in detonation calculation, Oran et al. (1979) proposed eight species and 43 elementary reactions in 1979. They also proposed eight species and 24 elementary reactions (1982) in 1982. Shepherd (1986) also proposed in 1986. He studied the relationship between calculated reaction zone length and measured cell size. This reaction model has a set of 23 reactions and 11 species (H_2, O_2, O, H, OH, H_2O, HO_2, H_2O_2, N_2, CO_2, CO).

Recent several detailed chemical kinetic mechanisms of hydrogen combustion have been developed and are being updated by many researchers (Petersen and Hanson 1999; Mueller et al. 1999; Li et al. 2004; O'Conaire et al. 2004; Konnov 2008: Davis et al. 2005; Saxena and Williams 2006; Shimzu et al. 2011). These models have been validated using a wide range of measurements and were generally found to agree with experimental data, including ignition delay times with shock tubes, reaction behavior in flow reactors, and laminar flame speeds. Rate constants and third-body efficiencies for many elementary reactions seem to be evident in hydrogen/oxygen systems. However, determining some rate constants characterized by high sensitivity at high pressures has remained a challenge.

Although predictions of those models agree quite well with each other and with the experimental data of ignition delay times and flame speeds at pressures lower than 10 atm, substantial differences are observed between recent experimental data of high-pressure mass burning rates and model predictions, as well as among the model predictions themselves. Different pressure dependencies of mass burning rates above 10 atm in different kinetic models result from using different rate constants in these models for HO_2 reactions, especially for $H + HO_2$ and $OH + HO_2$ reactions. The rate constants for the reaction $H + HO_2$ involving different product channels were found to be very important for the prediction of high-pressure combustion characteristics. In order to obtain better performance of the model prediction for the high-pressure combustion of H_2, UT-JAXA model (Shimzu et al. 2011) adopts more precise values of the rate constants for the following reactions:$H + OH + M = H_2O + M$, $O + OH + M = HO_2 + M$, channel-specific rate constants for $H + HO_2$, and the temperature dependence of the $OH + HO_2 = H_2O + O_2$ reaction.

The production rate of each species relevant to the two-body reaction, ω_i, in Eq. (5.1) is given by

$$\dot{\omega}_i = W_i \sum_{k=1}^{K} \left(v_{ik}^{''} - v_{ik}^{'} \right) \left\{ k_{f,k} \prod_{i=1}^{N} \left(c_{\chi_i,k} \right)^{v_{ik}^{'}} - k_{b,k} \prod_{i=1}^{N} \left(c_{\chi_i,k} \right)^{v_{ik}^{''}} \right\} \tag{5.18}$$

where χ_i is the chemical symbol for the ith species (e.g., hydrogen: $\chi = H_2$), $v_i^{'}$ is the stoichiometric coefficient of the reactants for the ith species, $v_{ik}^{''}$ is the stoichiometric coefficient of the products for the ith species, the subscript k is the index of the elementary reactions ($k = 1, 2, \ldots, K$), K is the total number of elementary reactions, $k_{f,k}$ is the forward specific reaction-rate constant of the kth elementary reaction, and $k_{b,k}$ is the backward specific reaction-rate constant for the kth elementary reaction, respectively. The forward specific reaction-rate constant of the kth elementary reaction is given by the modified Arrhenius rate law:

$$k_{f,k} = A_k T^{n_k} \exp\left(-\frac{E_{a,k}}{RT}\right) \quad (5.19)$$

where A_k is the frequency factor of the kth elementary reaction, n_k is the exponent that determines the power-law relationship between the temperature and the forward specific reaction-rate constant, and $E_{a,k}$ is the activation energy per unit mass of the kth elementary reaction in cal/mol. Values for A_k, n_k, and $E_{a,k}$ are provided by the detailed reaction model by UT-JAXA model (Shimzu et al. 2011). $k_{b,k}$, the backward reaction rate constant of kth reaction which is related with the forward reaction rate constant is given as follows

$$k_{b,k} = k_{f,k} / K_{c_k}, \quad (5.20)$$

where K_{c_k} is the concentration equilibrium constant of kth reaction. When K_{p_k} is the pressure equilibrium constant,

$$K_{c_k} = K_{p_k}\left(\frac{p_{atm}}{RT}\right)^{\sum_{i=1}^{N}(v_{ik}'' - v_{ik}')} \quad (5.21)$$

where $p_{atm} = 1(atm)$ and R is the universal gas constant with an unit of T/P_{atm}. K_{p_k} is shown in the next Eq. (5.22) which is related with enthalpy and standard entropy.

$$K_{p_k} = \exp\left[\sum_{i=1}^{N}\left\{(v_{ik}'' - v_{ik}')\frac{s_i^0}{R_i}\right\} - \sum_{i=1}^{N}\left\{(v_{ik}'' - v_{ik}')\frac{h_i}{R_i T}\right\}\right] \quad (5.22)$$

The frequency of three-body reactions is promoted by the fact that the third body (that is, the third molecule) absorbs energy from the other two molecules. By multiplying the production rate of the two-body reaction, given in Eq. (5.18), by the mole concentration of the third body, C_M, the production rate of the three-body reaction is obtained to be

$$\dot{\omega}_i = W_i C_M \sum_{k=1}^{K}(v_{ik}'' - v_{ik}')\left\{k_{f,k}\prod_{i=1}^{N}(c_{\chi_i,k})^{v_{ik}'} - k_{b,k}\prod_{i=1}^{N}(c_{\chi_i,k})^{v_{ik}''}\right\} \quad (5.23)$$

The mole concentration of the third body, C_M, is given by

$$C_M = \sum_{i=1}^{N}(\alpha_{ik} c_{\chi_i}), \quad (5.24)$$

where α_{ik} is the collision coefficient of the third body for the ith species. The values of these collision coefficients are indicated by the UT-JAXA detailed reaction model (Shimzu et al. 2011).

4 Numerical Methods

In the detonation simulation in this study, an explicit method is used to solve the convective term and a point implicit method is applied to solve the source term of the chemical reaction, respectively. For the time integration, a third-order total validation diminishing Runge-Kutta scheme (TVDRK) (Gottlieb and Shu 1998) is used. The convective term is calculated by the AUSMDV scheme using flux evaluation with conservative variables that are interpolated at high orders by the WCNS. In the source term, a Crank-Nicholson type point implicit method is used. The inversion of Jacobian matrix in the source term is carried out by Gauss-Jordan elimination (Odlyzko 1985). The details are shown as follows.

Time Integration

In order to solve the governing equations of reactive flow in Eq. (5.1), a splitting technique into two parts is applied to treat for the difference of chemical kinetic time scale and fluid dynamic time scale. The partial equation for the fluid dynamic is solved by the TVDRK and the ordinary differential equation for the chemical kinetics is solved by the point implicit method. The detail of the procedure is shown as follows:

$$\mathbf{Q}^{n+1} = L_{PDE}^{\Delta t} L_{ODE}^{\Delta t} \mathbf{Q}^n \tag{5.25}$$

$$L_{PDE}^{\Delta t} : \frac{\partial \mathbf{Q}}{\partial t} + \frac{\partial \mathbf{E}}{\partial x} + \frac{\partial \mathbf{F}}{\partial y} = 0 \tag{5.26}$$

$$L_{ODE}^{\Delta t} : \frac{d\mathbf{Q}}{dt} = \mathbf{S} \tag{5.27}$$

Because of the superposition of the solution obtained from each differential operator, Eq. (5.25) corresponds to solving the following equations:

$$\mathbf{Q}^* = \mathbf{Q}^n + \Delta t L_{PDE}^{\Delta t}\left(\mathbf{Q}^n\right) \tag{5.28}$$

$$\mathbf{Q}^{n+1} = \mathbf{Q}^* + \Delta t L_{ODE}^{\Delta t}\left(\mathbf{Q}^*\right)$$

Here, Eq. (5.26), which serves as the partial differential equation for fluids, is applied to a following three-stage third-order TVDRK:

$$\mathbf{Q}^{(1)} = \mathbf{Q}^n + \Delta t L\left(\mathbf{Q}^n\right)$$

$$\mathbf{Q}^{(2)} = \frac{3}{4}\mathbf{Q}^n + \frac{1}{4}\mathbf{Q}^{(1)} + \frac{1}{4}\Delta t L\left(\mathbf{Q}^{(1)}\right) \quad (5.29)$$

$$\mathbf{Q}^* = \frac{1}{3}\mathbf{Q}^n + \frac{2}{3}\mathbf{Q}^{(2)} + \frac{2}{3}\Delta t L\left(\mathbf{Q}^{(2)}\right)$$

These equations yield conservative variables that represent the solution of the differential equation for the fluid at $t + \Delta t$.

Spatial Discretization Method

The method of evaluating the spatial differential term ($\partial \mathbf{E}/\partial x$) in Eq. (5.26) is shown below along the computational procedure in the simulation code. In this study, the WCNS is implemented to provide nonlinear interpolation to improve the accuracy of the simulation, and a robust linear difference scheme suggested by Nonomura and Fujii (2013) is used. The present schemes are summarized in Table. 5.1.

Nonlinear Interpolation

Interpolation is a method of constructing new data points within the range of a discrete set of known data points. In this case, the conservative variables Q_j on the grid point j are used to calculate $Q^L_{j+1/2}$ and $Q^R_{j+1/2}$, i.e. on the left and right sides of the cell interface in Fig. 5.1.

The WCNS is the nonlinear interpolation method used to determine the value at the computational cell interface, which is needed to evaluate the numerical flux. The WCNS uses the value at the node point. A previous study of the WCNS (Nonomura et al. 2012) indicated that the interpolation by using primitive variables is robust against the oscillation in the vicinity of the interference, where gas species vary drastically, and that the interpolation by conservative variables is robust to the

Table 5.1 Simulation methods for present scheme

	Nonlinear interpolation	Flux evaluation	Linear Interpolation
MUSCL	2nd-order MUSCL	AUSMDV	2nd-order central difference with minmod limiter[49]
WCNS5MN6	5th-order RWCNS	AUSMDV	Eq. (5.47) with 6th-order coefs

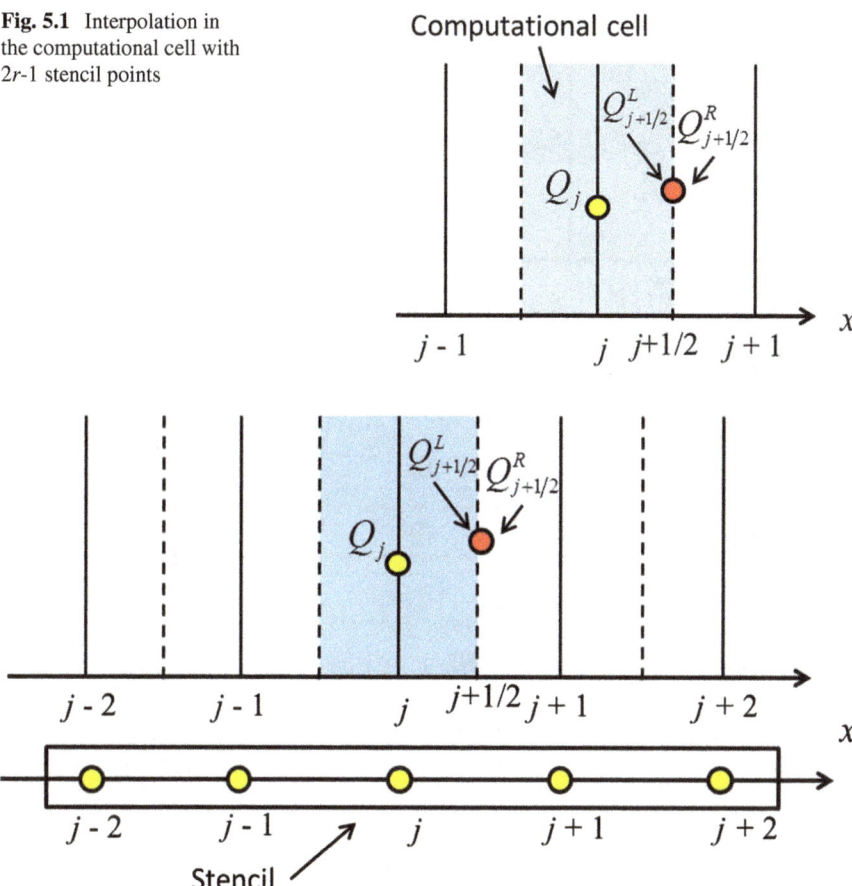

Fig. 5.1 Interpolation in the computational cell with 2r-1 stencil points

Fig. 5.2 Stencil at the point j for $r = 3$ (5th-order interpolation)

oscillation around shock waves. The detonation front structure is constructed by the complicated shock waves so that the interpolation by conservative variables is suitable for the simulation of detonation. From this perspective, the conservative variables are used in the interpolation of the WCNS.

In order to calculate the conservative variables at the computational cell interface, stencils $S_{j+1/2}$ of $2r - 1$ points are used to construct the conservative variables $Q_{j+1/2}^L$ and $Q_{j+1/2}^R$ with a $(2r - 1)$th-order interpolation:

$$S_{j+12} = \{x_{j-r+1},\ldots,x_j,\ldots,x_{j+r-1}\} \quad (5.30)$$

For example, if $r = 3$, five points $(j - 2, j - 1, j, j + 1, j + 2)$ are required and a 5th-order interpolation is performed in Fig. 5.2.

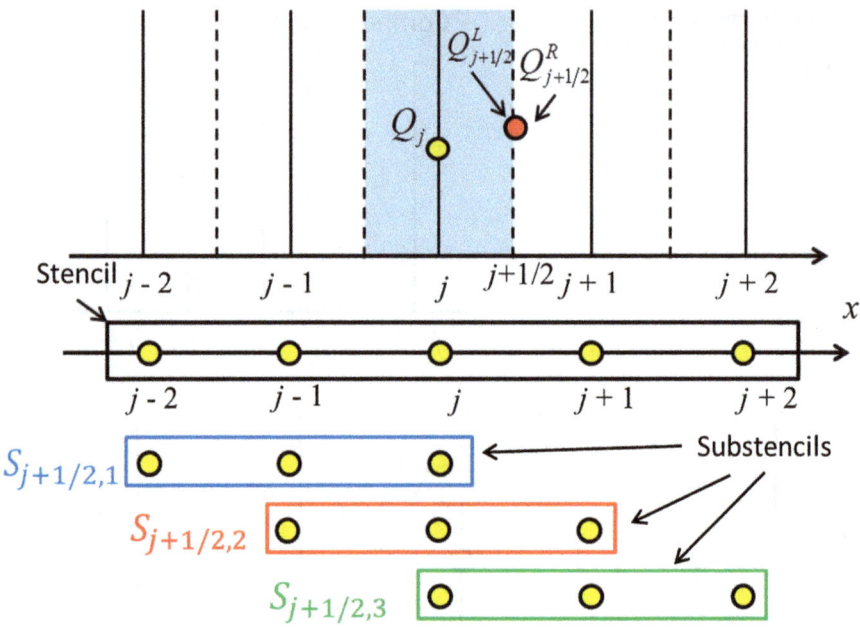

Fig. 5.3 Substencils for interpolation at the point j for $r = 3$ (5th-order interpolation)

The first step is to transform the conservative variables $\mathbf{Q}$ in the stencil to the characteristic variables $q_{j,m}$ by multiplying the left eigenvector, yielding the expression

$$q_{j,m} = \mathbf{l}_{j,m} \mathbf{Q}_j, \tag{5.31}$$

where the mth left eigenvectors of the Jacobian matrix $\partial \mathbf{E}/\partial \mathbf{Q}$ are $\mathbf{l}_{j,m}$ and the mth characteristic variables are $q_{j,m}$. The second step is to construct polynomials consisting of (r substencils. The r substencils are composed of r cell centers in Fig. 5.3. The kth substencil $S_{j+1/2,k}(k = 1, \cdots, r)$ is written as

$$S_{j+12,k} = \{x_{j+k-r},\ldots,\ldots,x_j,\ldots,\ldots,x_{j+k-1}\} \tag{5.32}$$

Then, points of each substencil are combined to calculate an $(2r - 1)$th-order interpolation in Fig. 5.4.

The characteristic variables in the kth polynomials $q^L_{j+1/2,k,m}$ are computed as

$$q^{(n)}_{j,k,m} = \sum_{l=1}^{r} c_{n,k,l} \cdot q_{j+k-r+l-1,m}. \tag{5.33}$$

5 Numerical Simulation on Rotating Detonation Engine…

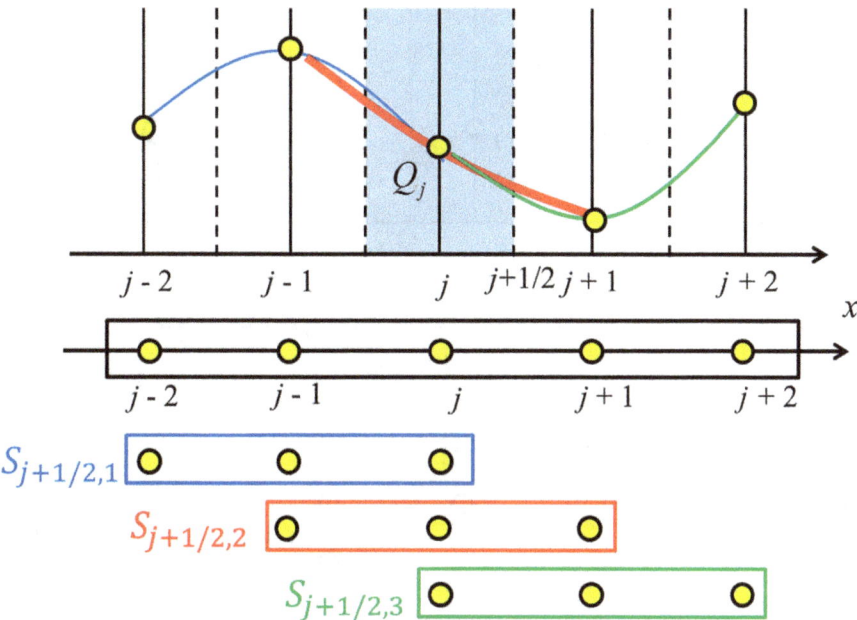

Fig. 5.4 5th-order interpolation by combination of each substencil ($r = 3$)

Here, the details of coefficients $c_{n,k,l}$ are shown in Nonomura and Fujii [5.33]. For example, the approximation of the 1st derivative of the mth characteristic variable at the 1st substencil ($n = 1, r = 3$) is expressed as:

$$q^{(1)}_{j,1,m} = c_{1,1,1} \cdot q_{j-2,m} + c_{1,1,2} \cdot q_{j-1,m} + c_{1,1,3} \cdot q_{j,m}. \tag{5.34}$$

By reproducing the same calculation for other kth substencils and other nth derivatives with their own c coefficients, we can obtain a full data set to calculate the weighted characteristic variables for each substencil at the cell interface as follows:

$$q^{L}_{j+1/2,k,m} = q_{j,m} + \sum_{n=1}^{r-1} \left(\frac{1}{n!}\right)\left(\frac{\Delta x}{2}\right)^n q^{(n)}_{j,k,m}. \tag{5.35}$$

This calculation is based on the Taylor expansion method which represents a function as an infinite sum of terms that are calculated from the values of the function's derivatives at a single point. In the case of $r = 3$ (5th-order), the mth weighted characteristic variable for the k substencils can be written as:

$$q^{L}_{j+1/2,k,m} = q_{j,m} + \frac{\Delta x}{2} q^{(1)}_{j,k,m} + \frac{1}{2}\frac{\Delta x^2}{4} q^{(2)}_{j,k,m}. \tag{5.36}$$

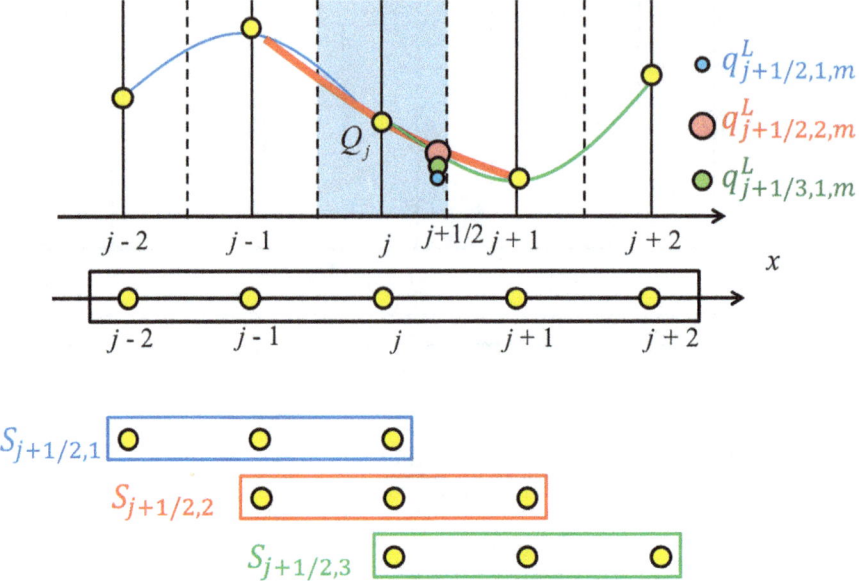

Fig. 5.5 Weighted characteristic variable at the left edge of the cell from each substencil ($r = 3$, 5th-order interpolation)

If we consider the 1st substencil, the above equation becomes:

$$q^L_{j+1/2,k,m} = q_{j,m} + \frac{\Delta x}{2}\left[c_{1,1,1} \cdot q_{j-2,m} + c_{1,1,2} \cdot q_{j-1,m} + c_{1,1,3} \cdot q_{j,m}\right]$$
$$+ \frac{\Delta x^2}{8}\left[c_{2,1,1} \cdot q_{j-2,m} + c_{2,1,2} \cdot q_{j-1,m} + c_{2,1,3} \cdot q_{j,m}\right]. \quad (5.37)$$

The idea is that a weight is assigned to each of the k substencils. It determines the contribution of the substencil to the final approximation of the cell-edge value $q_{j+1/2,m}$ in Fig. 5.5. The obtained variables in Eq. (5.37) are rth order cell-edge interpolated values.

The third step is to combine these characteristic variables $q^L_{j+1/2,k,m}$ together with a new final weight coefficient w_k to calculate a final $(2r-1)$th-order accurate interpolation value $q^L_{j+1/2,m}$ at the cell interface. This combination is done as follows:

$$q^L_{j+1/2,m} = \sum_{k=1}^{r} w_{j+1/2,k,m} \cdot q^L_{j+1/2,k,m}, \quad (5.38)$$

where

$$w_{j+1/2,k,m} = \frac{\alpha_{j+1/2,k,m}}{\sum_{l=1}^{r} \alpha_{j+1/2,l,m}}, \quad (5.39)$$

$$\alpha_{j+1/2,k,m} = \frac{C_k}{\left(IS_{j+1/2,k,m}\right)^2 + \varepsilon}, \quad (5.40)$$

$$IS_{j+1/2,k,m} = \sum_{l=1}^{r}\left(q^{(n)}_{j+1/2,k,m}\right)^2. \quad (5.41)$$

In Eq. (5.40), ε is a very small value $\varepsilon = 1 \times 10^{-6}$, which avoids the division by zero. The meaning of Eqs. (5.39), (5.40) and (5.41) is that if a polynomial of the form of Eq. (5.35) becomes smooth, i.e. in continuous regions, $w_{j+1/2,k,m}$ approaches to C_k. On the other hand, near discontinuities, $w_{j+1/2,k,m}$ decreases to prevent from numerical oscillations. This may result in a slight loss of accuracy but the stability of the calculation is improved. $w_{j+1/2,k,m}$ has to satisfy the following equation:

$$\sum_{k=1}^{r} w_{j+1/2,k,m} = 1. \quad (5.42)$$

Finally, multiplying Eq. (5.38) by the right eigenvectors of the matrix $\partial \mathbf{E}/\partial \mathbf{Q}$, $\mathbf{r}_{j,m}$, the conservative variables, $\mathbf{Q}_{j+1/2}$, can be calculated:

$$\mathbf{Q}^L_{j+1/2} = \sum_{m} q^L_{j+1/2,m} \cdot \mathbf{r}_{j,m}. \quad (5.43)$$

When $\mathbf{Q}^L_{j+1/2}$ is calculated, $\mathbf{Q}^R_{j+1/2}$ can be obtained symmetrically with a similar way.

Flux Evaluation

By using the conservative variables at the computational cell interface $\mathbf{Q}^L_{j+1/2}$ and $\mathbf{Q}^R_{j+1/2}$ in Eq. (5.43), the numerical flux at the computational cell interface, $\mathbf{E}_{j+1/2}$, is calculated. The equation used for the flux evaluation in the AUSMDV scheme is

$$\mathbf{E}_{j+12} = \frac{1}{2}\left\{(\rho u)_{12}\left(\Psi_R + \Psi_L\right) - \left|(\rho u)_{12}\right|\left(\Psi_R + \Psi_L\right)\right\} + p_{12}\mathbf{I}_2, \quad (5.44)$$

where

$$\Phi = \begin{pmatrix} \rho \\ \rho u \\ \rho v \\ \rho h \\ \rho_1 \\ \vdots \\ \rho_N \end{pmatrix}, \psi = \begin{pmatrix} 1 \\ u \\ v \\ h \\ Y_1 \\ \vdots \\ Y_N \end{pmatrix}, \mathbf{I}_2 = \begin{pmatrix} 0 \\ 1 \\ 0 \\ 0 \\ 0 \\ \vdots \\ 0 \end{pmatrix}. \quad (5.45)$$

The methods used to evaluate the pressure term, mass flux term, and momentum flux term are described in the original paper (Wada and Liou 1994).

Linear Difference

An approximation of flux derivative $\left(\dfrac{\partial \mathbf{E}}{\partial x}\right)_j$ is evaluated from $\tilde{\mathbf{E}}_{j+1/2}$. A general form of the explicit midpoint-to-node differencing (MD) scheme in Fig. 5.6 is written as follows:

$$\left(\frac{\partial \mathbf{E}}{\partial x}\right)_j = \frac{1}{\Delta x}\sum_{k=0}^{r-1} a_k \left(\tilde{\mathbf{E}}_{j+k+1/2} - \tilde{\mathbf{E}}_{j+k-1/2}\right). \tag{5.46}$$

On the other hand, the robust scheme for the RWCNS that Nonomura and Fujii (2013) improved as a midpoint-and-node-to-node difference (MND) in Fig. 5.7 is given by

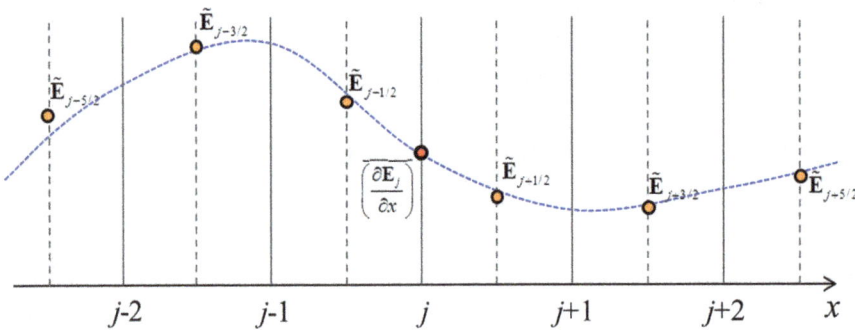

Fig. 5.6 Midpoint-to-node linear differentiation

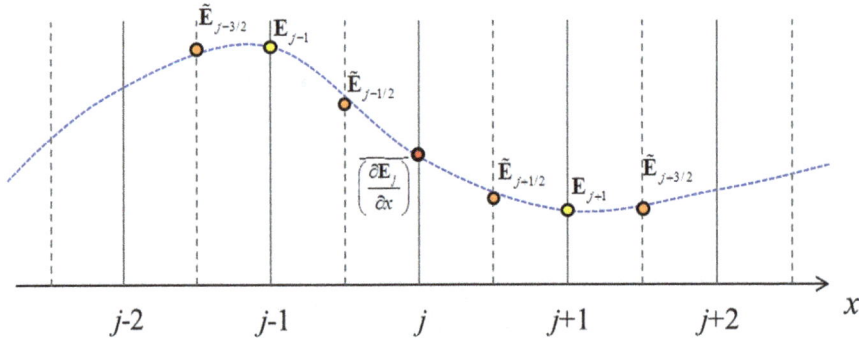

Fig. 5.7 Midpoint and node-to-node linear differentiation

$$\left(\frac{\partial \mathbf{E}}{\partial x}\right)_j = \frac{1}{\Delta x}\sum_{k=1}^{r}b_k\left(\tilde{\mathbf{E}}_{j+k/2}-\tilde{\mathbf{E}}_{j-k/2}\right), \qquad (5.47)$$

$$\bar{\mathbf{E}}_{j\pm k/2} = \begin{cases} \bar{\mathbf{E}}_{j\pm k/2}(k=2n-1) \\ \mathbf{E}_{j\pm k/2}(k=2n) \end{cases}, \qquad (5.48)$$

where the true flux at the node point is $\mathbf{E}_{j\pm k/2}$. The coefficients of a_k and b_k are shown in the original article by Nonomura and Fujii (2013). When Eq. (5.48) is substituted into Eq. (5.47), the true flux is interpolated into the numerical viscosity term.

This scheme in Eq. (5.47) prevents the negative numerical viscosity around the shock wave and improves stability more than the original scheme in Eq. (5.46). Thus, the difference between the WCNS and the RWCNS is the linear interpolation. The present research uses the nonlinear interpolation as 5th-order RWCNS and linear interpolation as Eq. (5.47) with 6th-order coefficients.

5 Simulation Conditions

The simulation conditions and grid system used in this study are presented in Table 5.2. The computational regions for the x- and y- directions are 3 mm × 6 mm.

The ambient conditions behind the exit are pressure p_e of 0.01 MPa and temperature of 300 K. The reason to use low ambient pressure is the comparison of I_{sp} between the present simulation and H_2/O_2 conventional rocket engine without nozzle under vacuum condition. The simulation conditions in the stagnation chamber are pressure p_0 of 2–6 MPa and temperature of 300 K. The micro-nozzle area ratios of the throat to nozzle exit at the injection port, A^*/A, are 0.02. The stoichiometric H_2/O_2 gas mixture is supplied through the micro-nozzles. The numerical setup of the present simulation and boundary conditions are shown in Fig. 5.8. The numerical

Table 5.2 Simulation conditions and computational grids (constant resolution)

p_0, MPa	L, mm (imax)	H, mm (jmax)	$\Delta x, \Delta y,$ μm [a]1	A^a/A
2	3.0 (1201)	6.0 (2401)	2.5	0.2
3	3.0 (1451)	6.0 (2901)	2.07	0.2
4	3.0 (1543)	6.0 (3087)	1.95	0.2
5	3.0 (1593)	6.0 (3187)	1.89	0.2

[a]1: approximately five grid points in H_2 half reaction length

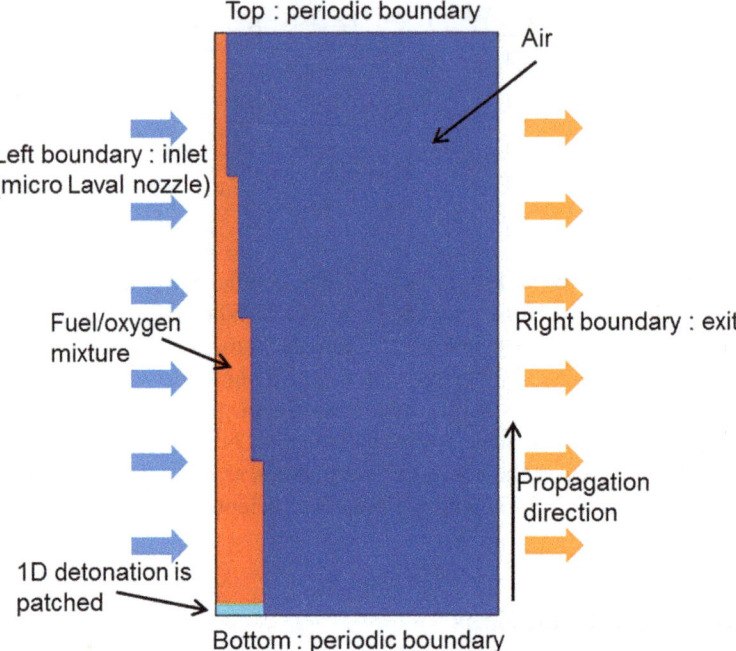

Fig. 5.8 Numerical setup and boundary conditions

result of the one-dimensional detonation is used for the ignition in the two-dimensional simulation. As the stagnation pressure is lower than 2 MPa, the rotating detonation becomes failure. Although the successful of the detonation ignition will depend on the stoichiometric ratio and stagnation temperature in the injection gas, the present simulations does not estimate the effect of them yet.

There are two boundary condition systems for the mixture injection: the supersonic and subsonic inlets. The supersonic inlet condition is used for most of the simulations because the inlet nozzles for premixed gas typically have a choked condition at the exits of small nozzles. However, a supersonic inlet condition exists in many real cases because of the high pressure in the combustion chamber. The subsonic inlet condition is discussed by Zhdan et al. (2007). The present calculations are performed based on the following four conditions:

1. The pressure is extrapolated, and velocity is set to zero when the inlet pressure is higher than the manifold pressure. This case means that the gas injection is impossible.
2. The pressure is extrapolated when the pressure just before the inlet is lower than the manifold pressure and higher than the pressure just behind the inlet. In this case, there is no choking at the throat with a subsonic gas injection.
3. The pressure is extrapolated when the inlet pressure is higher than the supersonic condition and lower than the subsonic condition. In this case, a normal shock wave is generated in the nozzle with the subsonic injection.

4. The pressure achieves a supersonic value and the premixed gas is accelerated due to isentropic expansion with supersonic injection when the inlet pressure is higher than the supersonic condition.

The outlet boundary conditions are given at the exit of the RDE by two patterns, but the flow cannot go backwards from the downstream to the upstream:

1. The exit pressure of the RDE is set to the ambient pressure when the exhaust gas speed is subsonic.
2. The exit pressure of the RDE is extrapolated from the values in the combustion chamber when the exhaust gas speed is supersonic.

The effects of the grid resolution on the MUSCL are discussed by Tsuboi et al. (2013, 2015, 2017). As the grid resolution increases in 2D and 3D RDE simulations, a cellular structure appears near the detonation head. However, the effects of the grid resolution on the I_{sp} is approximately a few seconds.

6 Results and Discussions

Basic Flow Structure

The comparison of the density gradient between the MUSCL and the WCNS5MN is presented in Fig. 5.9. The injection conditions are $A^*/A = 0.2$ and $p_0 = 2.0$ MPa. For both cases, some triple points appear near the detonation front. The rotating

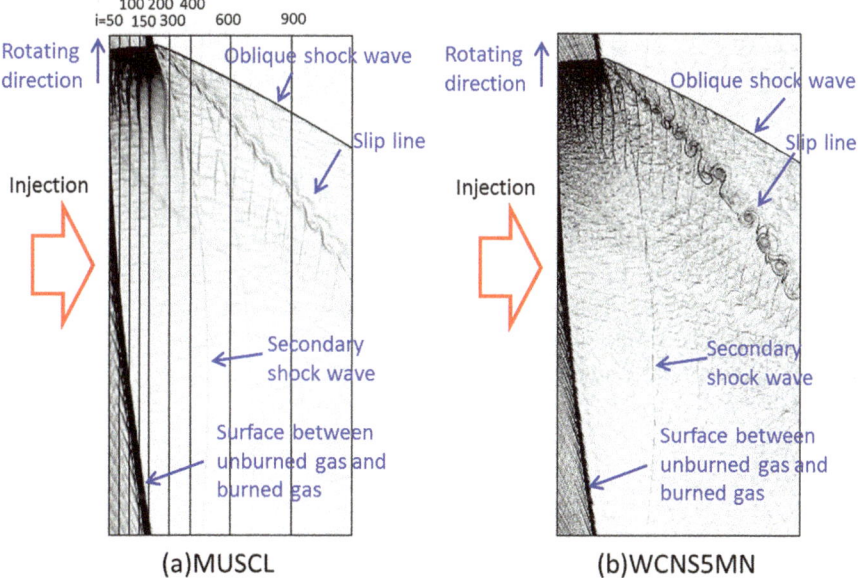

Fig. 5.9 Instantaneous density gradient contours and flow structure for $p_0 = 2$ MPa

Fig. 5.10 Instantaneous pressure profile along y-directions in Fig. 5.9 for p_0= 2 MPa

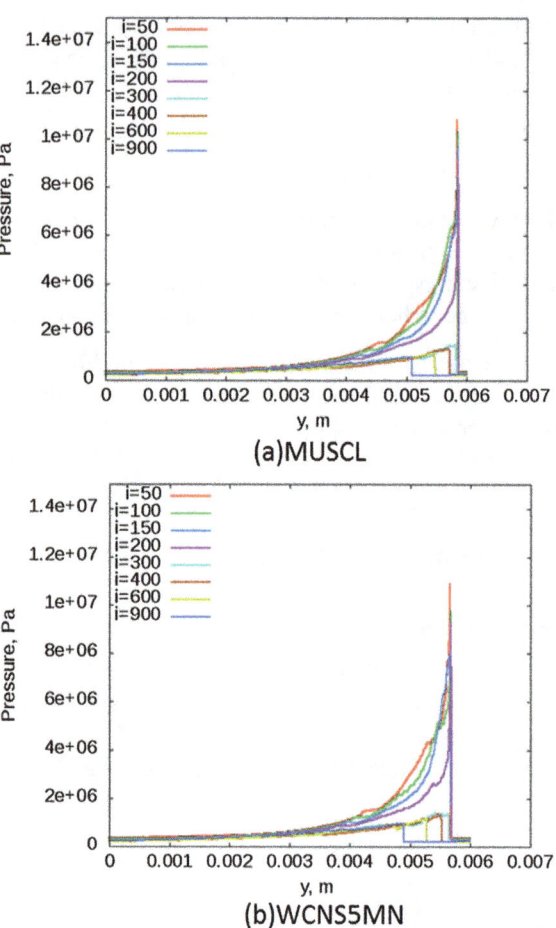

detonation head produces some shock waves and a slip line. The WCNS5MN can resolve the slip line with small vortices better than the MUSCL.

Figure 5.10 plots the pressure along the lines near the injection, as shown in Fig. 5.9. The pressure behind the rotating detonation along lines i = 50 − 900 decreases because of the strong expansion in the unconfined region. Both pressure profiles show similar feature.

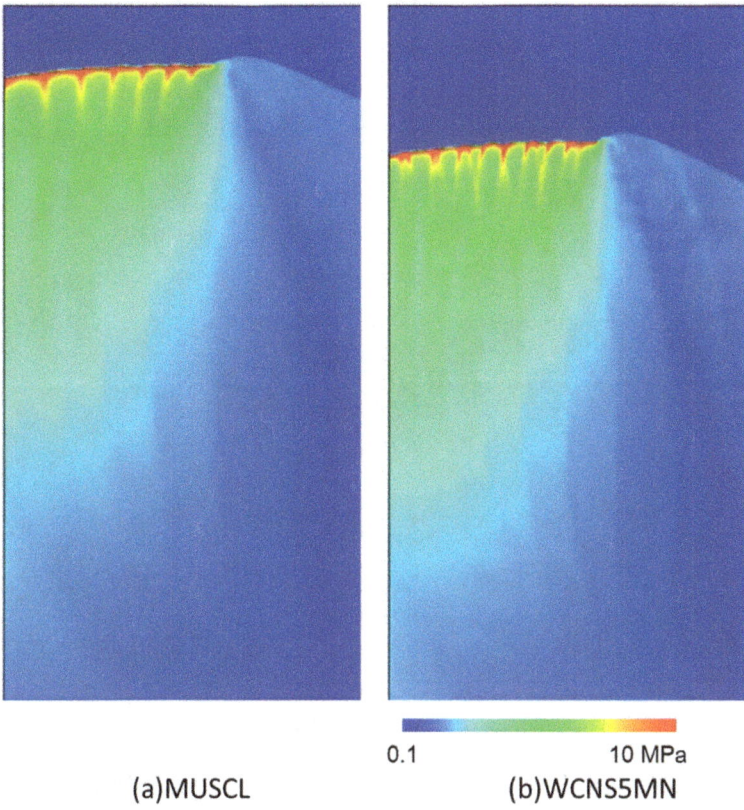

Fig. 5.11 Instantaneous pressure contours for p_0= 2 MPa

Close-Up View Near Detonation Front

This subsection discusses the effects of the high-resolution scheme near the rotating detonation front. Figures 5.11, 5.12, 5.13 and 5.14 show the instantaneous pressure, temperature, OH mass fraction, and local heat release contours near the rotating detonation front. Both schemes can capture some triple points near the rotating detonation front in the present grid resolution. Some disturbance in the temperature and OH mass fraction contours are captured behind the rotating detonation front for the WCNS5MN. The instantaneous profiles along $i = 50\sim300$ are shown in Figs. 5.15, 5.16 and 5.17. The WCNS5MN can capture the large discontinuity near the detonation front better than the MUSCL as shown in Fig. 5.15. However, there are small difference on temperature and OH mass fraction in Figs. 5.16 and 5.17 because the WCNS5MN improves spatial resolution for fluid and this scheme does not improve resolution for chemical reaction as shown in Eqs. (5.25), (5.26) and (5.27).

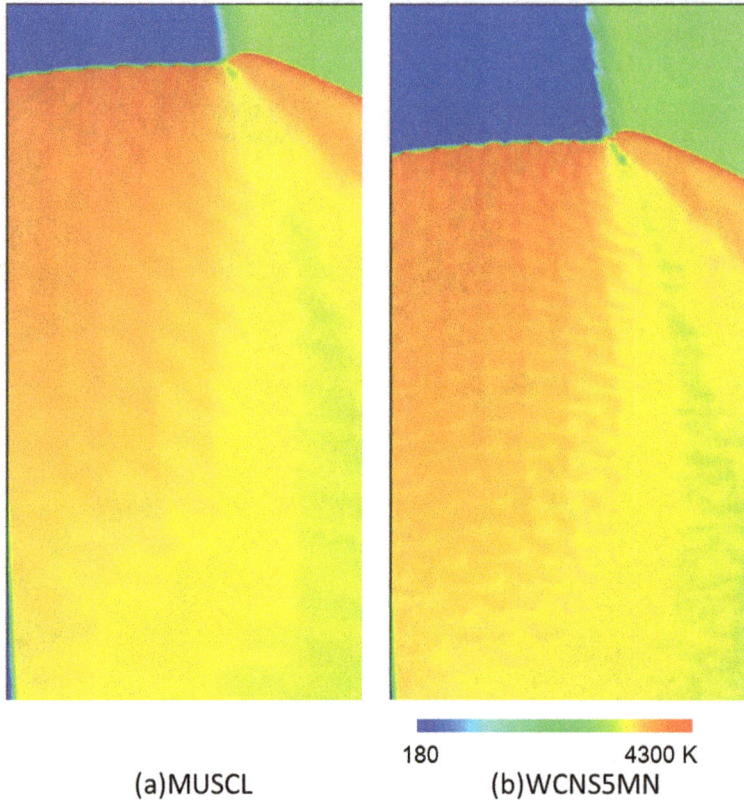

Fig. 5.12 Instantaneous temperature contours for p_0= 2 MPa

Detonation Velocity

Figure 5.18 shows the comparison of the rotating detonation velocity between the MUSCL and the WCNS5MN. The detonation velocity for both schemes is 2849 m/s and 97% of D_{CJ}. In the experimental data (Bykovskii et al. 2006), the rotating detonation velocities are 80–95% of D_{CJ} because of incomplete mixing and unconfined effects. The detonation velocity deficit in Hishida's 2D RDE simulation (Hishida et al. 2009) is approximately 5% because of the unconfined detonation effects. The present results agree well with those in Hishida's results (Hishida et al. 2009).

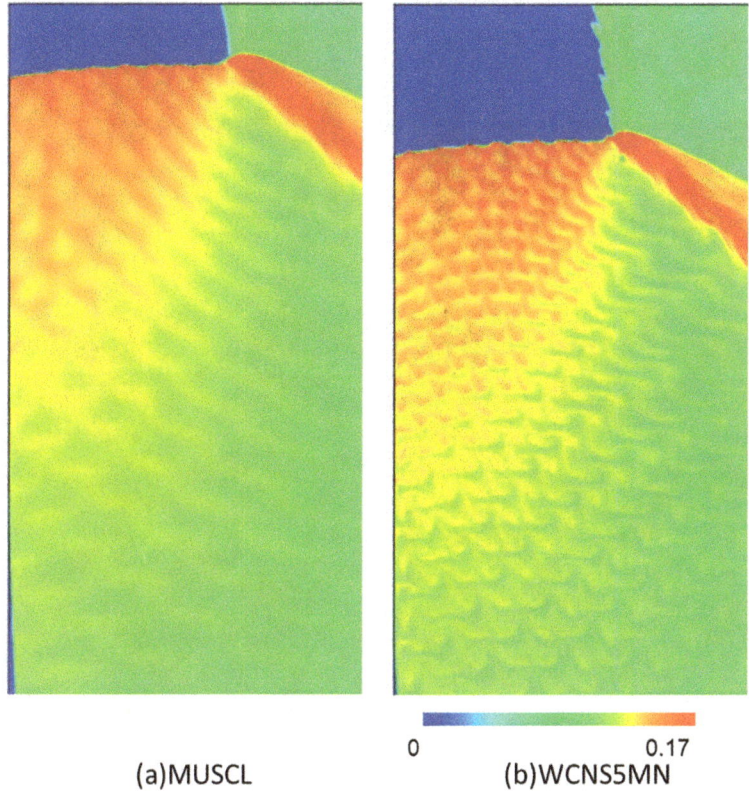

Fig. 5.13 Instantaneous OH mass fraction contours for p_0= 2 MPa

Thrust Performance

Figure 5.19 shows the comparison of I_{sp} between the MUSCL and the WCNS5MN. This figure also includes the calculated I_{sp} for a H_2/O_2 rocket engine without a diverging nozzle, assuming a chemical equilibrium state under a vacuum environment. This value is calculated using the Gordon and McBride method (Gordon and McBride 1971). This figure shows that I_{sp} for the WCNS5MN is approximately 5 sec larger than I_{sp} for the MUSCL. I_{sp} for both schemes is actually greater than I_{sp} of a conventional rocket engine.

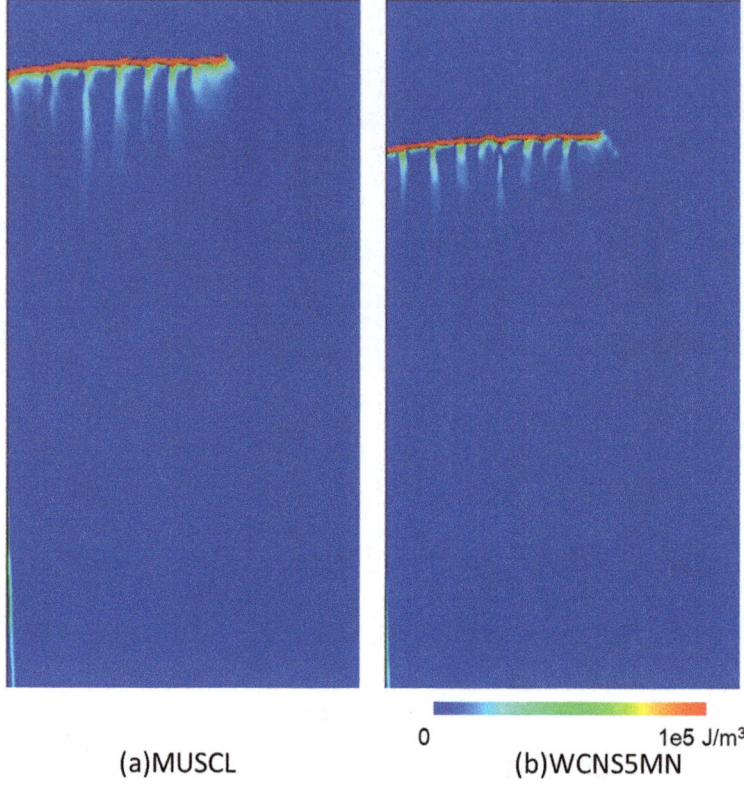

Fig. 5.14 Instantaneous heat release contours for p_0= 2 MPa

The time-averaged thrust per cycle for both schemes is also proportional to the mass flow rate, as shown in Fig. 5.20. Because the mass flow rates for the WCNS5MN are 2–4% smaller than those for the MUSCL, I_{sp} for the WCNS5MN is 5 s larger than I_{sp} for the MUSCL in Fig. 5.19.

Fig. 5.15 Instantaneous pressure profiles near rotating detonation front along y-direction in Fig. 5.11 for p_0= 2 MPa

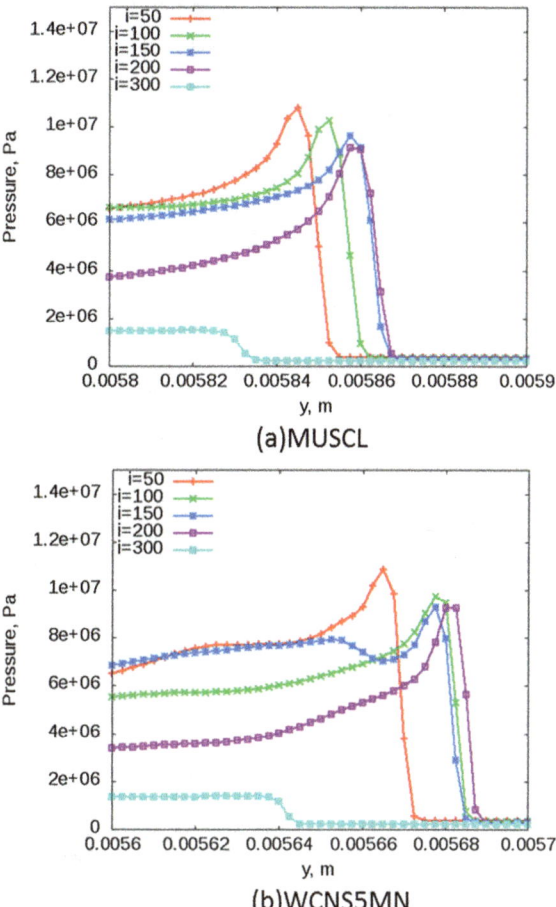

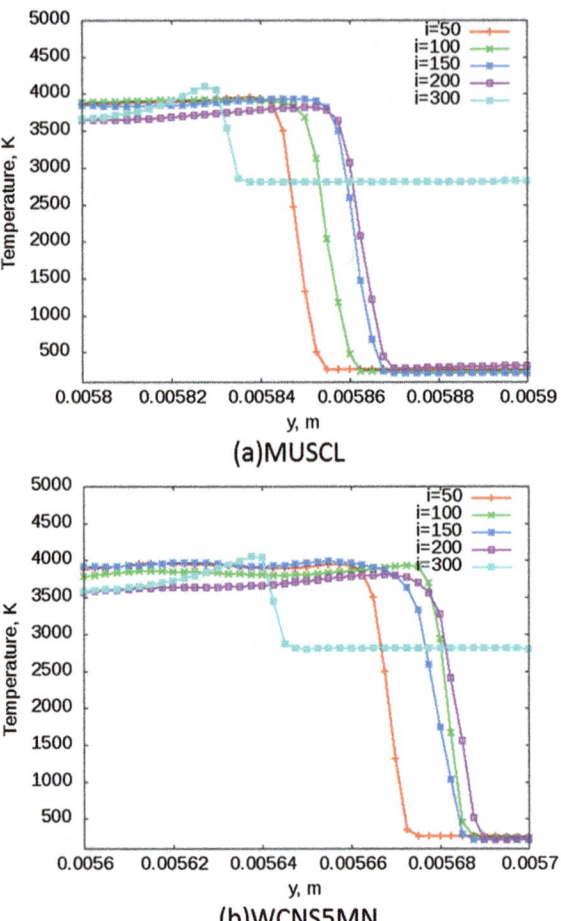

Fig. 5.16 Instantaneous temperature profiles near rotating detonation front along y-direction in Fig. 5.12 for $p_0 = 2$ MPa

Fig. 5.17 Instantaneous OH mass fraction profiles near rotating detonation front along y-direction in Fig. 5.13 for p_0= 2 MPa

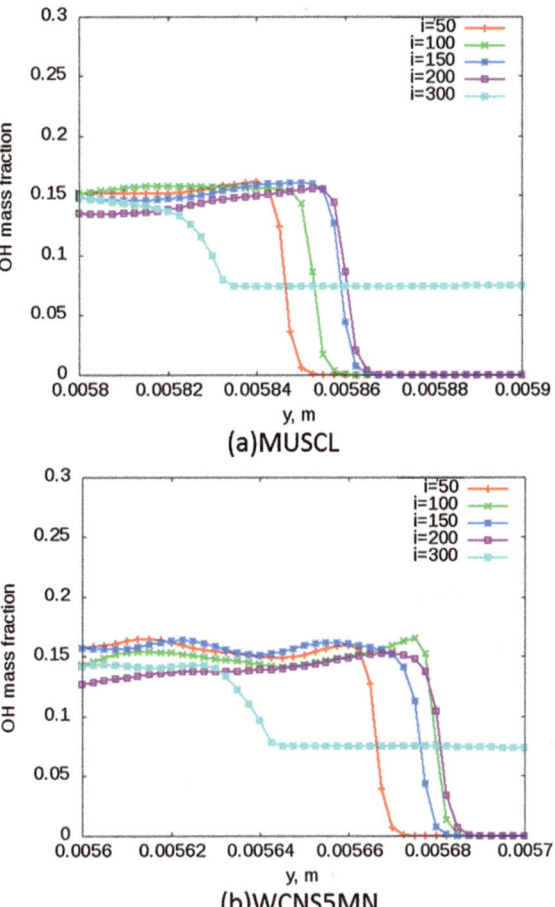

Fig. 5.18 Comparison of detonation velocities for p_0= 2 MPa

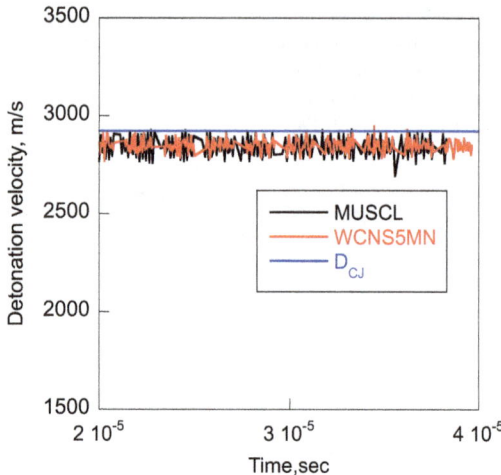

Fig. 5.19 Comparison of mixture-based specific impulse

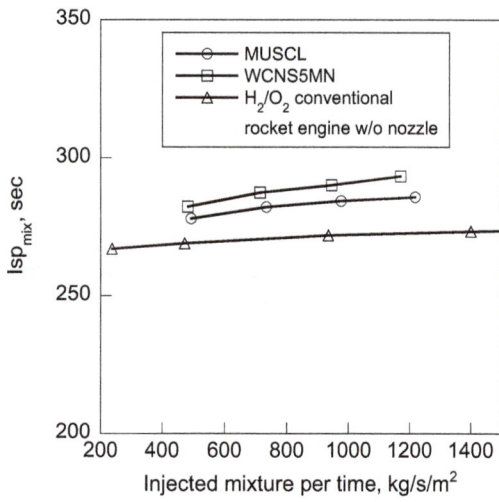

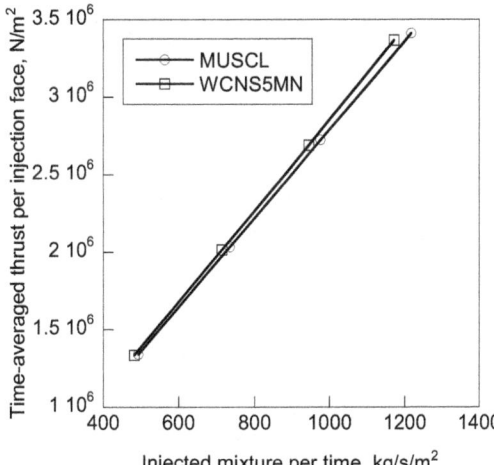

Fig. 5.20 Comparison of time-averaged thrust

7 Conclusions

The implementation and simulations of the robust weighted compact nonlinear scheme (RWCNS) for the two-dimensional rotating detonation engine are performed using the detailed chemistry model. The comparison of the MUSCL and the 5th-order RWCNS (WCNS5MN) indicates that the shock front and the contact surface for the WCNS5MN can be improved with the better resolution than those for the MUSCL and that both rotating velocities are approximately 97% of the CJ value. However, I_{sp} for the WCNS5MN is approximately 5 sec larger than I_{sp} for the MUSCL because the mass flow rates for the WCNS5MN are 2–4% smaller than those for the MUSCL.

Acknowledgements This research was done in collaboration with Cybermedia Center using the Osaka University supercomputer system.

References

Araki, T., Yoshida, K., Morii, Y., Tsuboi, N., & Hayashi, A. K. (2016). Numerical analyses on ethylene/oxygen detonation with multistep chemical reaction mechanisms: Grid resolution and chemical reaction model. *Combustion Science and Technology, 188*(3), 346–369.

Asahara, A., Hayashi, A. K., Yamada, E., & Tsuboi, N. (2012). Generation and dynamics of subtransverse wave of cylindrical detonation. *Combustion Science and Technology, 184*, 1568–1590.

Batten, P., Clarke, N., Lambert, C., & Causon, D. M. (1997). On the choice of wavespeeds for the HLLC Riemann solver. *SIAM Journal on Scientific Computing, 18*, 1553–1570.

Bykovskii, F. A., Zhdan, S. A., & Vedernikov, E. F. (2006). Continuous spin detonations. *Journal of Propulsion and Power, 22*(6), 1204–1216.

Davis, S. G., Joshi, A. V., Wang, H., & Egolfopoulos, F. (2005). An optimized kinetic model of H2/CO combustion. *Proceedings of the Combustion Institute, 30*(1), 1283–1292.

Deng, X. G., & Zhang, H. (2000). Developing high-order weighted compact nonlinear schemes. *Journal of Computational Physics, 165*, 22–44.

Einfeldt, B. (1988). On Godunov-type methods for gas dynamics. *SIAM Journal on Numerical Analysis, 25*, 294–318.

Eto, K., Tsuboi, N., & Hayashi, A. K. (2005). Numerical study on three-dimensional C-J detonation waves: Detailed propagating mechanism and existence of OH radical. *Proceedings of the Combustion Institute, 30*, 1907–1913.

Eto, S., Tsuboi, N., Takayuki, K., & Hayashi, A. K. (2016). Three-dimensional numerical simulation of a rotating detonation engine: Effects of the throat of a converging-diverging nozzle on engine performance. *Combustion Science and Technology, 188*(11–12), 2105–2116.

Godunov, S. K. (1959). A difference scheme for numerical solution of discontinuous solution of hydrodynamic equations. *Math. Sbornik, 43*, 271–306.

Gordon, S., & McBride, J.B. (1971). Computer program for calculation of complex chemical equilibrium compositions, rocket performance, incident and reflected shocks, and Chapman-Jouget detonations. NASA SP-273.

Gottlieb, S., & Shu, C. W. (1998). Total variation diminishing Runge-Kutta schemes. *Mathematics of Computation, 67*, 73–85.

Henrick, A. K., Aslam, T. D., & Powers, J. M. (2006). Simulations of pulsating one-dimensional detonations with true fifth order accuracy. *Journal of Computational Physics, 213*, 311–329.

Hishida, M., Fujiwara, T., & Wolanski, P. (2009). Fundamentals of rotating detonations. *Shock Waves, 19*(1), 1–10.

Hu, X. Y., Zhang, D. L., Khoo, B. C., & Jiang, Z. L. (2005). The structure and evolution of a twodimensional H2/O2/Ar cellular detonation. *Shock Waves, 14*, 37–44.

Iida, R., Asahara, M., Tsuboi, N., Hayashi, A. K., & Nonomura, T. (2014). Implementation of a robust weighted compact nonlinear scheme for modeling of hydrogen/air detonation. *Combustion Science and Technology, 186*(10–11), 1736–1757.

Jiang, G. S., & Shu, C. W. (1996). Efficient implementation of weighted ENO schemes. *Journal of Computational Physics, 126*, 200–212.

Konnov, A. A. (2008). Remaining uncertainties in the kinetic mechanism of hydrogen combustion. *Combustion and Flame, 152*(4), 507–528.

Korobenikov, V. P., Levin, V. A., Markov, V. V., & Chernyi, G. G. (1972). Propagation of blast wave in a combustible gas. *Astronautica Acta, 17*, 529–537.

Kurosaka, M., & Tsuboi, N. (2014). Spinning detonation, cross-currents, and the chapman-Jouguet velocity. *Journal of Fluid Mechanics, 756*, 728–757.

Li, J., Zhao, Z., Kazakov, A., & Dryer, F. L. (2004). An updated comprehensive kinetic model of hydrogen combustion. *International Journal of Chemical Kinetics, 36*(10), 566–575.

Mueller, M. A., Kim, T. J., Yetter, R. A., & Dryer, F. L. (1999). Flow reactor studies and kinetic modeling of the H2/O2 reaction. *International Journal of Chemical Kinetics, 31*(2), 113–125.

Niibo, T., Morii, Y., Ashahara, M., Tsuboi, N., & Hayashi, A. K. (2016). Numerical study on direct initiation of cylindrical detonation in H2/O2 mixtures: Effect of higher-order schemes on detonation propagation. *Combustion Science and Technology, 188*(11–12), 2044–2059.

Nonomura, T., & Fujii, K. (2009). Effects of difference scheme type in high-order weighted compact nonlinear schemes. *Journal of Computational Physics, 228*, 3533–3539.

Nonomura, T., & Fujii, K. (2013). Robust explicit formulation of weighted compact nonlinear scheme. *Computers and Fluids, 85*, 8–18.

Nonomura, T., Iizuka, N., & Fujii, K. (2010). Freestream and vortex preservation properties of highorder WENO and WCNS on curvilinear grids. *Computers and Fluids, 39*, 197–214.

Nonomura, T., Morizawa, S., Terashima, H., Obayashi, S., & Fujii, K. (2012). Numerical (error) issues on compressible multicomponent flows using a high-order differencing scheme: Weighted compact nonlinear scheme. *Journal of Computational Physics, 231*, 3181–3210.

O'Conaire, M. O., Curran, H. J., Simmie, J. M., Pitz, W. J., & Westbrook, C. K. (2004). A comprehensive modeling study of hydrogen oxidation. *International Journal of Chemical Kinetics, 36*(11), 603–622.

Odlyzko, A. M. (1985). Discrete logarithms in finite fields and their cryptographic significance. *Lecture Notes in Computer Science, 209*, 224–314.

Oran, E. S., Young, T., & Boris, J. P. (1979). Application of time-dependent numerical methods to the description of reactive shocks. *Seventeenth Symposium (International) on Combustion*, 43–54.

Petersen, E. L., & Hanson, R. K. (1999). Reduced kinetics mechanisms for ram accelerator combustion. *J. Propulsion Power, 15*, 591–600.

Saxena, P., & Williams, F. A. (2006). Testing a small detailed chemical-kinetic mechanism for the combustion of hydrogen and carbon monoxide. *Combustion and Flame, 145*(1–2), 316–323.

Schwer, D., & Kailasanath, K. (2013). Fluid dynamics of rotating detonation engines with hydrogen and hydrocarbon fuels. *Proceedings of the Combustion Institute, 34*, 1991–1998.

Shepherd, J. E. (1986). Chemical kinetics of H2-air-diluent detonations. In *Progr. in Astronautics and aeronautics* (Vol. 106, pp. 263–293).

Shimizu, K., Hibi, H., Koshi, M., Morii, M., & Tsuboi, N. (2011). Updated kinetic mechanism for high-pressure hydrogen combustion. *Journal of Propulsion and Power, 27*(2), 383–396.

Shuen, J. S. (1992). Upwind differencing and LU factorization for chemical non-equilibrium Navier-stokes equations. *Journal of Computational Physics, 909*, 233–250.

Stull, D. R., & Prophet, H. (1971). *JANAF Thermochemical Tables* (2nd ed.). NSRDS-NBS, 37.

Togashi, F., Lohner, R., & Tsuboi, N. (2009). Numerical simulation of H2/air detonation using unstructured mesh. *Shock Waves, 19*(2), 151–162.

Tsuboi, N., & Koichi Hayashi, A. (2007). Numerical study on spinning detonations. *Proceedings of the Combustion Institute, 31*, 2389–2396.

Tsuboi, N., Katoh, S., & Hayashi, A. K. (2002). Three-dimensional numerical simulation for hydrogen/air detonation: Rectangular and diagonal structures. *Proceedings of the Combustion Institute, 29*, 2783–2788.

Tsuboi, N., Eto, K., & Hayashi, A. K. (2007). Detailed structure of spinning detonation in a circular tube. *Combustion and Flame, 149*(1–2), 144–161.

Tsuboi, N., Asahara, M., Eto, K., & Hayashi, A. K. (2008a). Numerical simulation of spinning detonation in square tube. *Shock Waves, 18*(4), 329–344.

Tsuboi, N., Daimon, Y., & Hayashi, A. K. (2008b). Three-dimensional numerical simulation of detonations in coaxial tubes. *Shock Waves, 18*(5), 379–392.

Tsuboi, N., Hayashi, A. K., & Koshi, M. (2009). Energy release effect of mixture on single spinning detonation structure. *Proceedings of the Combustion Institute, 32*, 2405–2412.

Tsuboi, N., Morii, Y., & Koichi Hayashi, A. (2013). Two-dimensional numerical simulation on galloping detonation in a narrow channel. *Proceedings of the Combustion Institute, 34*(2), 1999–2007.

Tsuboi, N., Eto, S., Hayashi, A. K., & Kojima, T. (2017). Front cellular structure and thrust performance on hydrogen-oxygen rotating detonation engine. *Journal of Propulsion and Power, 33*(1), 100–111.

Tsuboi, N., Watanabe, Y., Kojima, T., & Hayashi, A. K. (2015). Numerical estimation of the thrust performance on a rotating detonation engine for a hydrogen-oxygen mixture. *Proceedings of the Combustion Institute, 35*(2), 2005–2013.

Wada, Y., & Liou, M. (1994). A flux splitting scheme with high resolution and robustness for discontinuities. *AIAA Paper, 94-0083*, 1994.

Wintenberger, E., & Shepherd, J. E. (2003). A model for the performance of air-breathing pulse detonation engines. *AIAA Paper, 03-4511*, 2003.

Yee, H. C. (1987). Upwind and symmetric shock-capturing schemes. NASA TM-89464.

Zhang, S., Jiang, S., & Shu, C. W. (2008). Development of nonlinear weighted compact schemes with increasingly higher order accuracy. *Journal of Computational Physics, 227*, 7294–7321.

Zhdan, S. A., Bykovskii, F. A., & Vedernikov, E. F. (2007). Mathematical modeling of a rotating detonation wave in a hydrogen-oxygen mixture. *Combustion, Explosion and Shock Waves, 43*(4), 449–459.

Zhou, R., & Wang, J. P. (2012). Numerical investigation of flow particle paths and thermodynamic performance of continuously rotating detonation engines. *Combustion and Flame, 159*, 3632–3645.

Chapter 6
Review on the Research Progresses in Rotating Detonation Engine

Mohammed Niyasdeen Nejaamtheen, Jung-Min Kim, and Jeong-Yeol Choi

Abstract Present paper focuses on the comprehensive survey of Rotating Detonation Engine (RDE) and their research from the basic to the advanced level. In this paper, an abridged archival background of Pulse/Rotating Detonation Engine (PDE/RDE) is briefed. This is followed by a short description of a Continuous Spin Detonation (CSD) and a few essential facts from the prior publications. Furthermore, a summarization of the Continuous Detonation Wave Rocket Engine (CDWRE) concepts is examined. At long last, a detailed numerical investigation and experiment work of RDE is also presented.

1 Introduction

Since the Detonative Propulsion Research has obtained its peak (Hoffmann 1940; Tsuboi et al. 2011; Braun et al. 2010a; Davidenko et al. 2011, 2009a; Zhou and Wang 2012; Eude and Davidenko 2011; Yi et al. 2009a; Naour et al. 2011; Endo et al. 2011), the practical application of such propulsive device is still a question mark (to author's insight). The current focus in utilizing Detonative Propulsion (DP) for air-breathing engines has come across a long journey from the Pulsed Mode (PM) to Continuous Rotating Mode (CRM). Contrary to the pulsed operation that is applicable to the flight Mach (M) number up to about 3–4, the concept of RDE is attractive for M of above 4. Moreover, the PDE designed for flight M exceeding 3–4 increases the complexity and are getting too expensive. By the same token, in use gas turbine engines (Jet Engine (JE)/Turboprop Engine (TpE)/Turbo shaft Engine (TsE)/Radial Gas Turbine (RGT)/Scale Jet Engine (SJE)/Micro-turbine (Mt)) are also having some disadvantages when compared to RDE in terms of both the engineering and materials standpoint. Noticeably, the gas

---

M.N. Nejaamtheen • J.-M. Kim • J.-Y. Choi (✉)
Department of Aerospace Engineering, Pusan National University, Busan, Republic of Korea
e-mail: aerochoi@pusan.ac.kr

© Springer International Publishing AG 2018
J.-M. Li et al. (eds.), *Detonation Control for Propulsion*, Shock Wave and High Pressure Phenomena, https://doi.org/10.1007/978-3-319-68906-7_6

turbines has reduced efficiency, increased cost and delayed response to changes in power settings.

The global interest in the development of RDE for aerospace propulsion has led to numerous studies in DP, particularly pertaining to its efficiency. This is evident from the formation of collaborative teams and groups by universities and industries worldwide. During the period of 2005–2008, the Lavrentyev Institute of Technology (LIT) and the Novosibirsk Institute of Hydrodynamics (NIH) in Russia have collaborated in the development of CSD. From 2006 to 2009, the France missile developer and manufacturing company so called Matra BAe Dynamics Aérospatiale (MBDA) joined together with the Lavrentyev Institute of Hydrodynamics of the Siberian Branch of the RAS in developing the CDWRE. The recent speedily numerical investigation of RDE was carried out in many places world-wide. Some of the noteworthy universities and collaborative include Warsaw University of Technology (WUT), Pusan National University (PNU), Aoyama Gakuin University (AGU), Japan Aerospace Exploration Agency (JAXA), N.N. Semenov Institute of Chemical Physics and collaborative work of Air Force Research Laboratory (AFRL) with Innovative Scientific Solution Incorporated (ISSI) and Air Force Institute of Technology (AFIT). The equal effort in numerical research is given to the experimental work of RDE, which is taken care by large number of research organizations in the US and many regions across the big blue marble. Some of them are Aerojet Rocketdyne (AR), Naval Research Lab (NRL), NASA Glenn Research Center, Collaborative work of University of Tsukuba (UoT) with Keio University (KU) and JAXA, AFRL, Peking University (PU), and also the Collaborative work of Nagoya University (NU) with University of Tsukuba and JAXA.

Recently, in 2013, the International Workshop on Detonation for Propulsion (IWDP) (International Workshop on Detonation for Propulsion 2013) was held in Taiwan with the participation of more than 20 experts. Likewise, the 14th International Detonation Symposium (IDS) (Heiser and Pratt 2002) has also contributed for DPR. As the result of these conferences and symposiums, a number of journals and papers have been published. The main goal of this review paper is to provide a full detailed review of RDE for those who are interested in recent accomplishments and thorough knowledge of RDE design concepts.

In order to use the advantages of rotating detonation concept in aircraft engines, there are number of challenges which include both fundamental and engineering problems are yet to be solved. These problems deal basically with low-cost achievement and control of successive detonations in a propulsion device. To ensure the practical use, one needs to improve the (1) thermodynamic scale of cycle efficiency; (2) to reduce the design complexities and geometrical constrains, geometry of the combustion chamber to promote detonation initiation and propagation at lowest possible pressure loss and to ensure high operation frequency; (3) Heat transfer rates and their losses; (4) As the structural components of RDE are subject to repeated high-frequency shock loading and thermal deformations, a considerable wear and tear can be expected within a relatively short period of operation. (5) low-energy source for detonation initiation to provide fast and reliable detonation onset; (6) cooling technique for rapid, preferably recuperative, heat removal from the walls of (Detonation Chamber) DC to ensure stable operation and avoid premature ignition of Fuel–Air Mixture (FAM) leading to detonation failure; Among the most challenging issue, is the problem of durability of the propulsion systems. The other additional problems include noise and vibration.

The paper is organized is such a way that the reader first gets acquainted with a basic idea of detonation based propulsion in pulsed mode, followed by the advantages and disadvantages of PDE (Sect. 2.1) which lead to the development of RDE. Then, based on the significance, a small description of the thermodynamic cycle analysis of RDE, their advantages and existing problems which restrict the technology to come into practical use is discussed (Sect. 2.2). With that clear knowledge of PDE/RDE, a brief historical background of detonation research is presented (Sect. 3). Then, some of the early Russian work of CSD is given (Sect. 4), which is continued by the France works of CDWRE, their advantages (Sect. 5.1), and their collaborative works are mentioned (Sects. 5.2 and 5.3). A turning point in detonative research by the French scientist is also elaborated in this section. Then a thorough survey of numerical investigation of RDE research around the world has been discussed from early 2007 to recent 2014 (Sects. 6.1, 6.2, 6.3, 6.4, 6.5, 6.6, 6.7 and 6.8). Besides, the experimental work of RDE research starting from the late 2011 to up-to-date 2014 has also been discussed (Sects. 7.1, 7.2, 7.3, 7.4, 7.5, 7.6, 7.7, 7.8 and 7.9). The investigation is presented in such a manner that the readers can comprehend the full research flow of RDE very easily. And conclusively, there are some novel concepts of utilizing detonative propulsion phenomenon in space application has been discussed (Sect. 8).

2 Fundamentals

Pulse Detonation Engine

Till last decade, PDEs are attractively considerable attention of research because they promised and provided hope for the performance improvements over existing air-breathing propulsion devices, especially at low flight Mach numbers (Heiser and Pratt 2002). Figure 6.1 shows the first pulsed detonation powered aircraft so called Rutan Long-EZ on Jan 31, 2008. PDE uses detonation waves to combust the fuel and oxidizer mixture. Pulsed mode of detonation engine operates in an intermittent cyclical manner that (1) propellant injected from the closed end, then (2) wave propagation and (3) transition in the tube, finally the (4) wave exist tube, (5) exhaust and (6) purge. Figure 6.2 shows the basic PDE cycle operation. Key issues for further PDE development includes efficient mixing of fuel/oxidizer, prevention of auto-ignition and integration with an inlet and nozzle.

Fig. 6.1 First pulsed detonation powered aircraft, Rutan Long-EZ on Jan 31, 2008 (Heiser and Pratt 2002)

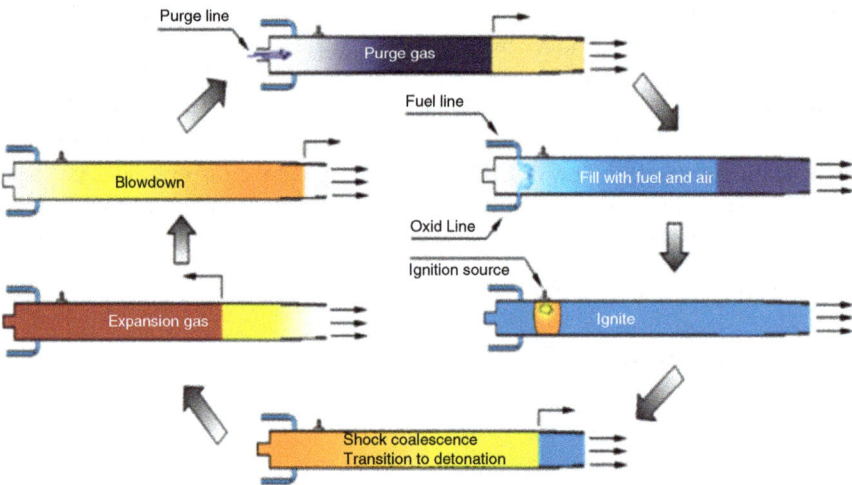

Fig. 6.2 PDE cycle

Historical Background

PDEs have been considered for propulsion for over 70 years. Merely, in the end of the 20th – the beginning of the twenty-first century, most of the Asian (china, Japan, Singapore and South Korea), European (France, Poland, Russia, Belarus) and North American (U.S.A) countries entangled in the research on PDE-based propulsion. PDE typically generate detonative combustion product in order to get high pressure, which in turn converted to thrust performance. It is acclaimed in the 1990's that the performance improvements of the PDE is relatively appreciable than the existing constant pressure conventional air-breathing propulsion engines (Eidelman et al. 1991, 1990a, b; Eidelman and Grossmann 1992; Helman et al. 1986; Lynch et al. 1994; Lynch and Edelman 1994). Further betterments in the thermodynamic efficiency of PDE cycle has been achieved in the late decades (Bussing and Pappas 1996; Kailasanath 1999). This enhancement insisted the researchers all over the globe to interfere intensely in the PDE research. As an outcome of action, the mechanical valve PDE concept (Bollay 1960; Winfree and Hunter 1999) has been renovated to valveless scheme of PDE's with continuous or intermittent airflow within a combined cycle concept (Roy et al. 2000a). In the same decennium, the two-step detonation initiation method the so-called Predetonator approach for PDE's Detonation Chamber (DC) was introduced (Roy et al. 2003a; Brophy et al. 2001). Concurrent studies were executed to enhance the Deflagration to Detonation Transmission (DDT) by including blockages to benefits the DDT to takes place within the distance of the detonation tube of PDE's (Yu 2001; Conrad et al. 2001). Nevertheless, rampant propellant injection in PDE's is a serious dispute which was rectified by stratified-charge phenomenon by Knappe and Edwards (2002, 2001). Further augmentation of detonation initiation in PDE's was brought by means of shock-booster method (Frolov et al. 2001; Roy et al. 2002a) and shock-implosion method (Roy et al. 2000b). Thrust production in PDE research gets appreciated when Pulse-reinitiation concept (Roy et al. 2000c) and Pulse-blasting concept (Borisov 2002) was acknowledged. Subsequently, much effort was made to examine the Multi-tube PDE's (Roy et al. 2002b, 2003b). Complete description of the research accomplished in detonation based propulsion is briefed and discussed in the survey documents that have been published by Wolanski (2013a) and Kailasanath (2011a). By the time mentioned, some of the expected and imminent perspectives of utilizing these detonative propulsion technologies in rockets/space applications were suggested by Roy et al. (2004), Falempin et al. (2006) and Davidenko et al. (2008, 2009b).

Thermodynamic Cycle Efficiency of PDE

The time independent, thermodynamic cycle analysis of the PDE was presented by W.H. Heiser et al. (2002). The most significant conclusion was drawn from the Heiser method is the thermal efficiency of the PDE cycle, called as 'Zel'dovich cycle' nowadays, or the fraction of the heating value of the fuel that is converted to work that can be used to generate thrust. The cycle thermal efficiency is then used to and all of the traditional propulsion performance measures. The benefits of this approach are the fundamental processes incorporated in PDEs are clarified, quantitative comparisons with other cycles are easily made, the influence of the entire ranges of the main parameters that influence PDE performance are easily explored, the ideal or upper limit of PDE performance capability is quantitatively established and that this analysis provides a basic building block for more complex Zel'dovich cycles. A comparison of cycle performance is made for ideal and real PDE, Brayton, and

Humphrey cycles, utilizing realistic component loss models (Heiser and Pratt 2002). The results show that the real Zel'dovich cycle has better performance than the real Brayton cycle only for flight Mach less than 3, or cycle static temperature ratios less than 3 (Heiser and Pratt 2002). For flight Mach greater than 3, the real Brayton cycle has better performance, and the real Humphrey cycle is an over optimistic surrogate for the real Zel'dovich cycle. The T – s diagram for the ideal Humphrey cycle is presented in Fig. 6.3, together with the Zel'dovich and Brayton cycles. (Heiser and Pratt 2002). It is apparent from a comparative inspection of the three ideal cycles displayed in Fig. 6.3 that the thermal efficiency of the ideal Humphrey cycle is close to, but always somewhat less than, that of the ideal Zel'dovich cycle. Figure 6.4 demonstrates that both the ideal Zel'dovich and Humphrey cycles enjoy a significant cycle thermal efficiency advantage over the ideal Brayton cycle. Significant papers on pulse detonation cycles by Heiser and Pratt and Dyer and Kaemming have been discussed and are often cited by researchers for the comparison of engine cycles (Heiser and Pratt 2002; Dyer and Kaemming 2002). These papers are significant to RDE research by establishing the Zel'dovich cycle as the basis for the PDE and introducing the effects of parasitic losses on the cycle.

Advantages/Disadvantages of PDE

Application of the pulsed detonation mode in the aircraft engines offers significant advantages over the existing conventional engines with constant pressure combustion. The compression effect by the detonation gets rid of the necessity of

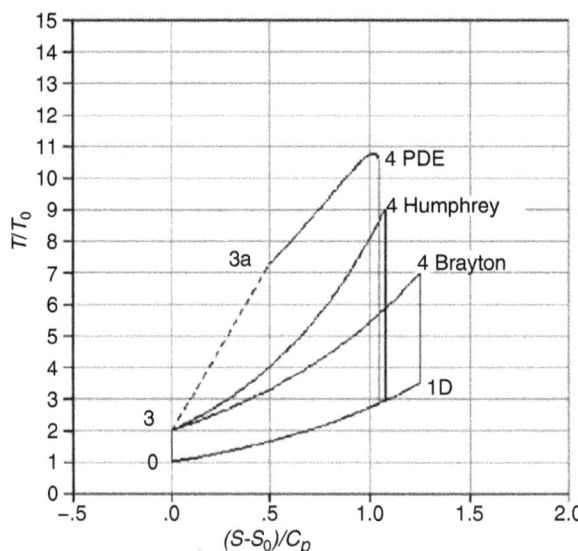

Fig. 6.3 Temperature-entropy diagrams for the ideal Zel'dovich, Brayton, and Humphrey cycles (Heiser and Pratt 2002).

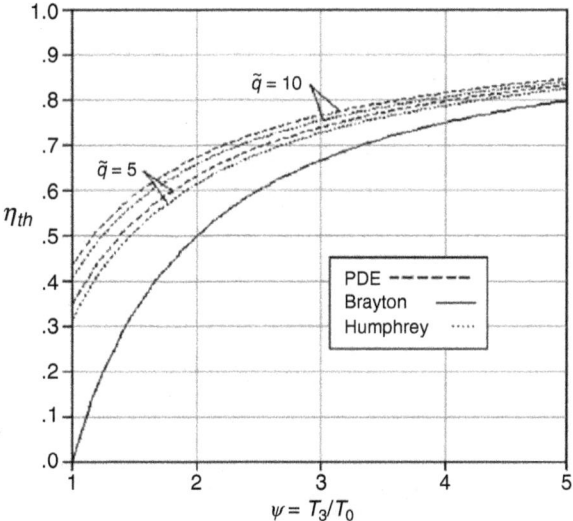

Fig. 6.4 Thermodynamic cycle of PDE (Heiser and Pratt 2002)

compressor as well as turbine. Moreover, the PDE's are scalable, light-weight with high Thrust/Weight (T/W) ratio. Because of no moving parts, PDE seems to be comparably low cost to acquire and operate with board operating range. In combinations with the existing compressor and turbine system, PDE is expected to provide additional thermodynamic efficiency by the compression effect. In the meanwhile, PDE pose some technical disadvantages. The intermittent operation of PDE at order of 100 Hz is the well-known disadvantage emitting high level of noise and vibration. The system integration of PDE exhibited design problem of unsteady turbine interacting with incoming detonation or shock wave. Also, the high temperature of detonation product gas makes the problem even worse. Since, the turbines rotating at high speed cannot withstand high temperature of the detonation gas, the heat transfer problem of turbine and the cooling of the detonation product gas increase the complexity of the PDE-turbine system.

To avoid the problems associated with the intermittent operation of PDE, they've sought for the continuous detonation operation for propulsion while preserving the potential of higher thermodynamic efficiency by using the detonation moving around the annual chamber, widely called as Rotating Detonation Engine (RDE).

Rotating Detonation Engine (RDE)

RDE, a form of continuous detonation-wave engine, is shown to have the potential to further increase the performance of air-breathing propulsion devices above pulsed or intermittent detonation wave engines. The principle of RDE is based on the

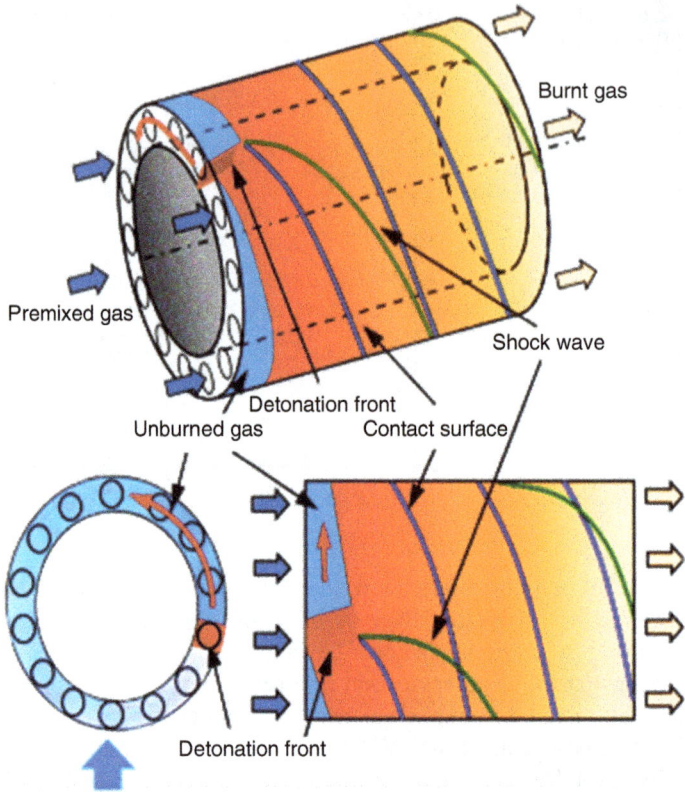

Fig. 6.5 Schematic of rotating detonation engines

creation of direction of tangent Detonation wave travels through disk like path within combustion chamber. Since the detonation wave is moving around the annular chamber at detonation speed the RDE's operation frequency is several kHz, which is same order of magnitude of vibration in ordinary gas turbine. Expanding detonation products from nozzle will produce the thrust. In RDE's, once the detonation is initiated into a combustion chamber containing propellant mixture, then the detonation propagates continuously without any intermission. It is noted, in the mid – 1950's, the efficiency of uninterrupted detonations are quiet impressible than the pulsed or intermittent detonations. From that juncture, rotating detonative propulsion research has drawn attention around the big blue marble. While the PDE poses the loss of thrust by the intermittent operation, RDE could provide better performance than PDE in the same supply conditions. Furthermore, the RDE's geometry is similar to the annular combustion chamber of a conventional gas turbine engine, which makes the system integration much easier with the existing configuration of gas turbine engines (Fig. 6.5).

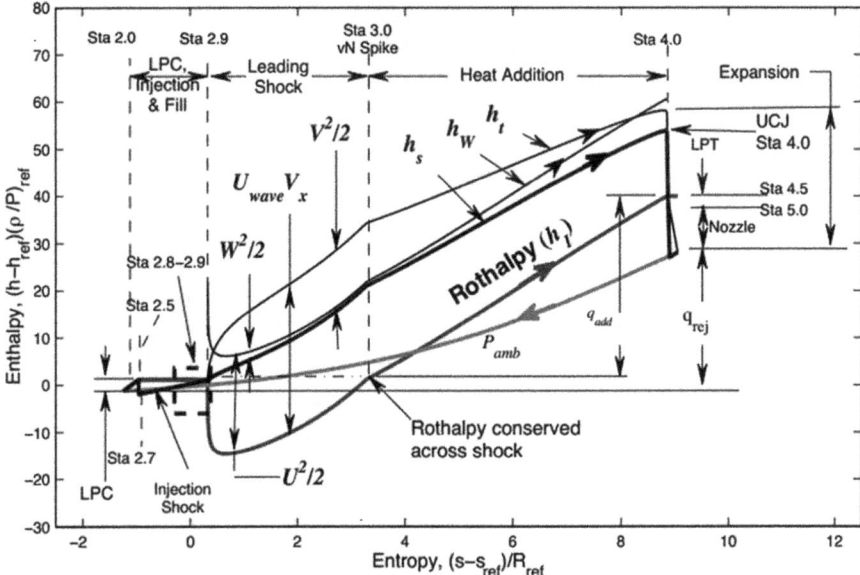

Fig. 6.6 Dimensionless h-s diagram of RDE analytical cycle model (Nordeen 2013)

Thermodynamic Cycle Efficiency of RDE

The thermodynamic efficiency of RDE was recently briefed by Craig A. Nordeen (2013). The general outline of a ZND based RDE cycle has emerged from the studies carried out by Schauer and Lu. Schauer has outlined an evolutionary understanding of the detonation cycle. Lu acknowledged the identification of the ZND cycle in the RDE and notes the differences between the ZND model and the Humphrey cycle (Lu et al. 2011a). The challenge of modeling a RDE is the realization that it is not a pure ZND cycle, but is comprised of a multiplicity of detonation streams, deflagrations and various incidental shocks (Nordeen et al. 2011). Kailasanath acknowledged the significance of multiple streams in the RDE cycle (Kailasanath 2011b).

Braun's approach to the RDE cycle is grounded in the construction of Brayton cycles by constructing the cycle as a single stream from assumed known processes (Braun et al. 2010b, c). The RDE is modeled as a supersonic ramjet. The inlet is isentropic. The significant role of injection losses due to mixing are recognized and modeled as a mass-averaged entropy increase. The detonation is modeled as a rotating wave traveling at the CJ velocity. The wave is assumed to be normal to the azimuthal direction. Expansion is assumed to be isentropic. Energy is used as an approach to analyze performance and identify trends. The performance gap between conventional jet propulsion and scramjet is identified and the RDE proposed as a possible solution. See Fig. 6.6 (Nordeen 2013).

The establishment of the thermodynamic cycle of the RDE as a modified ZND-type cycle is not new, but was disputed. The actual cycle contains multiple processes. Further studies are required to understand the full range of cycle behavior.

Comparison of Cycle Efficiencies of PDE & RDE

S.No	Pulse detonation engines	Rotating detonation engines
1.	PDE will have unsteady throttling and shock losses associated with the opening and closing of the valves	The RDE will have steady shock and viscous losses from flow through choked injectors
2.	Unsteadiness in the PDE flow has been shown to cause system losses	The RDE does see injection unsteadiness due to the passing detonation, which are expected to be of a small order
3.	The PDE detonation upstream flow velocity is zero	The RDE upstream velocity is non-zero. Non-zero inlet velocity is a source of system inefficiency because of expansion cooling of the inlet flow
4.	The upstream conditions of the PDE detonation are mostly isotropic	The RDE detonation upstream flow is exposed to a gradient of all thermodynamic properties
5.	The PDE exhaust undergoes a "blow down" process. (i.e.,) that is, pressure decreases as flow exits the fixed volume of the detonation tube. Each fluid parcel sees a different expansion process line and thus, generates a different amount of entropy	The RDE flow is continuous

Advantages of RDE

Advantages of application of the continuously rotating detonation (wave) combustion process in all jet engines will results in a very compact combustion chamber, and thus the engines will be shorter, simpler and, due to pressure increase in detonative combustion, will execute higher engine performance. It also seems that RDE will be superior over PDE, since they offer continuous thrust generation and can be applied basically to any Mach number. Also RDE will have a lower mass and will be less expensive (Wolanski 2013b). Such detonation engine has a tremendous improvement in the recent decade. The detailed historical background and briefed survey of numerical and experimental work of RDE is presented below.

3 Archival Background of RDE

Antecedent Effort at the Novosibirsk Institute of Hydrodynamics (NIH)

Pulse detonation has a very long history which was conscientiously surveyed by Wolanski (2013b) and Kailasanath (2011b). As it was done for PDE, specific numerical and experimental studies are being performed to assess some key points for the feasibility of an operational RDE. The first and foremost investigation of a Continuous Detonation Wave Engine (CDWE) was carried out in the mid-nineteenth century by Voitsekhovskii et al. (1957, 1958, 1959, 1960, 1963a, b, 1969) at the Novosibirsk Institute of Hydrodynamics (NIH) in Russia (Figs. 6.7 and 6.8).

Their idea of rotating detonation emerges during the investigation of the instability of the plane detonation wave which leads to the formation of intense transverse disturbances. Voitsekhovskii et al. records a rotating detonation with six heads using Dynafax camera in a disk-shaped combustion chamber. He found that in a circular tube, in the presence of single-head spin, the transverse velocity coincides closely with the value $1.84c$ predicted by the acoustic theory. He also mentioned that a region of very high temperature and pressure – a transverse wave – localized in the neighborhood of a detonation tube wall, is known to exist in spinning detonation.

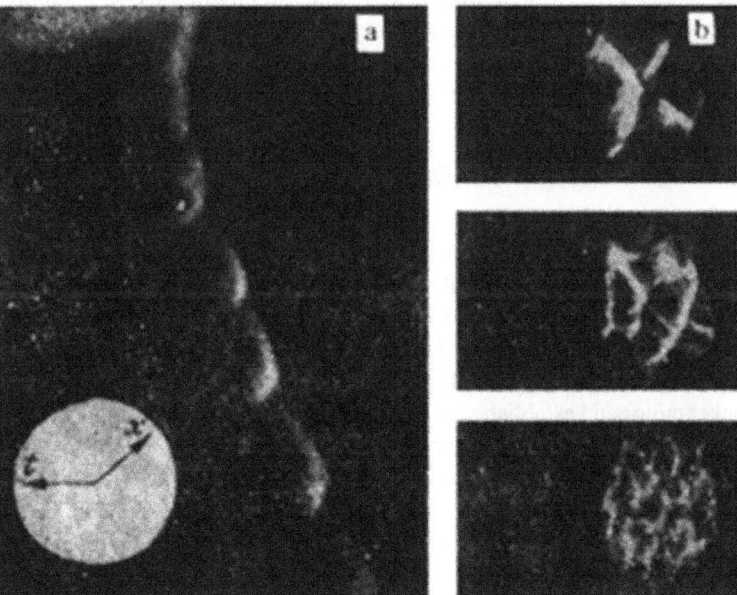

Fig. 6.7 (a) Transverse waves in a convergent cylindrical detonation recorded by the total compensation method (b) at the wave front in a circular tube (Voitsekhovskii et al. 1969)

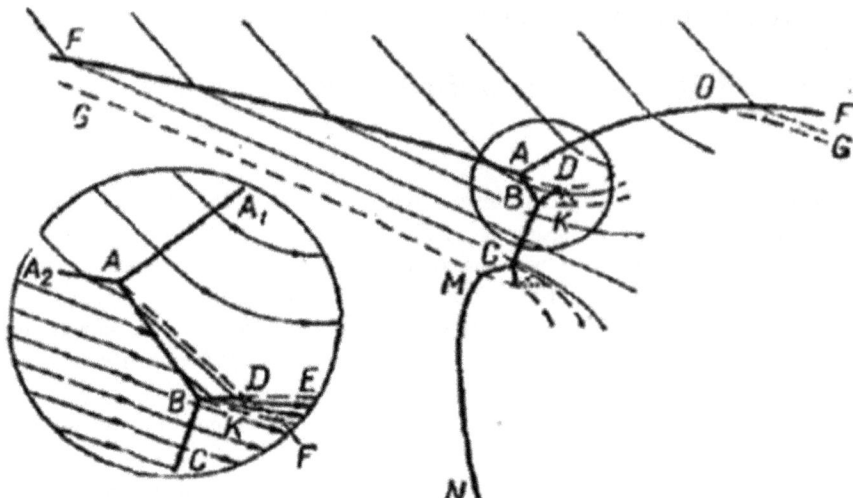

Fig. 6.8 Structure of the first: *BC* transverse wave, *AA1* jog in the primary front, *AA2* primary wave, *AB, BD* jumps coordinating the flow behind the transverse wave and the jog, *ADE, BK'F* contact (Voitsekhovskii et al. 1969)

The rotational frequency of the spinning head is the same as that of tangential acoustic oscillations. He also investigated the feasibility of a CDWE and favorably achieved stationary spinning detonation in the consecutive year (Voitsekhovskii et al. 1969).

Bi-directional Detonation Waves

Mikhailov's work was deliberated as the basics for studying the internal physics of continuous detonation in an annular channel (Voitsekhovskii et al. 1963b). This is followed by the researchers in the U.S. to develop a continuously rotating detonative propulsion engines for rocket applications. Nicholls et al. (1966, b, 1957) from the University of Michigan (UoM) was the first to undergone the feasibility studies of a rocket motor and established a simplex detonation wave. His theoretical examination was appreciated and lengthened additionally to bi-directional detonation waves which revealed the explicit notion of comprehending the RDE concept. Experiments were conducted in a small-scale annular rocket motor and he believed that the injector pattern with the consequent high turbulence level, premature burning, and limited mass flow characteristic was a major part of the difficulty.

The key factors affecting feasibility of the Rotating Detonation Wave Engine (RDWE) were deemed to be performance potential, heat transfer to the wall, the detonation characteristics at very low temperatures and elevated pressures.

Unfortunately, continuous operation of the model motor was never realized, so that important questions such as nozzle performance, actual over-all performance, and detonation in heterogeneous mixtures remain unanswered. A survey of propulsive detonation research studies in the late 1950's were briefed by Oppenheim et al. (1963). Further expedite evolution in RDE research were struggled by virtue of the existent heat transfer complications in rotating detonation wave however simultaneous research in heat transfer was bestowed by Sichel et al. (1998a) and Edwards et al. (1970) (Figs. 6.9, 6.10 and 6.11).

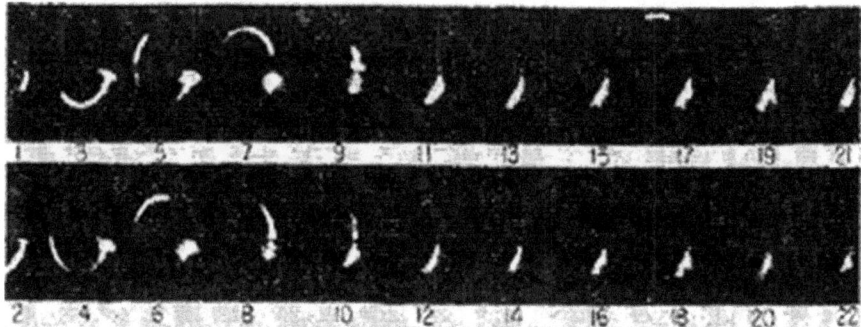

Fig. 6.9 Photographs from typical methane-O2 test using Beckman & Whitley camera (Nicholls et al. 1966)

Fig. 6.10 Cross-sectional view of rotating detonation wave motor (Oppenheim et al. 1963)

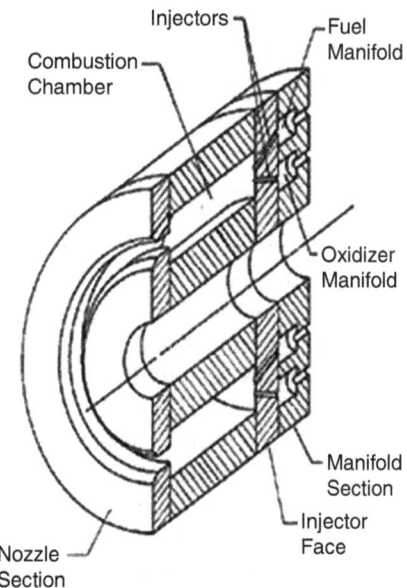

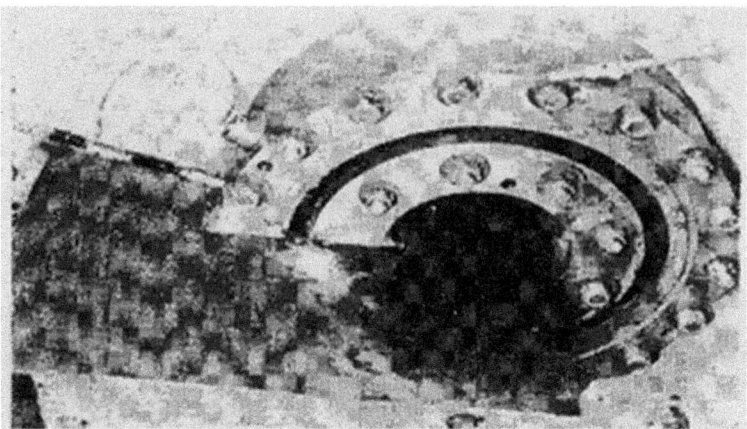

Fig. 6.11 Photograph showing diaphragm holder in annular motor (Nicholls et al. 1966)

Rotating Detonation in Liquid Rockets

Continues work of performance analysis of the rotating detonation waves in liquid rocket engines was carried out by Adamson et al. (1967). In 1973, theoretical calculation of a two-phase detonation in a rotating detonation in liquid rocket motor by Shen and Adamson (1973) is considered as the first and initial step in developing a theoretical analysis of a rotating detonation waves. Different research threads led to the RDE. In one instance, a type of combustion instability was found in rocket combustion chambers where waves rotated around the chambers cylindrical walls (Lu et al. 2011b). Clayton and Rogero (1965) found a rotating detonation like wave concept during a destructive liquid rocket resonant combustion mode. The pressure-wave-to-chamber wall intersection is found to curve in the direction of wave rotation with the nozzle end of the intersection leading the injector end substantially. The wave-to-injector-face intersection extends into the central area of the face, although the wave becomes weak in this area. Although the obtained results are not purported to provide the hypothesized detonation phenomenon, they added insight to the nature of the destructive wave. Figure 6.12 illustrates the triple wave pattern achieved by streak photography along with a tape record of pressure (Ar'kov et al. 1970). Edwards from the advanced research lab group, England has identified the main problem which has been successfully overcome by a gas dynamic valve system whereby differential feed pressures are used for the fuel and oxident to provide a buffer of fuel free oxident between the hot products of the previous wave and the incoming reactive mixture (Ar'kov et al. 1970). Ar'kov et al. discussed some few basic facts related to the effect of High Frequency Instability (HFI) on the combustion chamber, as well as to the nature of the phenomenon caused by the transverse detonation waves itself (Edwards 1977). They proposed that the overloading of a chamber by increasing the fuel atomization, rate of supply, etc. is tantamount to

Fig. 6.12 Symmetrical waves moving in liquid rocket motors (Edwards 1977)

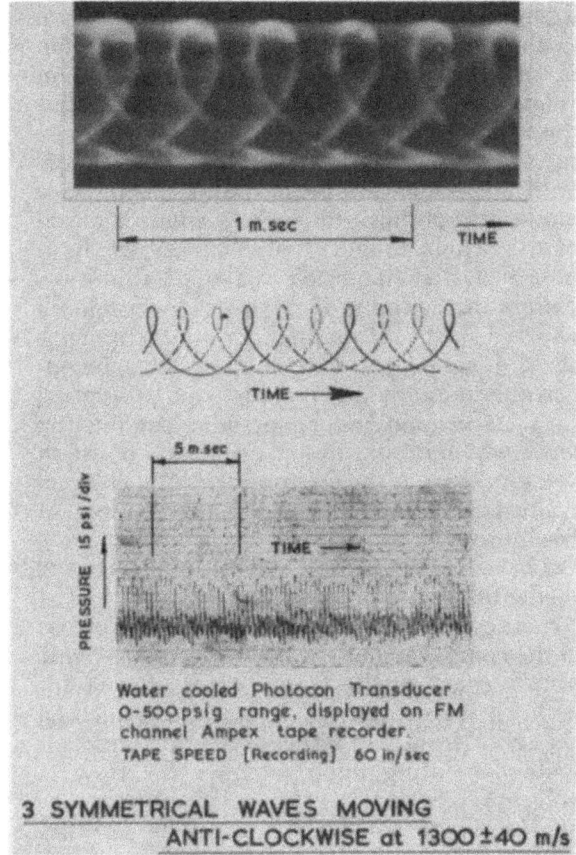

moving away from limit conditions. Moreover they have also mentioned that the most effective means would apparently be the use of transverse waves for burning of fuel. It could ensure a more complete combustion in chambers of reduced dimensions, since the shock waves produce further atomization of fuel droplets and reduce ignition arrests.

4 Continuous Spin Detonation (CSD)

First Mathematical Modeling of Spinning Detonation

In the very beginning of the twenty-first century, the Russian researchers at the Lavrentyev Institute of Technology (LIT) was working on the 2D unsteady mathematical modeling of a continuously spinning detonation wave in a supersonic flow

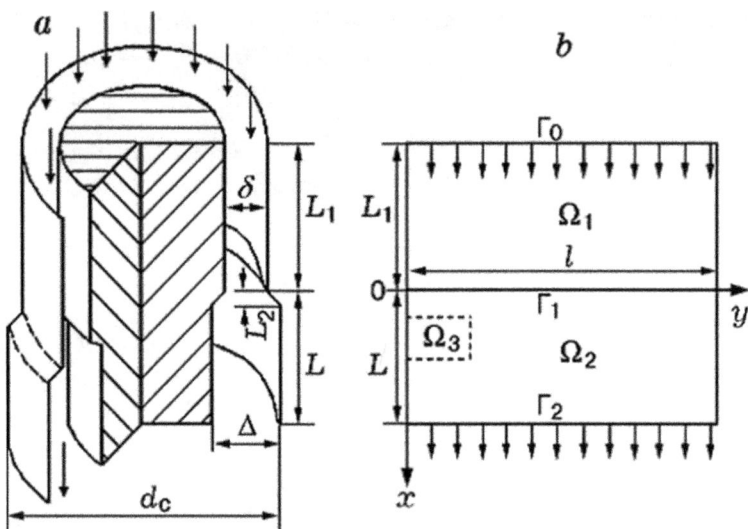

Fig. 6.13 Sketch of an annular detonation and domain of the numerical solution of the problem (Zhdan 2008)

in an annular cylindrical combustor. Their work was supported by the Russian Foundation for Basic Research and by the Foundation of the President of the Russian Federation for Supporting the Leading Scientific Schools. For the first time, the possibility of a continuous spin detonation with a supersonic flow velocity at the diffuser entrance is demonstrated numerically by Zhdan et al. (2008, 2007). It was found that the Mach number of the incoming supersonic flow is restricted for the case ($1 < Mo \leq 3$) of combustion of a H2–O2 mixture in an annular cylindrical combustor. Continuous spin detonation is not realized for Mo > 3. Figure 6.13a shows the sketch of an annular combustor (Fig. 6.13a; combustor diameter d_c, combustor length L, and annular channel width Δ ($\Delta > \delta$)) and Fig. 6.13b shows the domain of the numerical solution of the problem (cut section of the annular domain and unfold it into a rectangular domain $\Omega = \Omega 1 \cup \Omega 2$, which is shown in Fig. 6.13b.)

Dynamics of Transverse Detonation Waves

In Russia, the collaborative work of LIT with NIH has much contribution in the development of RDE (Bykovskii et al. 1980, 1994, 1997, 2004, 2005a, b, 2006a, b, 2008, 2009, 2011). The regime of CSD combustion of a $C_2H_2 - O_2$ mixture in a radial circular channel was employed for the first time by Bykovskii et al. in 2004. The detonation chamber was a coaxial channel; air was delivered into the chamber from a circular collector through a circular slit. At the same time, C_2H_2 was delivered into the chamber through a spray nozzle supplied with in-pair counter-flow channels. The entire process was photographed through the longitudinal windows of the detonation chamber by a photo-chronograph with a falling drum. Figure 6.14 shows the clear image of the transverse detonation waves in $C_2H_2 - O_2$ mixture (Bykovskii et al. 2004). Bykovskii et al. (2005a) studied the structure of Transverse Detonation Waves (TDW) by varying the flow rates of components of the mixture, width of the slot for oxidizer injection, point of fuel injection, and initial ambient pressure. The losses of the total pressure in the flow in oxygen-injection slots and in fuel-injector orifices are estimated. Bykovskii and Zhdan reported the controlled CSD of various fuels in liquid-propellant rocket motors. Later, a method for continuous photographic recording of the CD process with a microsecond resolution has been developed (Bykovskii et al. 2006a). It has been demonstrated experimentally, theoretically, and numerically that a transonic transition occurs in the flow in the CD process if the cross-sectional area of the chamber is unchanged. They concluded that in the case of high-quality mixing, the TDW velocity and structure are extremely stable in a wide range of the ratios of propellant components and in the examined range of pressures in the chamber. Figure 6.15 shows the photographic

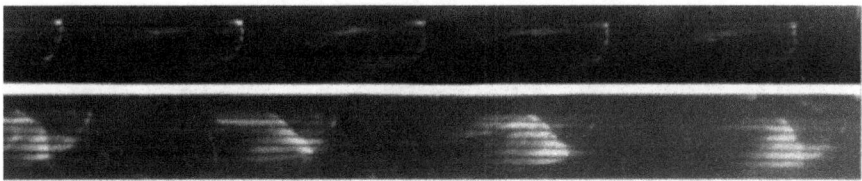

Fig. 6.14 show the clear image of the transverse detonation waves in C2H2-O2 mixture (Bykovskii et al. 2006a)

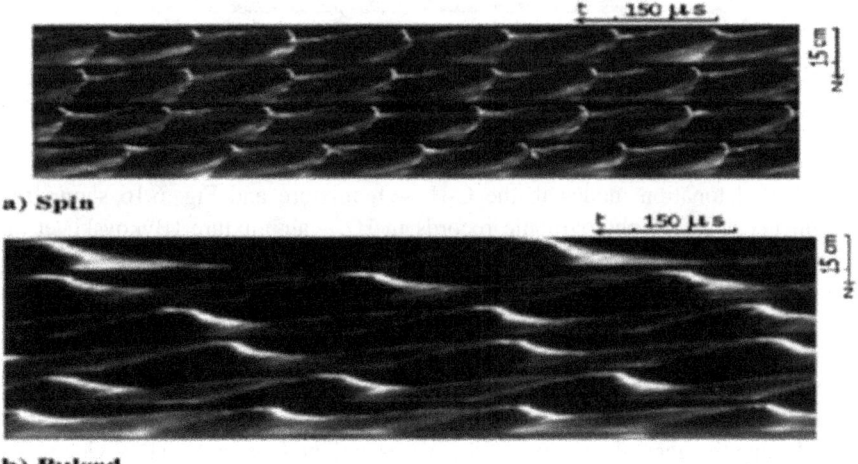

Fig. 6.15 Photographic record of detonation modes in the C2H2-O2 mixture between annular cylindircal (**a**) and expanded duct (**b**) chamber (Bykovskii et al. 2006a)

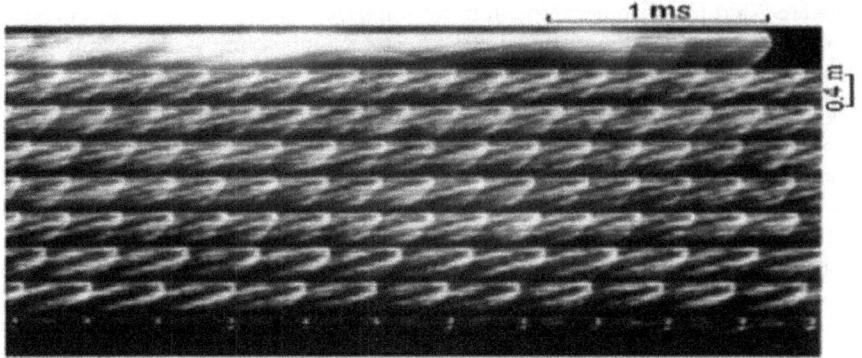

Fig. 6.16 Fragments of TDW photographic records in H2-air mixture (Bykovskii et al. 2006a)

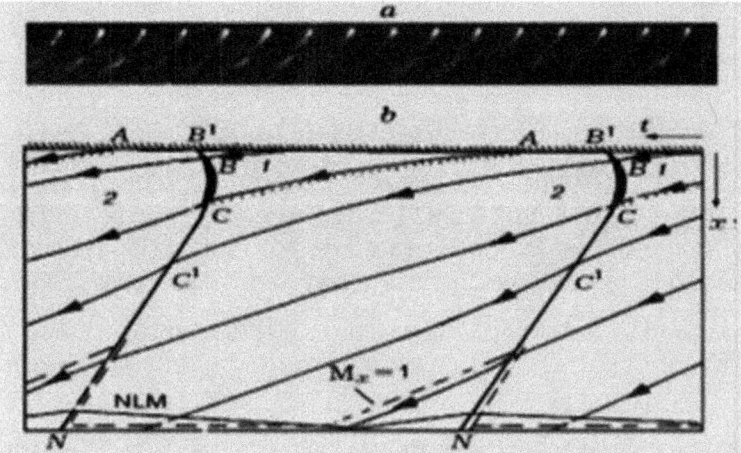

Fig. 6.17 Zoomed-in fragments of the TDW photoraphic records (**a**) and generalized structure of the waves and the flow in the TDW vicinity in the wave-fitted system (**b**): NML is the neutral Mach line (Bykovskii et al. 2008)

record of detonation modes in the C_2H_2 –O_2 mixture and Fig. 6.16 shows the Fragments of TDW photographic records in H2 – air mixture (Bykovskii et al. 2006a). Similarly, for the first time, the recent work includes a comprehensive numerical and experimental study of CSD of a H_2 – O_2 mixture in annular combustors by Bykovskii et al. (2009). A comparison with experiments reveals reasonable agreement in terms of the detonation velocity and pressure in the combustor. The calculated size and shape of detonation fronts are substantially different from the experimental data. A zoomed-in fragment of the TDW records is shown in Fig. 6.17a ($\varphi = 1$), and the general structure of the waves and the flow in the TDW vicinity in the wave-fitted system is illustrated in Fig. 6.17b.

5 Continuous Detonation Wave Rocket Engines (CDWREs)

What Is CDWRE

The main principle difference between CDWREs and currently existing rocket engines is an annular shape instead of a cylindrical combustion chamber. The working of CDWREs (Lentsch et al. 2005) includes, (i) fuel is injected continuously from one end; (ii) a detonation wave is initiated, which keeps rotating and turns into a multi-wave regime; (iii) finally, the detonation products are expelled on the opposite side of the injector plate. The conceptual sketch of such an engine is shown in Fig. 6.18 (Adamson and Olsson 1967). The primary advantages of CDWRE is significantly shorter combustion chamber without throat leading to a lower engine mass, higher chamber lifetime, increase in specific impulse, reduced sensibility, only starting ignition and higher chamber reliability (Lentsch et al. 2005).

First Experimental Verification of CDWRE

In the very beginning of the twenty-first century, the Laboratoire de Combustion et de Détonique (LCD) in Poitiers, financed by Centre National d'Etudes Spatiales (CNES) and Centre National de la Recherche Scientifique (CNRS), started the experimental research to evaluate the feasibility of rotating detonation rocket engines. This is considered as the first ever step in experimental verification of CDWRE in Western Europe and around the big-blue marble (Lentsch et al. 2005). Even though, only preliminary information's and qualitative data was obtained in the start, the main goal of the CDWRE study is to fully demonstrate the new engine

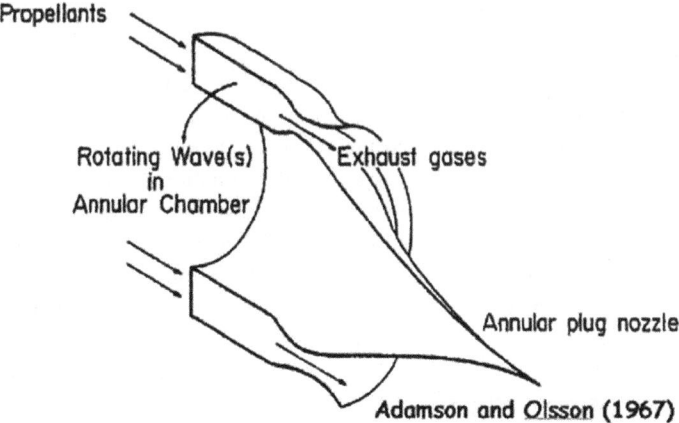

Fig. 6.18 Concept of CDWRE (Adamson and Olsson 1967)

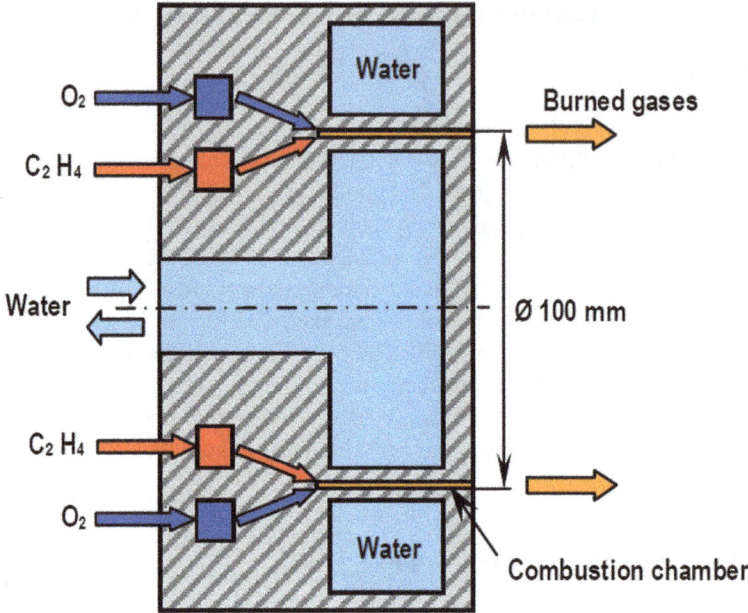

Fig. 6.19 First experimental work in LCD, France (Lentsch et al. 2005)

concept with sub-systems such as the propellant injection system, ignition system, etc. Figure 6.19 (Lentsch et al. 2005) shows a schematic cross section of the annular combustion chamber.

Collaborative Studies of MBDA with Lavrentyev Institute of Hydrodynamics (LIH)

Recently in French, both Falempin group in MBDA & Bykovskii group in LIH worked together in designing an operational CDWRE usable for the upper stage of a space launcher has been performed taking into account all engine/airframe integration issues in order to optimize the benefit of detonation wave engine (Falempin et al. 2006). Then, specific experimental works have been undertaken to address some key issues like noise generated by a CDWE operating at several kilo-Hertz, heat fluxes (intensity, areas) and cooling strategies, composite materials (Carbon/Silicon Carbide) compatibility, engine thrust vectoring capability (Falempin et al. 2006). Some preliminary simulations were performed using FLUCEPA code and found that this code was sufficiently robust to deal with the large pressure and velocity gradients found in a detonation front without the need to use mesh adaptation.

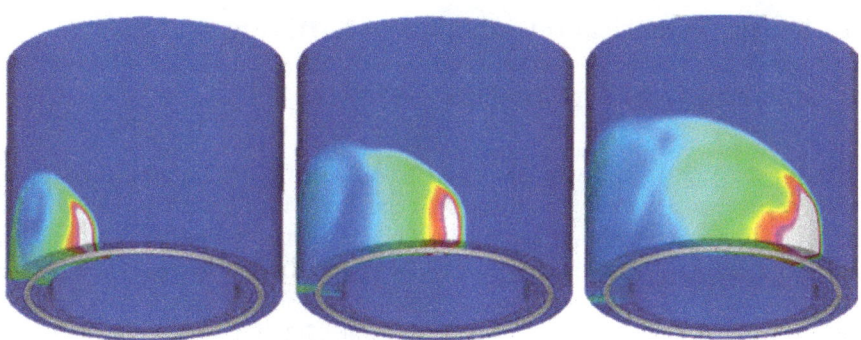

Fig. 6.20 Detonatio propagation in an annular chamber 10, 20 and 30 ms after ignition (Falempin et al. 2006)

Falempin et al. (2008) has proposed that compared to a PDE, the CDWE design allows an easier operation in reduced-pressure environment and an increase in engine Mass Flow Rate (MFR) and T/W ratio. Figure 6.20 shows the detonation propagation in an annular chamber 10, 20 and 30 ms after ignition. It was found, increasing specific MFR increase both lean & rich ignition limits of the engine and it was possible to start TDW at low combustion chamber pressure in a wide range of Equivalence Ratio (ER) (Falempin et al. 2006). One of the peculiarities of a CDWE is that the number of detonation waves inside the chamber is not constant and is a function of the combustible mixture, the combustion chamber geometry and the MFR (Falempin et al. 2006).

Meticulous research of both numerical and experimental studies of the CDWRE for space application was carried out by Davidenko et al. (2008, 2009b, 2011). He simulated a 2D Euler model of a CDWRE combustion chamber to investigate the effect of geometrical and injection parameters on the internal process and performance (Davidenko et al. 2009b). Furthermore he also carried out experimental studies to investigate the liquid oxygen breakup and vaporization in a helium flow as well as detonation initiation and propagation in a spray of liquid oxygen/gaseous hydrogen.

Some of the late existing challenges in experimental and numerical modeling of the rotating detonation wave propulsion and engine concepts were briefed by Lu et al. (2011c). Lu et al. have been simulated a more realistic environment of a rotating detonation in an annular chamber without any major drop in performance. The structure and cooling capability of the engine itself appears to have limited many of the tests by creating auto-ignition conditions and causing the rotating detonation to fail. Many recent investigations such as turbulence, non-uniform mixing, wall curvature, have addressed and there has been steady progress with understanding how they relate to an efficient RDE design (Lu et al. 2011c).

6 Recent Numerical Investigation of RDE (Fig. 6.21)

Stability and Sensitivity Analysis

In Japan, the Nagoya University (NU) and the Aoyama Gakuin University (AGU) in collaboration with Warsaw University of Technology (WUT) and Japan Aerospace Exploration Agency (JAXA) have involved in the sensitivity analysis of RDE with a detailed chemical kinetics model. Hayashi et al. (2009) proposed that the flow properties such as inlet pressure, Mach and T have a significant effect on rotating detonation device performance. Milanowski et al. (2007) have numerically stabilized a rotating detonation propagating at nearly C–J velocity on a 2D simple chemistry flow-model. Under purely axial injection of a combustible mixture, the rotating detonation is proven to give no average angular momentum at any cross section. Figure 6.22a, b shows the pressure profiles at 300 K and at two reservoir pressures; (a) 1.5 MPa and (b) 1.3 MPa. Figure 6.22c, d shows the T profiles in a RDE. In the Fig. 6.22, the T profiles give the rotating detonation structure more drastically and when the initial reservoir pressure is lower than the critical value, the detonation collapses that its front leaves from the inlet wall (Hayashi et al. 2009). Both experimental and theoretical analysis of the stability of RDE with different velocities consisting of two coaxial cylinders was carried out by Fujiwara et al. (2009). Further optimization study of RDE systems was undergone by Yamada et al. (2010) in 2010. The main aspects of their detonation study includes rotating detonation limits and specific impulse in doubled computational area and increased ignition energy, and they proposed that the lower detonation limit gets the computational area effect, but the upper one does not (Yamada et al. 2010).

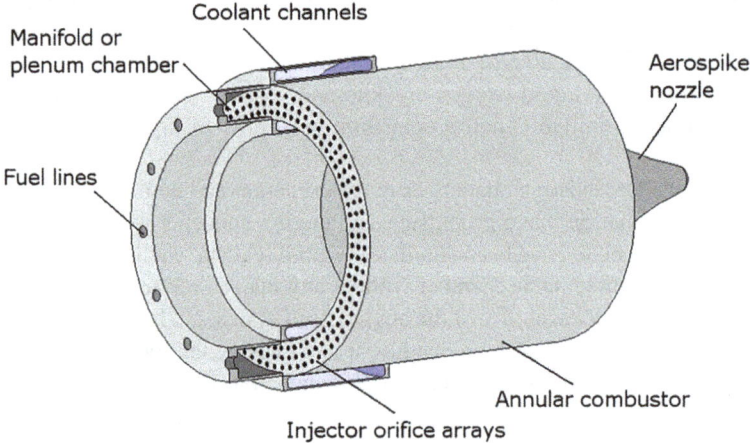

Fig. 6.21 Numerical modeling of the rotating detonation wave propulsion (Lu et al. 2011c)

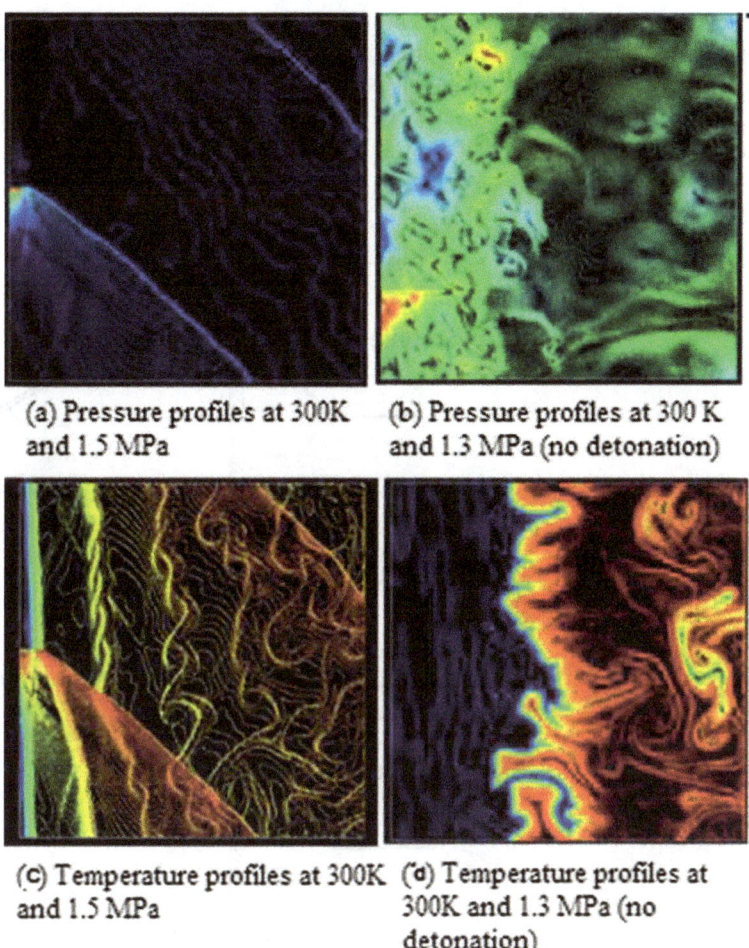

Fig. 6.22 (**a**, **b**) show the pressure profiles at 300 K and at two reservoir pressures; (**a**) 1.5 MPa and (**b**) 1.3 MPa. (**c**, **d**) show the temperature profiles in a RDE (Hayashi et al. 2009; Milanowski et al. 2007; Fujiwara et al. 2009; Yamada et al. 2010)

Propulsive Performance Analysis

There have been some attempts to experimentally and numerically analyze the propulsive performance of RDE by Daniau et al. (2005), who experimentally measured the thrust in a small rocket chamber. Another performance analysis study was carried out by Hishida et al. (2009), who numerically studied 2D RDE with a 70% argon-diluted hydrogen and oxygen mixture. Recently, propulsive performance of a rotating detonation based propulsion systems was carried out by Yi et al. (2009b, 2011). It was found that the propulsive performance is strongly dependent on the mass flow rate of an injected mixture but it is weakly dependent on the axial chamber length and the number of detonation waves. However, the effect of geometric parameters is insignificant on the engine performance (Yi et al. 2009b).

The injection system is modeled with three different injection conditions, depending on the left-wall pressure as follows. In Fig. 6.23a, the detonation wave is

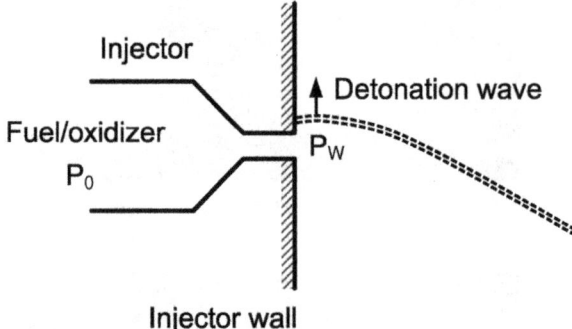

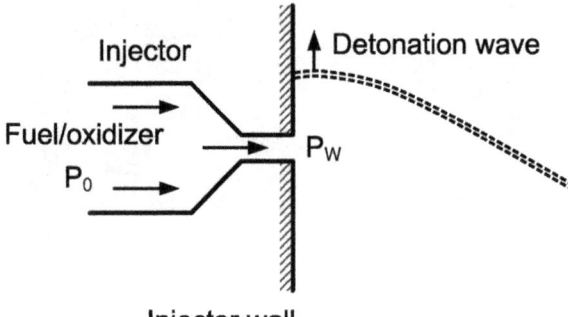

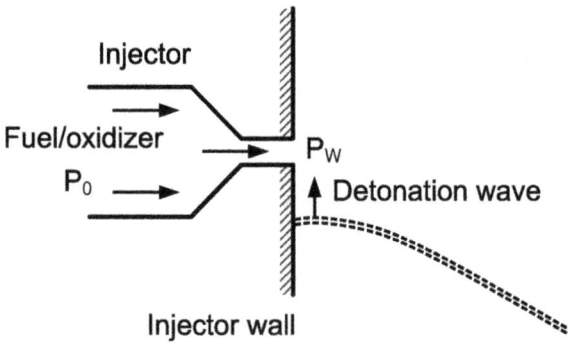

Fig. 6.23 Injector conditions (Yi et al. 2009b)

a) **No injection:** $p_w \geq p_0$

b) **Subsonic injection:** $p_0 > p_w > p_{cr}$

c) **Sonic injection:** $p_w \leq p_{cr}$

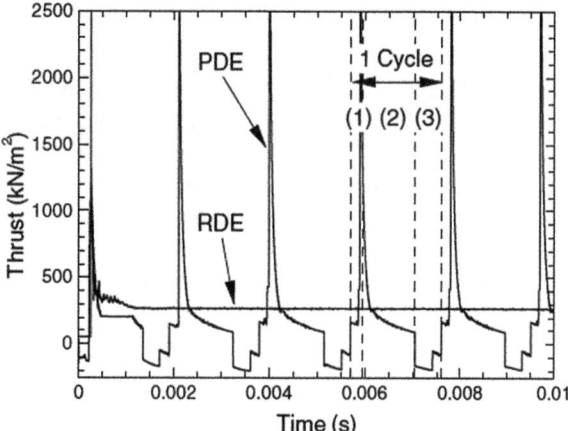

Fig. 6.24 Temporal evolution of thrust for performance calculation (Yi et al. 2009b)

located just in the front of an injector so that the wall pressure (p_w) is greater than the injection total pressure. In Fig. 6.23b, the wall pressure is less than the injection total pressure, but greater than the critical pressure (p_{cr}) obtained at a choking condition. In Fig. 6.23c where the detonation wave propagates far from or before the injector, here the p_w is less than the p_{cr}. It was demonstrated that the CRDE produces a total impulse higher than that of a single-tube PDE and the nozzle effect is not significant, compared with that of PDE (Yi et al. 2009b). Further effects of nozzle shape, angle and length on the propulsive performance of CRDE is numerically investigated in a 3D annular chamber with the one-step chemical kinetics of a H2/air mixture by Yi et al. (2011). They concluded that the RDE has the negligible nozzle effect on the propulsive performance, compared with that for PDE. Figure 6.24 gives the temporal evolution of thrust per unit area, indicating that because of purging and filling processes in the PDE operation, it is difficult to obtain high operating frequency, resulting in lower thrust. Figure 6.25 gives the temporal evolution of total impulse per unit area, indicating that the total impulse of the RDE linearly increases with time, but the PDE shows the discontinuous behavior in the total impulse, which is caused by a sudden increase in the thrust.

Heat Transfer Analysis

Recently, the Semenov ICP (Frolov et al. 2012) in Moscow has developed an efficient computational tool for transient 3D numerical simulation of RDC (Rotating Detonation Chamber) which has the capacity to solve various RDE issues such as chamber design, thermal management, operation control and conjugate heat transfer and fluid – structure interaction.

Resolution Study of RDE

Recently, Tsuboi et al. (2013) have involved in simulating 2D and 3D RDE by using a recent H2/O2 full chemistry model to estimate (1) the shock structure (2) Thrust performance and (3) Effect of resolution. They concluded that the Specific Impulse (Isp) with the diverging nozzle increases more than that without the nozzle, and also the time-averaged thrust with the diverging nozzle also increases more than that without the nozzle. Figure 6.26 gives the two different resolution study of a flow filed by using 2nd order MUSCL scheme.

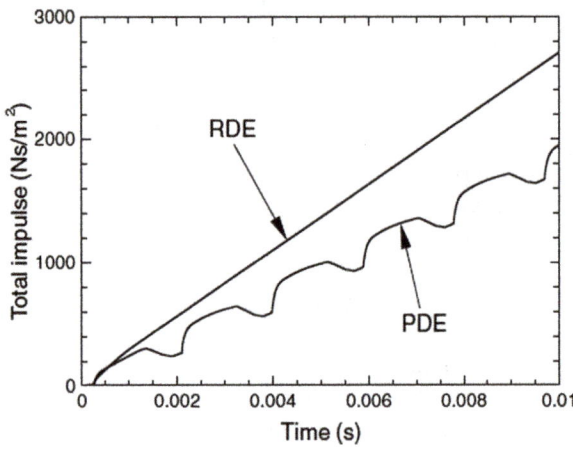

Fig. 6.25 Temporal evolution of impulse for performance calculation (Yi et al. 2009b)

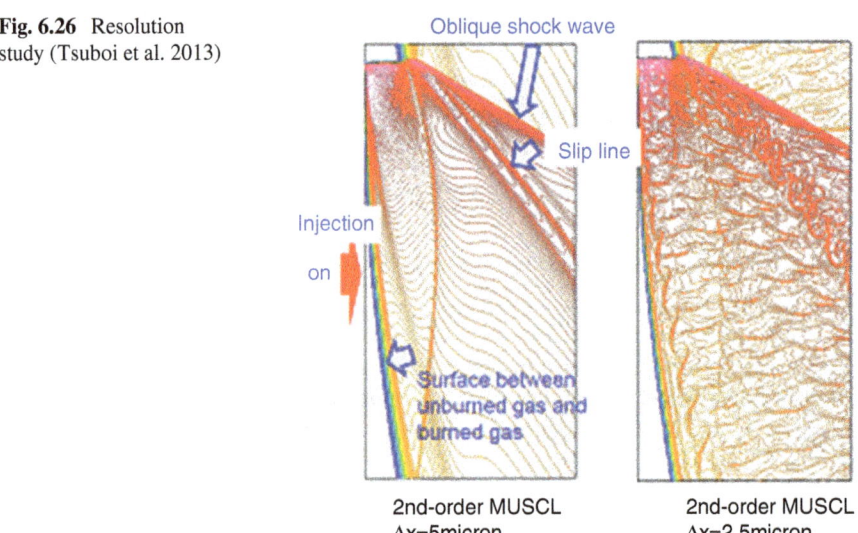

Fig. 6.26 Resolution study (Tsuboi et al. 2013)

Particle Tracking Analysis of RDE

In Peking University (PU), Zhou and Wang (2012) carried out a 2D and 3D particle tracking analysis and concluded that the particles' paths only have a small fluctuation along the azimuthal direction (About 8% of circumference of chamber) (Zhou and Wang 2012). Flow particles are injected into the combustion chamber, burned by the detonation wave or the deflagration wave, and then rapidly ejected almost along the axial direction to generate great thrust. When the flow particles encounter the detonation wave or the oblique shock wave, their paths are deflected. When the flow particles encounter the deflagration wave or the contact surface, their paths are not deflected. 3D path's fluctuant trend along the azimuthal direction is coincident with 2D path; 3D path's deflection is gentler than the 2D path. Figure 6.27 shows

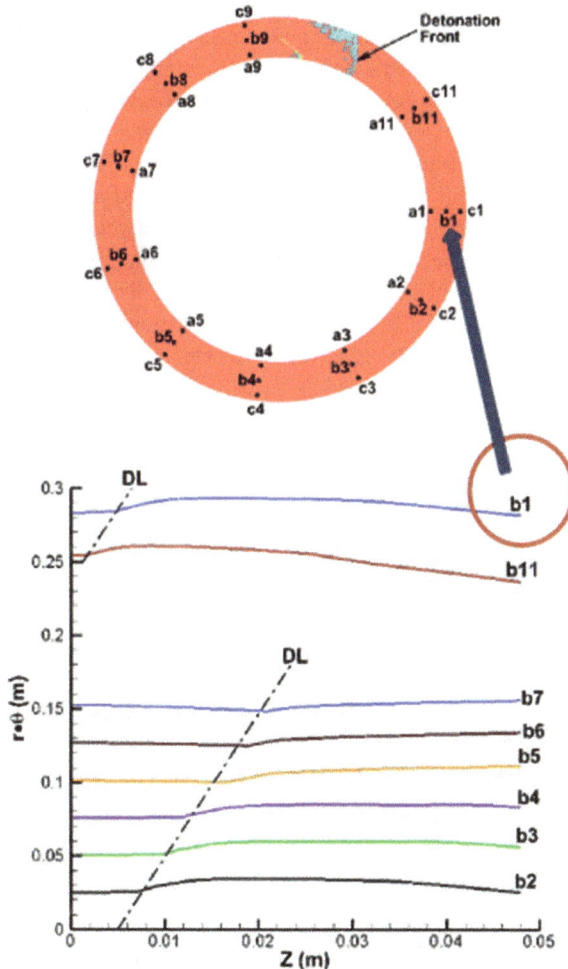

Fig. 6.27 Particle Tracking Analysis of RDE (Zhou and Wang 2012)

the 3D particle tracking technique utilized for particle path analysis. Moreover in thermodynamic point of view, Zhou and Wang (2012) have also proposed that the (1) 3D's p-v and T-s diagrams are consistent with 2D results; (2) they are all consistent with the Ideal ZND Model perfectly; (3) the maximum pressure in the 3D p-v diagram is higher than the 2D p-v diagram; (4) entropy increment in the 3D T-s diagram is slightly smaller than the 2D; (5) The mass flux and specific impulse of RDE increase with the stagnation pressure p_0 of inlet flow.

Recent Numerical Investigation at NRL

Very recently, time-accurate calculations of 2D and 3D RDE models was studied to using time-step-splitting and FCT (Flux-Corrected Transport) algorithm by Schwer and Kailasanath (Schwer and Kailasanath 2010, Schwer et al. 2011; Schwer and Kailasanath 2011; Kailasanath and Schwer 2013) at Naval Research Laboratory in U.S.A. It was found that the specific impulse was dependent on pressure ratio, whereas the mass flow and propulsive force were primarily dependent on the stagnation properties of the inlet micro-nozzles. Figure 6.28 shows a typical wave-structure within the RDE. In Fig.6.28, (A) Detonation Wave, (B) oblique shock wave, (C) slip line between freshly detonated products and older products, (D) secondary shock wave, (E) mixing region between fresh premixed fuel-air gases and detonated gases, (F) region with blocked injection micro-nozzles, and (G) unreacted pre-mixed fuel-air mixture injected from the micro-nozzles. In NRL, investigation on the effect of different engine sizing parameters on MFR, performance, and thrust of RDE is carried out. Some of the clear-cut understanding of the internal physical behavior of the combustion chamber of RDE is ongoing in the NRL. Recent RDE studies includes flow-field description, stagnation/back pressure effect, engine sizing effect, 3D effect, simulation of specific rigs, injection/inflow effects and preliminary fuel-air effects. Results indicate that for many of the above parameters, the

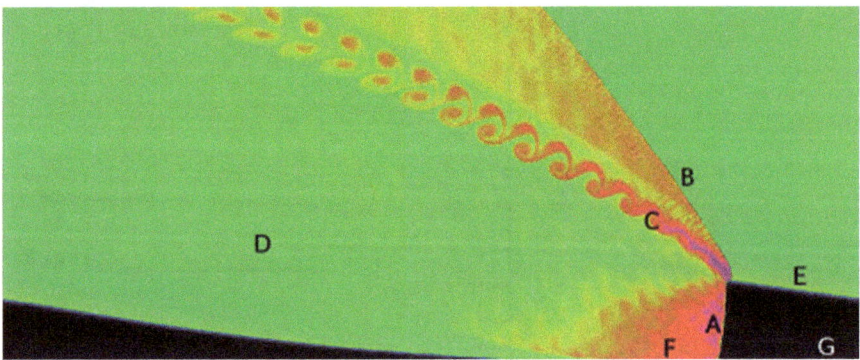

Fig. 6.28 Temp. solution as chamber length is increased (Schwer and Kailasanath 2010)

characteristics of the engine scale in predictable ways for high plenum pressures. A subsequent examination by Nordeen et al. (2014) confirms the operation of the Euler equation in a 3D numerical simulation and is consistent with the modified ZND model.

The Euler constant of integration or rothalpy (Relative total enthalpy) and the wave velocity were extracted from the simulation data by means of a linear regression. Figure 6.29 shows the 2-D time-accurate simulation temperature with time-averaged streamline and Fig. 6.30 shows the time-averaged rothalpy. A complex relationship is suggested by regression fit with area ratio squared. But, a 2nd order relationship is not inconsistent with physics of RDE flow.

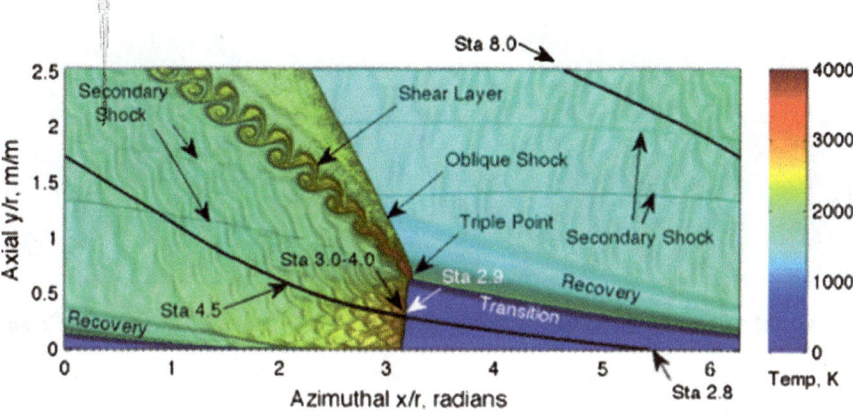

Fig. 6.29 2-D Time-accurate simulation temperature with time-averaged streamline 18 (Tsuboi et al. 2013)

Fig. 6.30 Time-averaged rothalpy (Tsuboi et al. 2013)

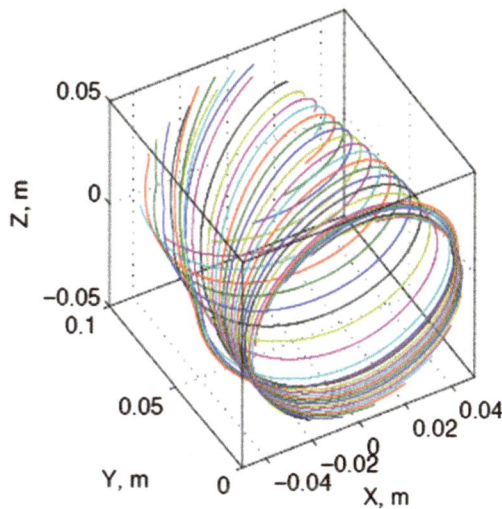

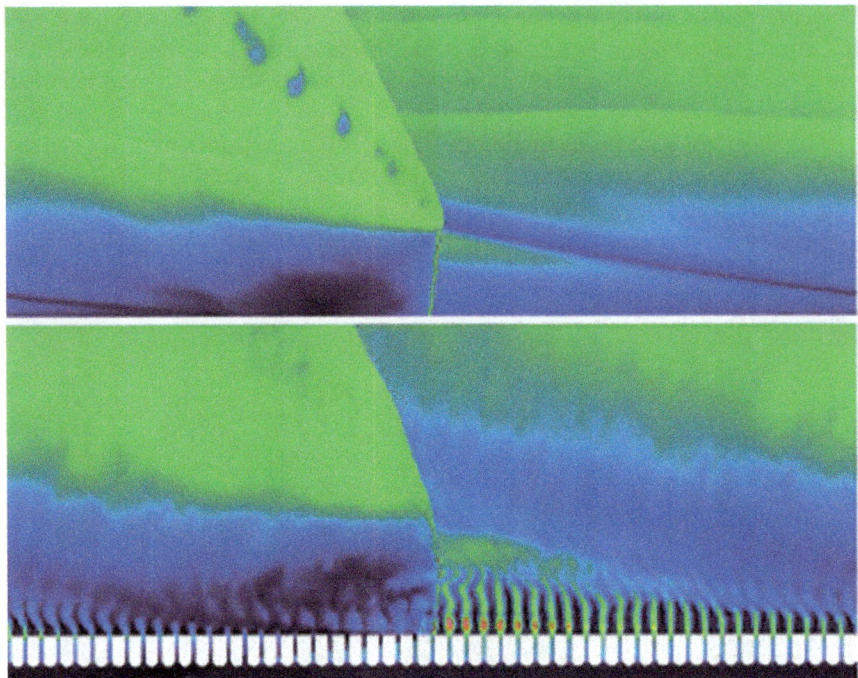

Fig. 6.31 Mach number comparison for ideal injectors (*top*), and 2D injectors (*bottom*) (Kailasanath and Schwer 2011)

RDE injection process is critically important to the stability and performance of the RDE. This has allowed the researchers in NRL to do a variety of parametric studies on the effect of plenum pressure, back pressure, and engine geometric parameters (Kailasanath and Schwer 2011). Figure 6.31 shows the Mach number comparison for ideal injectors (top), and 2D injectors (bottom). The results show that the RDE simulation is very sensitive to how these injectors are modeled; but, the more stable ideal injection approximation still provides valuable information on the influence of different parameters on overall performance. The current work examines the flow field, exhaust flow characteristics for low pressure RDEs and include 3D effects & effects of an exhaust plenum (Kailasanath and Schwer 2014). Figure 6.32 shows the Maximum pressure over one cycle for 3D RDE simulation with no exhaust plenum. Further investigation should also be done to find if an appropriate boundary condition fix can be made to 2D simulation results to bring them into agreement with the much expensive 3D + EP results. In addition to the milestones, Schwer et al. (2014) took the efforts to develop a new code, Propel, for simulating complex engine designs. Their work also compares 2D & 3D solutions using Propel for a detonation tube and baseline RDE with current RDE simulation

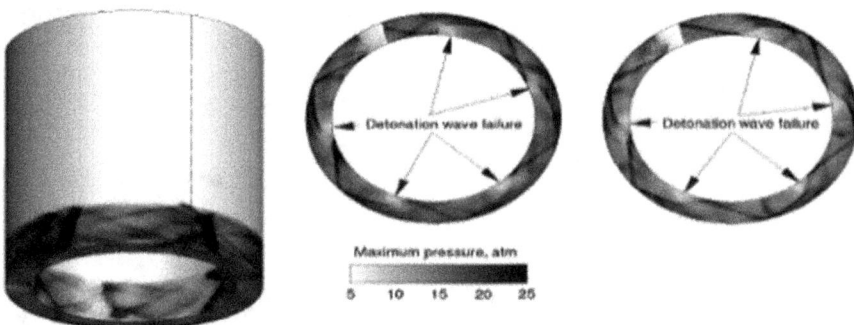

Fig. 6.32 Maximum pressure over one cycle for 3D RDE simulation with no exhaust plenum (Kailasanath and Schwer 2011)

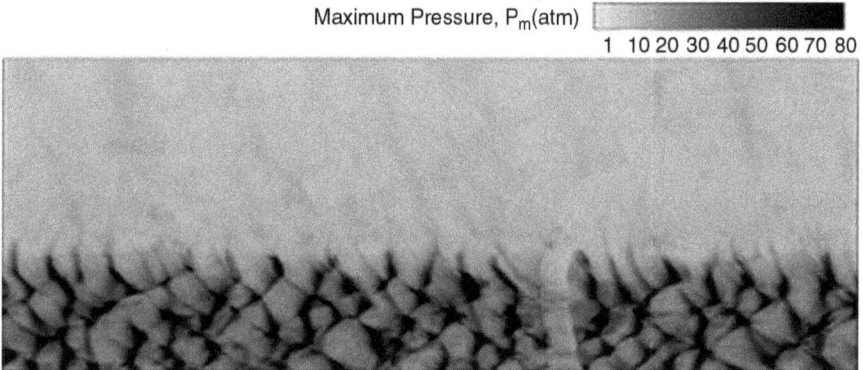

Fig. 6.33 Unrolled pressure maximum plot for inner walls with 2D propel (Schwer et al. 2014)

tool. Figure 6.33 shows the unrolled pressure maximum plot for inner walls with 2D Propel and Fig. 6.34 shows the unrolled pressure maximum plot for outer walls with 2D Propel.

In pursuit of greater thermal and propulsive efficiencies in rockets or gas turbines, a 1D thermodynamic model of RDE is compared to a numerical simulation model with good results by Nordeen et al. (2011). A ZND detonation model is modified to include stagnation properties and account for the velocity vectors that occur upstream of the detonation. Figure 6.35 shows the Static P-V diagram of streamlines & ZND cycle; Fig. 6.36 shows the static H-S diagram with streamlines & ZND cycle and Fig. 6.37 shows the Stagnation P-V Diagram of streamlines & modified ZND cycle. Velocity triangles, commonly used in the gas turbine industry, are shown to be an effective tool for understanding energy transfer in RDE's.

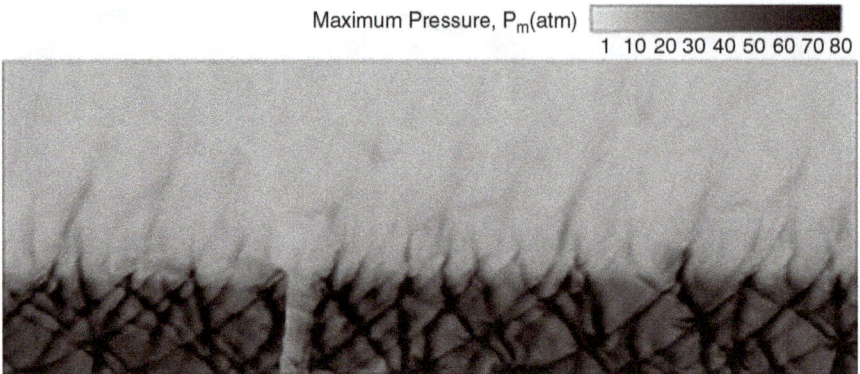

Fig. 6.34 Unrolled pressure maximum plot for outer walls with 2D propel (Schwer et al. 2014)

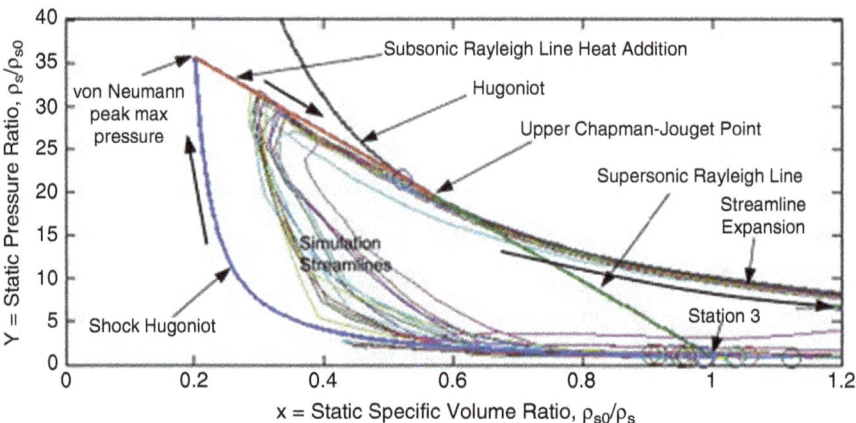

Fig. 6.35 Static P-V diagram of streamlines & ZND cycle (Nordeen et al. 2011)

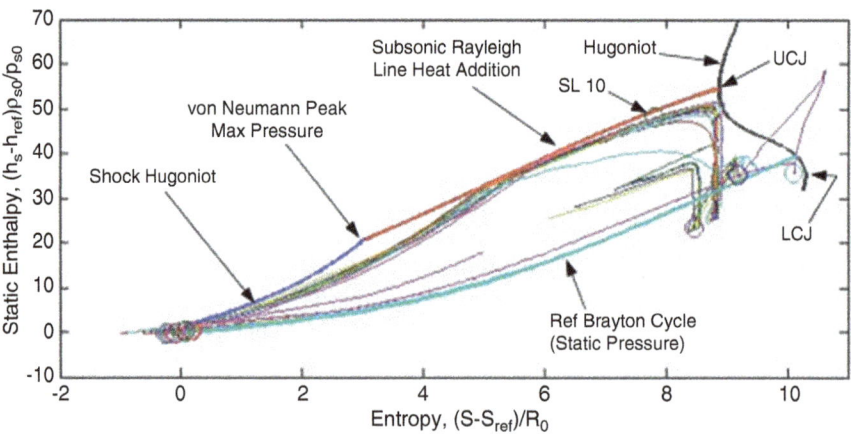

Fig. 6.36 Stagnation P-V Diagram of streamlines & modified ZND cycle (Nordeen et al. 2011)

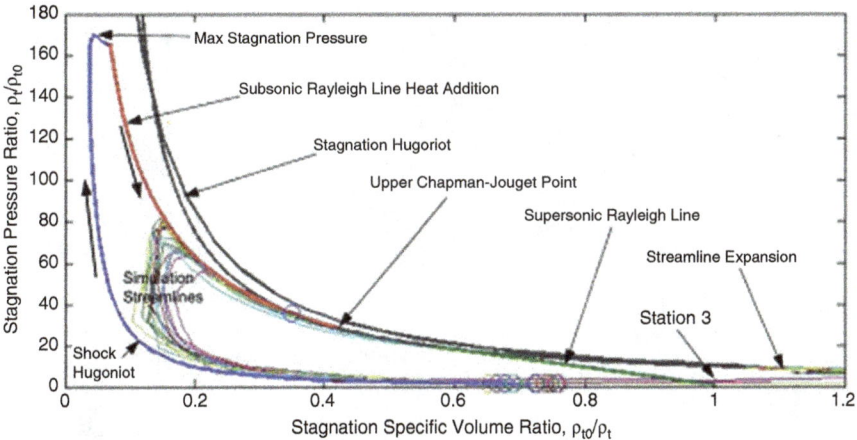

Fig. 6.37 Static H-S diagram with streamlines & ZND cycle (Nordeen et al. 2011)

RDE Injection and Design Criteria

Premixed operation of an RDE has proven difficult, with the flame igniting the supply upstream of the RDE. As a consequence, much energy has been put toward the design of a mixing system that occurs near or at the inlet of the RDE. Designs utilizing many thin 2D slots have shown some promise, however ignition of the incoming flow due to dead zones behind the plates has occurred.

The current study by Gutmark (2014) aims to investigate more complex three dimensional RDE injection and mixing designs and to reduce the ability to flame-hold, and ensure fewer hot dead zones in the post-detonation refill. Figure 6.38 shows the injection concept (left) with mesh showing refinement near throat (right). Arrows show hydrogen injection direction. Figure 6.39 shows the top and front views of H2 contours from steady operation for 3 different early injection designs during steady operation. The effect of various 3D geometries are studied, including central pintel injection and side wall simple constriction injection, along with modifications to a currently viable experimental design. The simulations have shown a novel RDE design which shows promise, with modification, of successfully repeating a detonation over it with reduced risk of flame-holding, which may result in RDE failure.

RDE Work at NASA

In recent times, NASA has described a semi-idealized 2D RDE simulation which operates in a detonation frame of reference (Paxson 2014) and utilizes a relatively coarse grid such that only the essential primary flow field structure is captured. This construction yields rapidly converging steady solutions and found to be in reasonable agreement with the results of a more complex and refined code. Figure 6.40

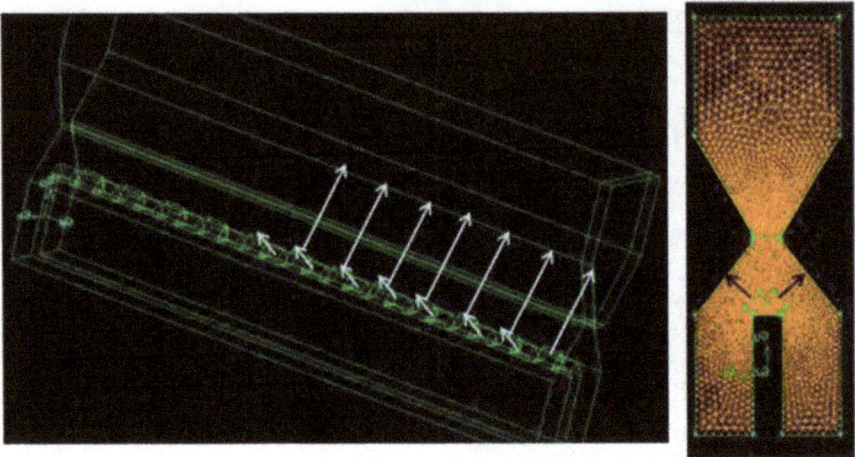

Fig. 6.38 Injection concept (*left*) with mesh showing refinement near throat (*right*). *Arrows* show hydrogen injection direction (Gutmark 2014)

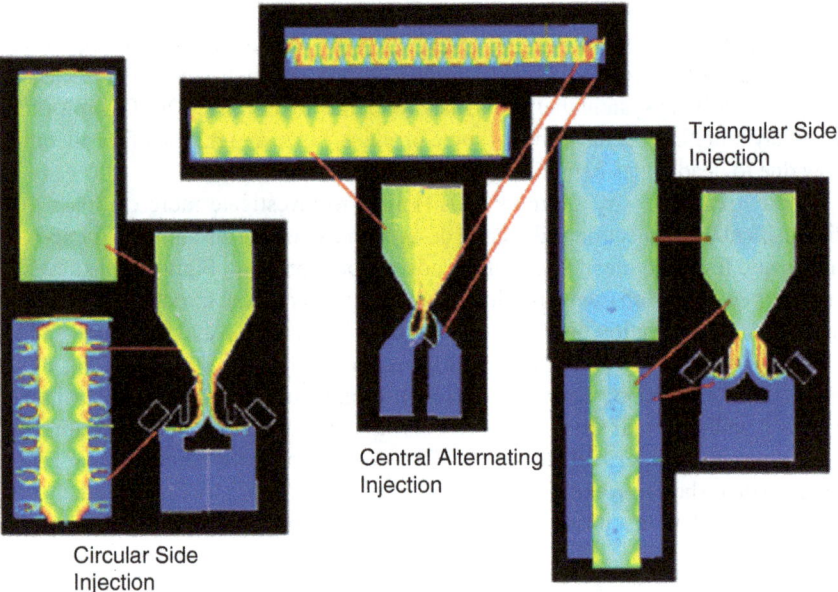

Fig. 6.39 *Top* and *front views* of H2 contours from steady operation for 3 different early injection designs during steady operation (Gutmark 2014)

shows the computed contours of temperature and vortices throughout the annulus of an 'unwrapped' RDE, at steady state. Further, the performance impacts of several RDE design parameters are examined and finally, it is found that the direct performance comparison can be made with a straight-tube PDE and for a particular RDE configuration (Paxson 2014).

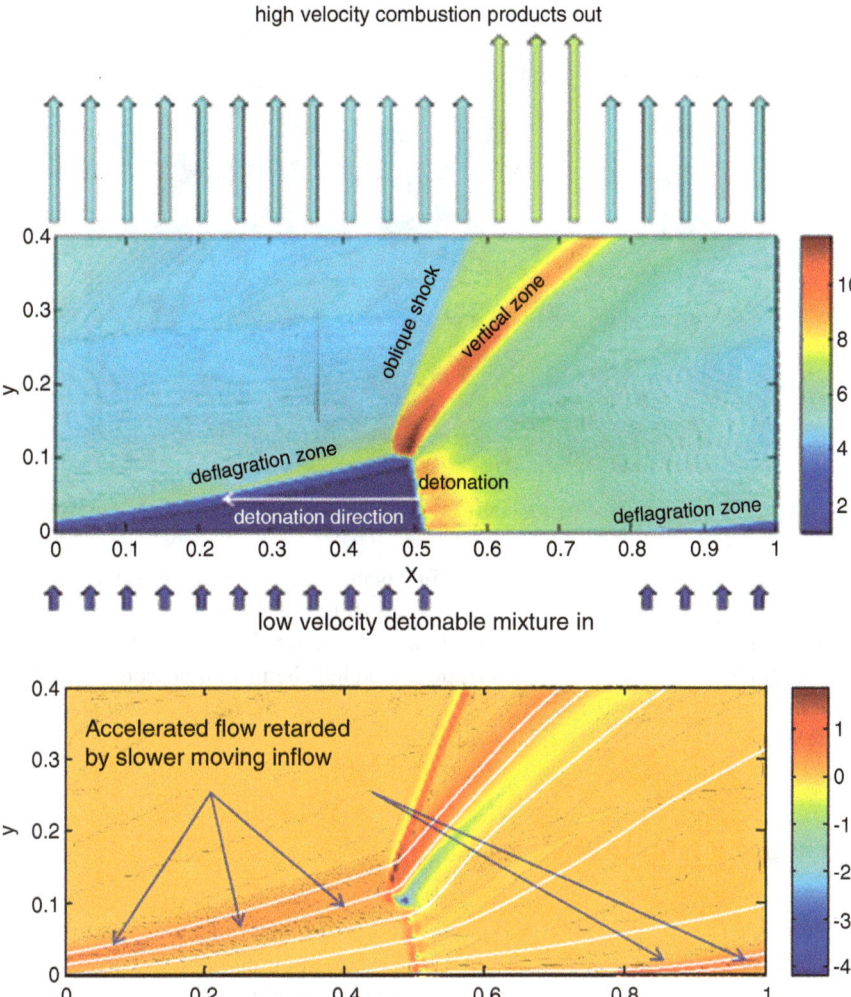

Fig. 6.40 Computed contours of temperature and vorticity throughout the annulus of an 'unwrapped' RDE, at steady state (Paxson 2014)

7 Experimental Work of RDE

Optically Accessible RDE's

Experimental RDE research in AFRL (Air Force Research Laboratory) includes optically accessible RDE's with quartz outer wall having aft end ignition utilizing H_2/O_2 detonation and rig design based on Nicholls and Edwards's research. Figure 6.41 shows the optically accessible RDE manufactured in AFRL (Naples 2011). Moreover, the comparison of experimental result with the obtained

Fig. 6.41 Optically accessible RDE(a) (Naples 2011)

numerical result as shown in Fig. 6.41 (Fievisohn 2012) was made to measure the angles of rough flow structure. They proposed RDE bulk flow looks much steadier to upstream and downstream turbo-machinery – essential for incorporation into turbine engines High speed flow visualization so called chemi-luminescence and RDE heat flux calculations are done in order to conclude that the heat flux increases with increasing wall conductivity (Fig. 6.42).

Integrated RDE's

Uncomplicated design criteria, uninterrupted propellant injection, performance increase and efficiency improvements of the detonation engines so called RDE's enforces the scientists to integrate the detonation technology with air-breathing engines. Recent experimental research on combined small gas turbine with rotating detonation combustion chamber was carried out at the Institute of Heat Energy in Poland by J. Kindracki (2009, 2012). Such a system is shown in Fig. 6.43. Furthermore, AFRL is presently developing a new peculiar engine by implementing Pressure Gain Combustion (PGC) or Constant Volume Engine Cycle (CVEC) approach via RDE. The researchers briefed that the Quasi-steady state outcome from the rotating detonation combustion chamber to the turbine and enormous amount of power density are the substantial advantages of integrating RDE's with gas turbines. However, the bottle-neck challenges of merging RDE's include immense supply of thermal loading, efficient starting, operability betterment, and valving difficulties. Scrupulous investigations are ongoing with the hope to use integrated RDE's in commercial airplanes in nearby future.

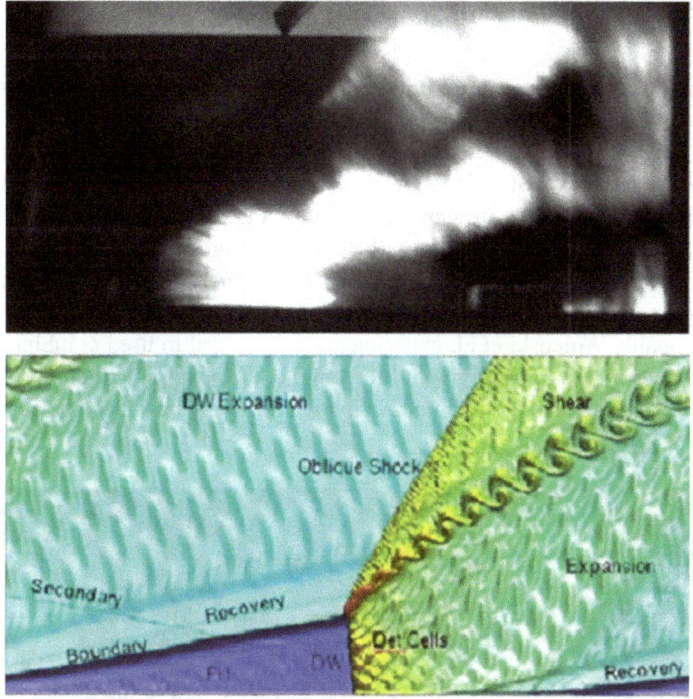

Fig. 6.42 Optically accessible RDE(b) (Fievisohn 2012)

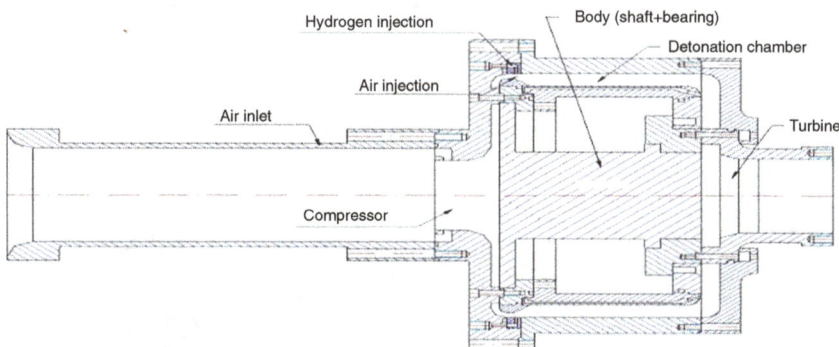

Fig. 6.43 Experimental research on small turbine engine with Rotating Detonation Combustion Chamber (RBCC) (Kindracki et al. 2009)

RDE Research at Poland

A detailed in-depth experimental study of rotating detonation in small rocket engine models were carried out by Wolanski (2011a, b, 2013b) at the Warsaw University of Technology (WUT) in Poland. The experiments clearly showed a possibility of the continuous propagation of rotating detonation in different cylindrical chambers in a wide range of fuel-lean hydrogen air mixtures. Figure 6.44 shows the test stand model of the experimental rocket with RDE and Fig. 6.45 shows the picture of chamber operating under conditions of continuously rotating detonation for hydrogen-air mixture.

Additionally, Wolanski et al. experimental researched to used combined engine between RDE and GTD-350 gas turbine engine. The experiment showed that the continuous rotating detonation could be applied in jet engines. It can be achieved by

Fig. 6.44 Experimental rocket with RDE on test stand (Wolanski 2011a, b, 2013c)

Fig. 6.45 Picture of chamber operating under conditions of continuously rotating detonation for hydrogen-air mixture (Wolanski 2011b)

increasing their efficiency and reducing the weight of the combustion chamber. It was found that it is much easier to get the rotating detonation for fuel-lean and stoichiometric mixtures than for fuel-rich mixtures.

Cooling System of RDE's

Vivid heating of the continuously rotating combustion chamber affects the performance characteristics of the engine as well as leads to structural failure. This kind of heat transfer problems was already discussed by the researchers in 1960's (Siechel and David 1998b; Edwards et al. 1970). However, the recent advancement and solution for novel cooling systems in CRDE were given by the researchers all over the globe. In late 2013, wall cooling and heat release of the wall interaction (Roy 2013) was carried out by the U.S. defense agency. Similarly heat transfer rate in a semi-infinite heat transfer problem and heat flux was given by Wen (2013) in Hong Kong. Wolanski (2013c) in Poland calculated the heat flux distribution in a 16 cooling channel configuration. Figure 6.46 shows the temperature distribution in a cooling channel. Some of the clear-cut understanding of the internal physical behavior of the combustion chamber of RDE was ongoing in the Naval Research Laboratory (NRL) in U.S. Kailasanath and Schwer (2013) in NRL is researching about the flow-field behavior, Stagnation/back pressure effects, injection/inflow effects, exhaust flow effects and preliminary Fuel-Air misting studies of the RDE's.

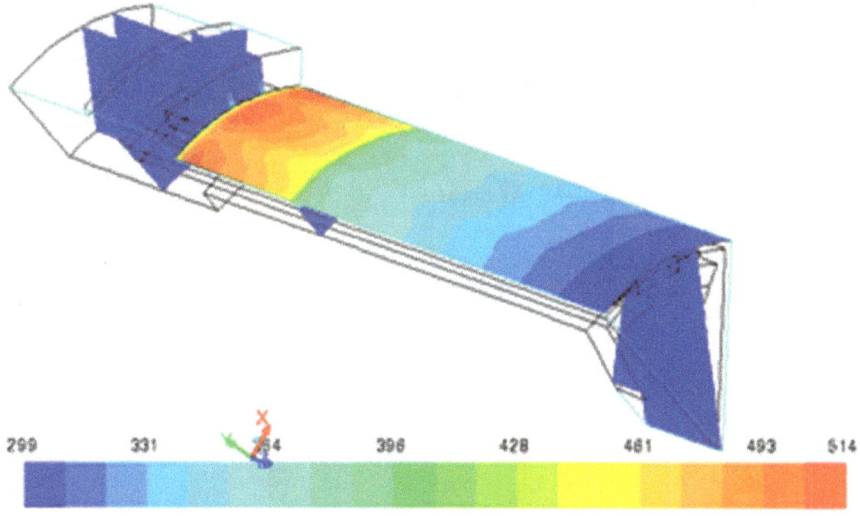

Fig. 6.46 Temperature distributions in a 16 cooling channel configuration (Kindracki et al. 2009)

Visualization and Measurement Techniques of RDE's

The visualization of the flow behaviors and measurement of detonation velocities in RDE's are still remains as a hot topic in detonation research area. Recent study on detonation engine momentum and thrust loss measurement by ballistic pendulum and laser displacement method was presented by Kasahara et al. (Ashida and Kasahara 2014) in Japan. Wolanski (2013c) at the institute of Aviation has briefed a novel measurement technique for investigation on heat transfer characteristics of the gaseous RDE's. Figure 6.47 shows the optical measurement technique developed by Chung (2013). High speed shadow-graphy (Kuznetsov 2013) method was used to optically visualize the inhomogeneity in transparent media and refraction of light rays due to the different in density. Moreover, durability testing system was used recently in Aerospace Science and Technology Research Centre (ASTRC) by Chung (2013) for calibration of flow measurement at high pressure, temperature and compact nozzle conditions.

Long Time Operation Experimental Work of Aerojet Rocketdyne (AR)

Since 2010, AR has been investigating the behavior of a RDE using modular hardware. In 2010, The Power Innovations Group (PIG) at AR successfully conducted over 200 tests of a 4 and 10 cm diameter RDE at Purdue University (PU) and Penn State University (PSU). In 2011, a novel plasma augmentation integrated system (RDE and Turbine engines) (Claflin 2013) was developed by Aerojet Rocketdyne (AR) in California to improve the efficiency of the engine by continuing the air-breathing operation without supplemental oxygen and also to increase the detonation

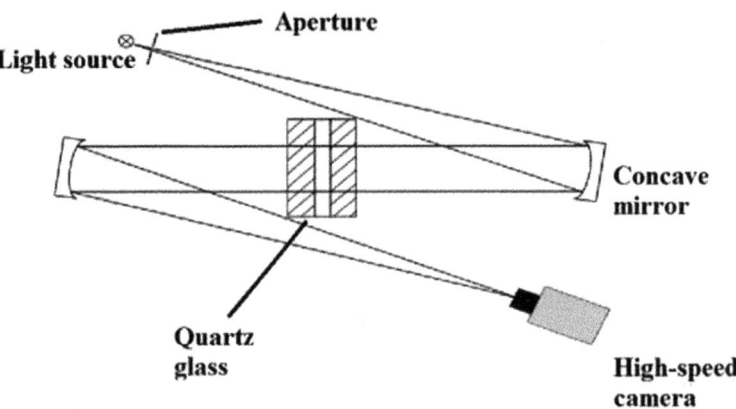

Fig. 6.47 Optical measurement technique (Chung 2013)

velocity. AR is investigating the behavior of RDE for a very long period with lot of experimental testing's using modular hardware, multiple propellant combinations, exchangeable injectors, nozzles and annulus. Figure 6.48 shows the Plasma augmented system developed by AR. In the very recent year it joined with the Defense Advanced Research Projects Agency (DARPA) for developing the high efficient Vulcan exhaust probe (Claflin 2013) to use that for CFD models to generate continuous strong detonation waves and also to sustain the detonation waves in real world scenarios.

High Performance Computing, Incorporated (HyPerComp, Inc.) is currently working on building theoretical and computational model of RDE's. They concluded the continuous detonation behavior of rotating engines is highly dependent on injector behaviors. Figure 6.49 shows the DARPA's recent collaborative projects progress with AR.

In 2013, HyPerComp, Inc. joined with DARPA so called Hypercomp/DARPA program (Claflin 2013) to explore the characteristics of RDE using liquid fuels. The forthcoming work of AR/DARPA includes the optimization of efficient energy conversion and scaling.

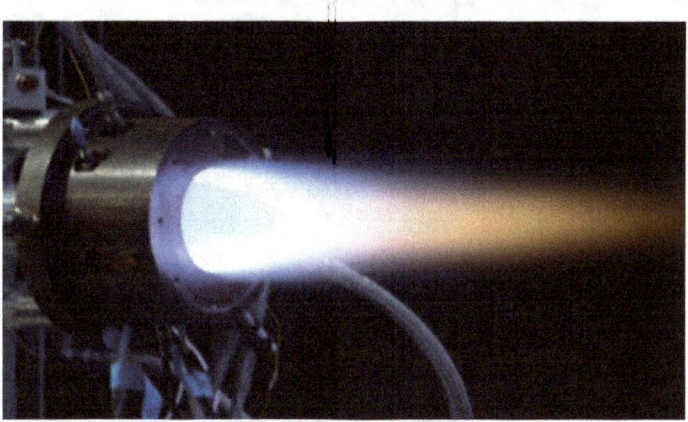

Fig. 6.48 Experimental work at Aerojet Rocketdyne (Claflin 2013)

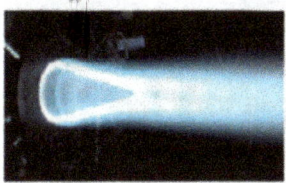 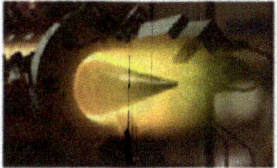 

DARPA/Hypercomp Contract Testing   **DARPA Vulcan Exhaust Probes RDE Program**   **DARPA/Hypercomp Liquid Fueled RDE Testing.**

Fig. 6.49 Recent progress in Rotating Detonation Engine development at Aerojet Rocketdyne (Claflin 2013)

RDE's in Space

Recently, Gawahara et al. (2013) taking the advantage of RDE benefits, application to the reaction control system of a spacecraft is expected. Figure 6.50 shows the full schematic photograph of the research stand used for the RDE developmental research and Fig. 6.51 shows the sequence of photographs while the wave front to go around from the incidence of blast. In order to perform internal visualization RDE, and produced a combustor having a straight section the flow channel, and combustion experiments was conducted, and it was confirmed basic combustion characteristics acquisition, the propagation of detonation.

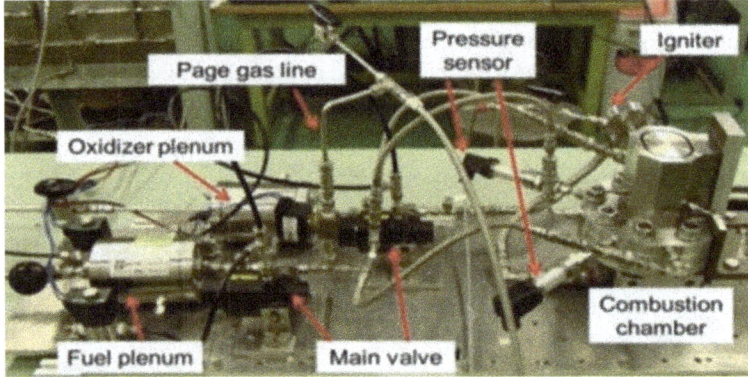

Fig. 6.50 The schematic photograph of the research stand (Gawahara et al. 2013)

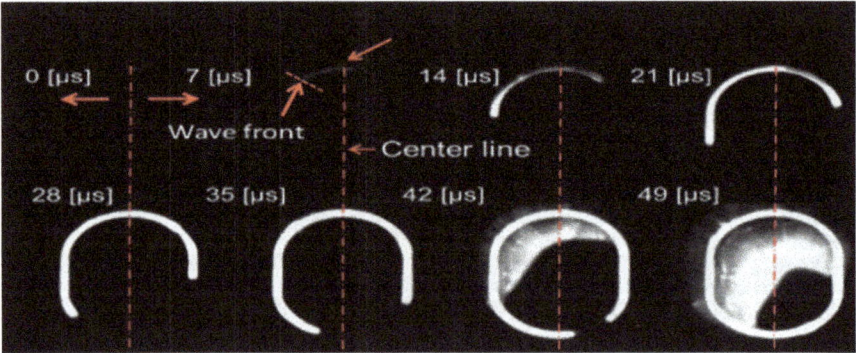

Fig. 6.51 Sequence of photographs while the wave front to go around from the incidence of blast (Gawahara et al. 2013)

RDE Exhaust

Exhaust gas analysis of two RDE's (vertical and horizontal) using Time-Division-Multiplexed Tunable Diode Laser Absorption Spectroscopy (TDM-TDLAS) was carried out by McMahan et al. (2014). These experiments were analyzed to obtain the temperature and H_2O concentration as well as the velocity of the exhaust of the RDE. The comparative result shows that after an initial startup region, a steady state operating condition was reached relatively quicker in vertical RDE's then Horizontal RDE's. This may be caused by the difference in flow rates, or it may be that the temperature stabilizes faster than the velocity. Figures 6.52 and 6.53 shows the vertical and horizontal RDE's respectively. The relatively small standard deviation of the temperature and concentration for the steady state region showed that not only the RDE remained steady, but the system taking the measurements could take them repeatedly.

RDE Thrust Measurement

Towards the practical use of RDE's in Rocket Engine, the evaluation of thrust efficiency and the knowledge of combustion under high acceleration or multiple degrees of freedom are essential. Kato et al. (2014) was conducted the RDE thrust measurement experiments by using sled model test in Nagoya University. In order

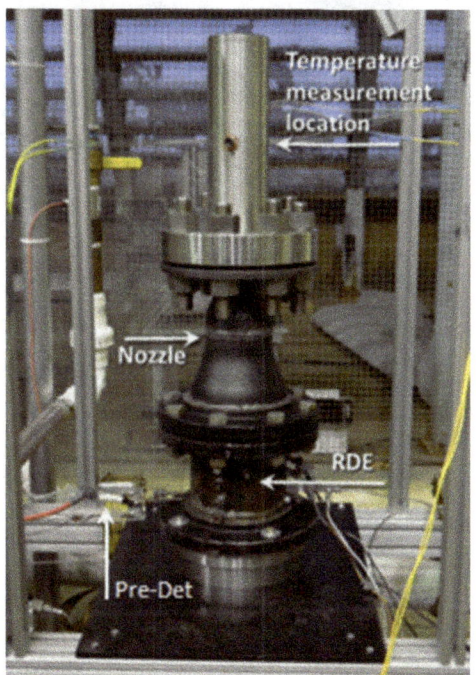

Fig. 6.52 Vertical RDE (McGahan et al. 2014)

Fig. 6.53 Horizontal RDE (McGahan et al. 2014)

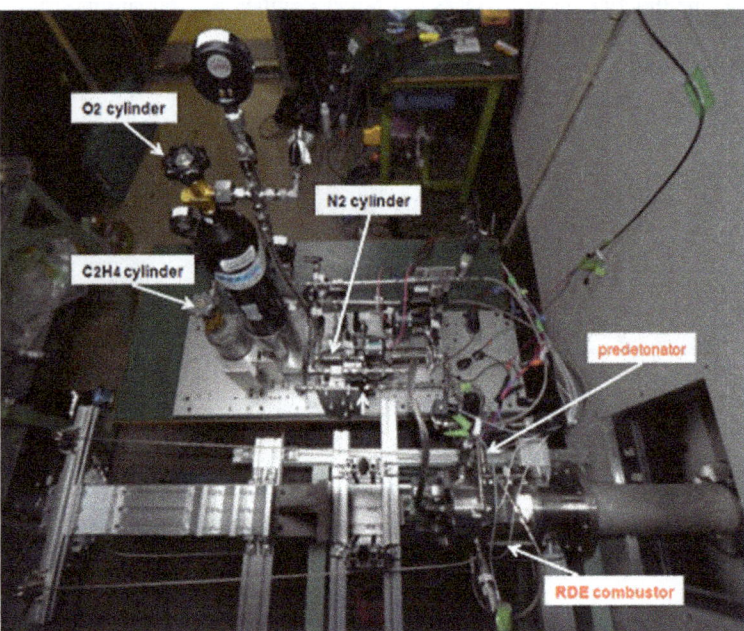

Fig. 6.54 RDE thrust measurement (Kato et al. 2014)

to measure thrust, the test stand with the slide mechanism for noise reduction and exhaust of burned gas was developed. Because of the exhaust chamber pressure, there is a possibility that output large thrust measured by the load cell. Figure 6.54 shows the full photograph of the test stand.

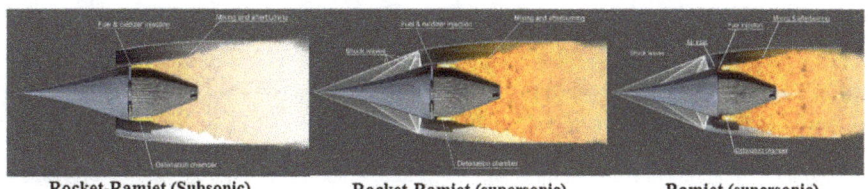

| Rocket-Ramjet (Subsonic) | Rocket-Ramjet (supersonic) | Ramjet (supersonic) |

Fig. 6.55 Novel RDE rocket (Wolanski 2013b)

8 Novel Concepts of RDE

Now-a-days among the detonative researchers, liquid propellants are getting quit famous and demanding because of their improved engine performance (Wolanski 2013b; Claflin 2013; Wei 2013). Recently, Wang et al... carried a numerical study of RDE for a circular cylinder without center-body similar to the liquid rocket combustor. (Wang 2012). Integrated RDE based air-breathing propulsion studies are in the stage of advancement. Researchers from Poland (Wolanski 2013b) and U.S. (Kailasanath and Schwer 2013) are progressing to development a small gas turbines with Rocket Booster Combustor Chamber (RBCC) and PGC via RDE's respectively. A suggestion of implementing rotating detonation based engines to space propulsion system has been in ongoing research for several years, since detonative propulsion engines provide a significant rise in engine efficiency over the conventional gas turbine engines.

Poland researchers moved one step forward in integrating RDE's with Rocket engines for space access. Wolanski (2011c) proposed a novel concept of rotating detonation in RBCC. They proposed that the general RDE RBCC engine configuration was not necessarily to be circular. RDE can wrap around any cross section. Figure 6.55 shows the integration of detonative propulsion technology in Rocket-Ramjet engines.

9 Conclusion

A detailed survey of RDE technology is carried out to provide a clear understanding. The recent promising advancement in experimental, theoretical and computational studies of Pulsed or intermittent detonation-wave engines and continuous rotating detonation-wave engines to improve the performance characteristic and cycle efficiency of in-practical gas turbines are also discussed. Moreover, some of the existing crucial problems in detonation technology which holds this propulsion system to enter into practical use are also briefed. Base on the efforts worldwide, detonation based engines would be effectively utilized in both air-breathing and rocket propulsion systems in nearby future.

References

Adamson, T. C., & Olsson, G. R. (1967). Performance analysis of a rotating detonation wave rocket engine. *Acta Astronautica, 13*, 405–415.

Ar'kov, O. F., et al. (1970). On the spinning-detonation-like properties of high frequency tangential oscillations in combustion chambers of liquid fuel rocket engines. *Journal of Applied Mechanics and Technical Physics, 11*(1), 159–161.

Ashida, T., & Kasahara, J. (2014). Study on detonation engine momentum and thrust loss measurement by ballistic pendulum and laser displacement methods. In *52nd Aerospace Science Meeting*, AIAA 2014–1016.

Bollay, W. (1960). Pulse detonation jet propulsion. *US Patent 2,942,412*, 28 June 1960.

Borisov, A. A. (2002). *Proceeding of the 15th ONR Propulsion Meeting*, University of Maryland.

Braun, E. M., et al. (2010a). Air-breathing rotating detonation wave engine cycle analysis. In *46th AIAA/ASME/SAE/ASEE Joint Propulsion Conference and Exhibit*, AIAA 2010–7039, 25–28 July 2010.

Braun, E. M., et al. (2010b). Air-breathing rotating detonation wave engine cycle analysis. In *46th AIAA/ASME/SAE/ASEE Joint Propulsion Conference and Exhibit*, AIAA 2010–7039, 25–28 July 2010.

Braun, E. M., et al. (2010c). Detonation engine performance comparison using first and second law analyses. In *46th AIAA/ASME/SAE/ASEE Joint Propulsion Conference and Exhibit*, AIAA 2010–7040, 25 – 28 July 2010.

Brophy, C., Sinibaldi, J., & Netzer, D. (2001). *Proceeding of the 14th ONR Propulsion Meeting*, Chicago.

Bussing, T. R. A., & Pappas, G. (1996). Pulse detonation engine theory and concepts. In S. N. B. Murthy & E. T. Curran (Eds.), *Developments in high-speed vehicle propulsion systems, AIAA* (pp. 421–472).

Bykovskii, F. A., et al. (1980). Detonation combustion of a gas mixture in a cylindrical chamber. *Combustion, Explosion and Shock Waves, 16*(5), 570–578.

Bykovskii, F. A., et al. (1994). Explosive combustion of a gas mixture in radial annular chambers. *Combustion, Explosion and Shock Waves, 30*(4), 510–516.

Bykovskii, F. A., et al. (1997). Continuous detonation combustion of fuel-air mixtures. *Combustion, Explosion and Shock Waves, 33*(3), 344–353.

Bykovskii, F. A., et al. (2004). Continuous spin detonation in ducted annular combustors. In G. Roy & S. Frolov (Eds.), *Application of detonation to propulsion*. Moscow: Torus Press.

Bykovskii, F. A., et al. (2005a). Spin detonation of fuel–air mixtures in a cylindrical combustor. *Doklady Physics, 50*(1), 56–58.

Bykovskii, F. A., et al. (2005b). Continuous spin detonation in annular combustors. *Combustion, Explosion and Shock Waves, 41*(4), 449–459.

Bykovskii, F. A., et al. (2006a). Continuous spin detonation. *Journal of Propulsion and Power, 22*(6), 1204–1216.

Bykovskii, F. A., et al. (2006b). Continuous spin detonation of fuel-air mixtures combustion. *Combustion, Explosion and Shock Waves, 42*(4), 463–471.

Bykovskii, F. A., et al. (2008). Continuous spin detonation of hydrogen–oxygen mixtures. 1. Annular cylindrical combustors. *Combustion, Explosion and Shock Waves, 44*(2), 150–162.

Bykovskii, F. A., et al. (2009). Continuous spin and pulse detonation of hydrogen–air mixtures in a supersonic flow generated by a detonation wave. In *22nd International Colloquium on the Dynamics of Explosions and Reactive Systems*, 27–31 July 2009.

Bykovskii, F. A., et al. (2011). Continuous spin detonation of a hydrogen–air mixture in the air ejection mode. In *Detonation Wave Propulsion Workshop*, France, 11–13 July 2011.

Chung, K. M. (2013). Activities of aerospace science and technology research center. In *International Workshop on Detonation for Propulsion*, Taiwan, 26–28 July 2013.

Claflin, S. (2013). Recent progress in rotating detonation engine development at aerojet rocketdyne. In *International Workshop on Detonation for Propulsion*, Taiwan, 26–28 July 2013.

Clayton, R. M., & Rogero, R. S. (1965). *Experimental measurements on a rotating detonation-like wave observed during liquid rocket resonant combustion*. Technical Report 32–788, Jet Propulsion Laboratory.

Conrad, C., et al. (2001). *Proceeding of the 14th ONR Propulsion Meeting*, Chicago.

Daniau, E., et al. (2005). Pulsed and rotating detonation propulsion systems: First step toward operational engines. In *AIAA/CIRA 13th International Space Planes and Hypersonic Systems and Technologies Conference*, AIAA 2005–3233, Italy, 16–20 May 2005.

Davidenko, D. M., et al. (2008). Numerical study of the continuous detonation wave rocket engine. In *15th AIAA International Space Planes and Hypersonic Systems and Technologies Conference*, AIAA 2008–2680, 28 April – 1 May 2008.

Davidenko, D. M., et al. (2009a). Continuous detonation wave engine studies for space application. Progress in propulsion. *Physics, 1*, 353–366.

Davidenko D. M., et al. (2009b) Continuous detonation wave engine studies for space application. Progress in Propulsion Physics, 2009.

Davidenko, D., et al. (2011). Theoretical and numerical studies on continuous detonation wave engines. In *17th AIAA International Space Planes and Hypersonic Systems and Technologies Conference*, AIAA 2011–2334, 11–14 April 2011.

Dyer, R. S., & Kaemming, T. A. (2002). The thermodynamic Basis of pulsed detonation engine Thrust production. In *38th AIAA/ASME/SAE/ASEE Joint Propulsion Conference and Exhibit*, AIAA 2002–4072, 7–10 July 2002.

Edwards, B. D. (1977). Maintained detonation waves in an annular channel: A hypothesis which provides the link between classical acoustic combustion instability and detonation waves. International. *Symposium on Combustion, 16*(1), 1611–1618.

Edwards, D. H., et al. (1970). The influence of wall heat transfer on the expansion following a C-J detonation wave. *Journal of Physics D: Applied Physics, 3*(3), 365–376.

Eidelman, S., & Grossmann, W. (1992). Pulsed detonation engine: experimental and theoretical review. In *28th Joint Propulsion Conference and Exhibit*, AIAA 1992–3168, U.S.A.

Eidelman, S., et al. (1990a). Air-breathing pulsed detonation engine concept: a numerical study. In *26th Join Propulsion Conference*. AIAA 1990–2420, U.S.A.

Eidelman, S., et al. (1990b). Computational analysis of pulsed detonation engines and applications. In *28th Aerospace Sciences Meeting*, AIAA 1990–460, U.S.A.

Eidelman, S., et al. (1991). Review of propulsion applications and numerical simulations of the pulse detonation engine concept. *Journal of Propulsion and Power, 7*(6), 857–865.

Endo, T., et al. (2011). Development of pulse-detonation technology in valve-less mode and its application to turbine-drive experiments. In *International Workshop on Detonation for Propulsion*, Korea, 14–15 November 2011.

Eude, Y., & Davidenko, D. (2011). *Numerical simulation and analysis of a 3D continuous detonation under rocket engine conditions*. France: Detonation Wave Propulsion Workshop. 11–13 July 2011.

Falempin, F., et al. (2006). Toward a continuous detonation wave rocket engine demonstrator. In *14th AIAA/AHI Space Planes and Hypersonic Systems and Technologies Conference*, AIAA 2006–7956, Australia.

Falempin, F., et al. (2008). A contribution to the development of actual continuous detonation wave engine. In *15th AIAA International Space Planes and Hypersonic Systems and Technologies Conference*, 28 April – 1 May 2008.

Fievisohn, R. (2012). Rotating detonation engine research at AFRL. In *International Workshop on Detonation for Propulsion*, Japan, 3–5 September 2012.

Frolov, S. M., Basevich, V.Ya., & Aksenov, V. S. (2001). *Proceeding of the 14th ONR Propulsion Meeting*, Chicago.

Frolov, S., Dubrovskii, A., & Ivanov, V. (2012). Three-dimensional numerical simulation of the operation of a rotating-detonation chamber with separate supply of fuel and oxidizer. *Russian Journal of Physical Chemistry B, 7*(1), 35–43.

Fujiwara, et al. (2009). Stabilization of detonation for any incoming mach numbers. *Combustion, Explosion and Shock Waves, 45*(5), 603–605.
Gawahara, K., et al. (2013). Detonation engine development for reaction control systems of a spacecraft. In *49th AIAA/ASME/SAE/ASEE Joint Propulsion Conference*, AIAA 2013–3721.
Gutmark, E. J. (2014). Comparative numerical study of RDE injection designs. In *52nd Aerospace Sciences Meeting*, AIAA 2014–0285.
Hayashi, et al. (2009). Sensitivity analysis of rotating detonation engine with a detailed reaction model. In *47th AIAA Aerospace Sciences Meeting including the New Horizons Forum and Aerospace Exposition*, AIAA 2009–633, 5–8 January 2009.
Heiser, W. H., & Pratt, D. T. (2002). Thermodynamic cycle analysis of pulse detonation engines. *Journal of Propulsion and Power, 18*(1), 68–76.
Helman, D., et al. (1986). Detonation pulse engine. In *22nd Join Propulsion Conference*, AIAA 1986–1683, U.S.A.
Hishida, et al. (2009). Fundamentals of rotating detonations. *Shock Waves, 19*(1), 1–10.
Hoffmann, N. (1940). Reaction propulsion by intermittent detonative combustion (trans: Volkenrode). German ministry of supply, AI152365.
International Workshop on Detonation for Propulsion. (2013). http://conf.ncku.edu.tw/iwdp2013/. Accessed 26 July 2013.
Kailasanath, K. (1999). Applications of detonations to propulsion: A review. In *37th aerospace Sciences Meeting and Exhibit*, AIAA 1999–1067, U.S.A.
Kailasanath, K. (2011a). The rotating-detonation-wave engine concept: A brief status report. In *49th AIAA Aerospace Science Meeting including the New Horizons Forum and Aerospace Exposition*, AIAA 2011–580, 4–7 January 2011.
Kailasanath, K. (2011b). The rotating-detonation-wave engine concept: A brief status report. In *49th AIAA Aerospace Science Meeting including the New Horizons Forum and Aerospace Exposition*, AIAA 2011–580, 4–7 January 2011.
Kailasanath, K., & Schwer, D. A. (2011). Effect of inlet on fill region and performance of rotating detonation engines. In *47th AIAA/ASME/SAE/ASEE Joint Propulsion Conference and Exhibit*, AIAA 2011–6044, 31 July – 03 August 2011.
Kailasanath, K., & Schwer, D. A. (2013). Rotating detonation engine research at NRL. In *International Workshop on Detonation for Propulsion*, Taiwan, 26–28 July 2013.
Kailasanath, K., & Schwer, D. A. (2014). Effect of low pressure ratio on exhaust plumes of rotating detonation engines. In *50th AIAA/ASME/SAE/ASEE Joint Propulsion Conference and Exhibit*, AIAA 2014–3901.
Kato, Y., et al. (2014). Thrust measurement of rotating detonation engine by sled test. In *50th AIAA/ASME/SAE/ASEE Joint Propulsion Conference*, AIAA 2014–4034.
Kindracki, J., Institute of Heat Engineering, WUT. (2012). project no. UMO-2012/05/D/ST8/02308, granted by National Science Centre, Poland.
Kindracki, J., et al. (2009). Experimental and numerical research on rotating detonation in small rocket engine model. Combustion Engines, 2009.
Knappe, B. M., & Edwards, C. F. (2001). *Proceeding of the 14th ONR Propulsion Meeting*, Chicago.
Knappe, B. M., & Edwards, C. F. (2002). *Proceeding of the 15th ONR Propulsion Meeting*, University of Maryland.
Kuznetsov, M. (2013). Flame acceleration and DDT in linear and circular geometries. In International Workshop on Detonation for Propulsion, Taiwan, 26–28 July 2013.
Lentsch, A., et al. (2005). Overview of current French activities on PDRE and continuous detonation wave rocket engines. In *AIAA/CIRA 13th International Space Planes and Hypersonic Systems and Technologies Conference*, AIAA 2005–3232, Italy, 16–20 May 2005.
Lu, F. K., et al. (2011a). Rotating detonation wave propulsion: Experimental challenges, modeling, and engine concepts (invited). In *47th AIAA/ASME/SAE/ASEE Joint Propulsion Conference and Exhibit*, 31 July – 03 August 2011.

Lu, F. K., et al. (2011b). Rotating detonation wave propulsion: Experimental challenges, modeling, and engine concepts (invited). In *47th AIAA/ASME/SAE/ASEE Joint Propulsion Conference and Exhibit*, 31 July – 03 August 2011.

Lu, et al. (2011c). Rotating detonation wave propulsion: Experimental challenges, modeling, and engine concepts. In *47th AIAA/ASME/SAE/ASEE Joint Propulsion Conference and Exhibit*, AIAA 2011–6043, 31 July – 03 August 2011.

Lynch, E. D., & Edelman, R. B. (1994). Analysis of flow processes in the pulse detonation wave engine. In *30th Joint Propulsion Conference and Exhibit*, AIAA 1994–3222, U.S.A.

Lynch, E. D., et al. (1994). Computational fluid dynamic analysis of the pulse detonation engine concept. In *32nd Aerospace Sciences Meeting and Exhibit*, AIAA 1994–264, U.S.A.

McGahan, C. J., et al. (2014). Exhaust gas analysis of a rotating detonation engine using tunable diode laser absorption spectroscopy. In *52nd Aerospace Science Meeting*, AIAA 2014–0391.

Milanowski, et al. (2007). Numerical simulation of rotating detonation in cylindrical channel. In *21st International Colloquium on the Dynamics of Explosions and Reactive Systems*, 23–27 July 2007.

Naour, B. L., Falempin, F., & Miquel, F. (2011). *Recent experimental results obtained on continuous detonation wave engine*. France: Detonation Wave Propulsion Workshop. 11–13 July 2011.

Naples, A. (2011). Recent progress in detonation. In *International Workshop on Detonation for Propulsion*, Korea, 14–15 November 2011.

Nicholls, J. A., Wilkinson, H. R., & Morrison, R. B. (1957). Intermittent detonation as a thrust-producing mechanism. *Journal of Jet Propulsion, 27*(5), 534–541.

Nicholls, J. A., Cullen, R. E., & Ragland, K. W. (1966). Feasibility studies of a rotating detonation wave rocket motor. *Journal of Spacecraft and Rockets, 3*(6), 893–898.

Nordeen, C. A. (2013). Thermodynamics of a rotating detonation engine (Doctoral Dissertations, University of Connecticut Graduate School).

Nordeen, C. A., et al. (2011). Thermodynamic modeling of a rotating detonation engine. In *49th AIAA Aerospace Science Meeting including the New Horizons Forum and Aerospace Exposition*, AIAA 2011–803, 4 – 7 January 2011.

Nordeen, C. A., et al. (2014). Area effects on rotating detonation engine performance. In *50th AIAA/ASME/SAE/ASEE Joint Propulsion Conference*, AIAA 2014–3900.

Oppenheim, A. K., Manson, N., & Wagner, H. G. (1963). Recent progress in detonation research. *AIAA Journal, 1*(10), 2243–2252.

Paxson, D. E. (2014). Numerical analysis of a rotating detonation engine in the relative reference frame. In *52nd Aerospace Sciences Meeting*, AIAA 2014–0284.

Roy, G. D. (2013). Propulsion and detonation engines-a navy perspective. In *International Workshop on Detonation for Propulsion*, Taiwan, 26–28 July 2013.

Roy, G. D., et al. (Eds.). (2000a). *High-speed deflagration and detonation: Fundamentals and control*. Moscow: Elex-KM Publishers.

Roy, G. D., et al. (Eds.). (2000b). *High-speed deflagration and detonation: Fundamentals and control* (pp. 289–302). Moscow: Elex-KM Publishers.

Roy, G. D., et al. (Eds.). (2002a). *Advances in confined detonations* (pp. 150–157). Moscow: Torus Press.

Roy, G. D., et al. (Eds.). (2002b). *Advances in confined detonations* (pp. 231–234). Moscow: Torus Press.

Roy, G. D., et al. (Eds.). (2003a). *Confined detonations and pulse detonation engines* (pp. 59–72). Moscow: Torus Press.

Roy, G. D., et al. (Eds.). (2003b). *Confined detonations and pulse detonation engines* (pp. 219–234). Moscow: Torus Press.

Roy, G. D., et al. (2004). Pulse detonation propulsion: Challenges, current status, and future perspective. *Progress in Energy and Combustion Science, 30*(6), 545–672.

Schwer, D. A., & Kailasanath, K. (2010). Numerical investigation of rotating detonation engines. In *46th AIAA/ASME/SAE/ASEE Joint Propulsion Conference and Exhibit*, AIAA 2010–6880, 25–28 July 2010.

Schwer, D. A., & Kailasanath, K. (2011). Numerical study of the effects of engine size on rotating detonation engines. In *49th AIAA Aerospace Sciences Meeting including the New Horizons Forum and Aerospace Exposition*, AIAA 2011–581, 4–7 January 2011.

Schwer, D. A., et al. (2011). Numerical investigation of the physics of rotating-detonation-engines. *Proceedings of the Combustion Institute, 33*(2), 2195–2202.

Schwer, D. A., et al. (2014). Towards efficient, unsteady, three-dimensional rotating detonation engine simulations. In *52nd Aerospace Sciences Meeting*, AIAA 2014–1014.

Shen, I., & Adamson, T. C. (1973). *Theoretical analysis of a rotating two phase detonation in a rocket motor*. Technical Report, NASA CR 121194.

Siechel, M., & David, T. S. (1998a). Transfer behind detonations in H2-O2 mixtures. NASA technical. *Notes, 4*(6), 1098–1099.

Siechel, M., & David, T. S. (1998b). Transfer behind detonations in H2-O2 mixtures. *NASA Technical Notes, 4*(6), 1098–1099.

Tsuboi, N., et al. (2011). Three-dimensional simulation on a rotating detonation engine: Three-dimensional shock structure. In *4th International Symposium on Energy Materials and their Applications*, Japan, 16–18 November 2011.

Tsuboi, N., Hayashi, A., & Kojima, T. (2013). Numerical study on a rotating detonation engine at KIT. In *International Workshop on Detonation for Propulsion*, Taiwan, 26–28 July 2013.

Voitsekhovskii, B. V. (1957). About spinning detonation. *Doklady Akademii Nauk SSSR, 114*(4), 717–720.

Voitsekhovskii, B. V. (1959). Stationary detonation. *Doklady Akad. Nauk. USSSR, 129*(6), 1254–1256.

Voitsekhovskii, B. V. (1960). J. Of app. Mech. And. *Technical Physics, 3*, 157–164.

Voitsekhovskii, B. V., & Kotov, B. E. (1958). Optical investigation of the front of spinning detonation wave. *Izv. Sibirsk. Otd. Akad. Nauk SSSR, 4*, 79.

Voitsekhovskii, B. V., Mitrofanov, V. V., & Topchiyan, M. E. (1963a). *Structure of detonation front in gases*. Novosibirsk: Siberian Branch USSR Academy Science.

Voitsekhovskii, B. V., Mitrofanov, V. V., & Topchiyan, M. E. (1963b). *Structure of the detonation front in gases*. Novosibirsk: Izd-vo SO AN SSSR.

Voitsekhovskii, B. V., Mitrofanov, V. V., & Topchiyan, M. E. (1969). Structure of the detonation front in gases (survey). *Fizika Goreniya i Vzryva, 5*(3), 385–395.

Wang, J. P. (2012). Numerical and experimental study on continuously rotating detonation engine at Peking university. In *International Workshop on Detonation for Propulsion*, Japan, 3–5 September 2012.

Wei, F. A. N. (2013). Efforts to increase the operating frequency of two-phase pulse detonation rocket engines. In *International Workshop on Detonation for Propulsion*, Taiwan, 26–28 July 2013.

Wen, C. Y. (2013). Experimental study on self-ignition of a pre-heated H2 transverse jet in a supersonic free-stream. In *International Workshop on Detonation for Propulsion*, Taiwan, 26–28 July 2013.

Winfree, D. D., & Hunter, L. G. (1999). Pulse detonation igniter for pulse detonation chambers. *US Patent 5,937,635*, 17 August 1999.

Wolanski, P. (2011a). Graphics processors as a tool for rotating detonation simulations. In *23rd International Colloquium on the Dynamics of Explosions and Reactive Systems*, 24–29 July 2011.

Wolanski, P. (2011b). Rotating detonation wave stability. In *23rd International Colloquium on the Dynamics of Explosions and Reactive Systems*, 24–29 July 2011.

Wolanski, P. (2011c). Detonation propulsion. *Journal of KONES Powertrain and Transport, 18*(3), 515–521.

Wolanski, P. (2013a). Detonative propulsion. *Proceedings of the Combustion Institute, 34*(1), 125–158.

Wolanski, P. (2013b). Research on rotating detonation engine. In *Detonation Wave Propulsion Workshop*, 2013.

Wolanski, P. (2013c). Research on RDE in POLAND. In *International Workshop on Detonation for Propulsion*, Taiwan, 26–28 July 2013.

Yamada, et al. (2010). Numerical analysis of threshold of limit detonation in rotating detonation engine. In *48th AIAA Aerospace Sciences Meeting including the New Horizons Forum and Aerospace Exposition*, AIAA 2010–153, 4–7 January 2010.

Yi, T. H., et al. (2009a). Propulsive performance of a continuously rotating detonation-based propulsion system. In *22nd International Colloquium on the Dynamics of Explosions and Reactive System*, Minsk, 27–31 July 2009.

Yi, T. H., et al. (2009b). Propulsive performance of a continuously rotating detonation-based propulsion system. In *22nd International Colloquium on the Dynamics of Explosions and Reactive System*, Minsk, 27–31 July 2009.

Yi, T. H., et al. (2011). Propulsive performance study of continuously rotational detonation engine. In *International Workshop on Detonation for Propulsion*, Korea, 14–15 November 2011.

Yu, S. T. J. (2001). *Proceeding of the 14th ONR Propulsion Meeting*, Chicago.

Zhdan, S. A. (2008). Mathematical model of continuous detonation in an annular combustor with a supersonic flow velocity. Combustion, explosion and. *Shock Waves, 44*(6), 690–697.

Zhdan, S. A., et al. (2007). Mathematical modeling of a rotating detonation wave in a hydrogen oxygen mixture. *Combustion, Explosion and Shock Waves, 43*(4), 449–459.

Zhou, R., & Wang, J.-P. (2012). Numerical investigation of flow particle paths and thermodynamic performance of continuously rotating detonation engines. *Combustion and Flame, 159*(12), 3632–3645.

Chapter 7
Continuous Detonation Engine Researches at Peking University

Jian-Ping Wang, Song-Bai Yao, and Xu-Dong Han

Abstract In this chapter we reviewed the research of the continuous detonation engine (CDE) performed at Peking University. The research team at Peking University was the first to conduct numerical and experimental research of the CDE in China. We designed several types of CDE combustion chambers and carried out experiments to verify its feasibility. In addition, we have performed a series of two- and three-dimensional simulations of the CDE. Numerical studies covered many aspects of the CDE, including the detailed flow structure, fuel injection, nozzle design, viscous effect, propulsive performance, initiation method, particle path, thermodynamic performance, shock wave reflections near the head-wall, spontaneously formation of multiple detonation waves, etc. In this chapter, we also discussed several recent examples of progress and accomplishments.

1 Introduction

The use of detonation combustion is of significant interest as a means of improving the current propulsion systems. One of the detonation-based devices has attracted an increasing interest – the continuous detonation engine (CDE), which is also called the rotating detonation engine (RDE), over the past decade. The CDE requires only a single ignition and the frequency of the high-speed rotating detonation wave is in the kHz range. Therefore, the CDE can produce an almost continuous thrust. The simple structure of the CDE also means a higher thrust-to-weight ratio (Lu and Braun 2014). Voitsekhovskii (1959) first realized continuous detonations. Following his early attempts, Bykovskii et al. (2006) made great progress towards feasible CDEs. Currently, there are many other research teams working on CDE experiments in Russia, Poland, USA, France, Japan, and China. The research team in Center for Combustion and Propulsion at Peking University was the first to perform the CDE research in China (Wang et al. 2010, 2012, 2014; Liu et al. 2013). In

---

J.-P. Wang (✉) • S.-B. Yao • X.-D. Han
Center for Combustion and Propulsion, CAPT & SKLTCS, Department of Mechanics and Engineering Science, College of Engineering, Peking University, Beijing, China
e-mail: wangjp@pku.edu.cn

© Springer International Publishing AG 2018
J.-M. Li et al. (eds.), *Detonation Control for Propulsion*, Shock Wave and High Pressure Phenomena, https://doi.org/10.1007/978-3-319-68906-7_7

November 2009, we successfully achieved hydrogen/oxygen rotating detonations, and calculated the propagating velocity of the detonation wave, which could reach 2041 m/s. In the experiment, hydrogen/air and hydrogen/oxygen rotating detonations could run for more than 2 s until fuel supply was cut. The detonation wave rotated about 6000 cycles per second in the hydrogen/oxygen detonation cases. Since then, we have carried out extensive studies of the CDE and achieved good progress. In addition, we have conducted a series of numerical simulations to investigate the fundamental physics of the flow field in the CDE and its engineering applications. Specific topics are discussed in the following sections.

2 Experimental Research

Figures 7.1 and 7.2 show the schematic and photograph of the experimental system which includes five major parts: the gas supply system, combustion chamber, ignition system, exhaust system, and data acquisition system.

The outer and inner diameters of the channel are 79 mm and 59 mm respectively, resulting in a width of 20 mm. The length of the chamber is 100 mm. Pressure transducers are positioned at different locations to record pressure histories. Non-premixed fuel and oxidizer are injected separately into the combustor from the head-wall.

A PCB® piezoelectric ICP dynamic pressure transducer is used to measure the pressure variation in the combustor during experiments. The inside crystal of this kind of PCB transducer possesses a maximum sensitivity frequency up to 200 kHz, which is sufficient to capture the characteristics of the pressure oscillation in the detonation flow. A high-speed camera is used to observe the propagation of the detonation wave.

The reactive mixture in the channel is ignited by a pre-detonator connected to the chamber tangentially (Fig. 7.3). The pre-detonator is also filled with the reactive mixture ignited by a spark plug. Deflagration-to-detonation-transition (DDT) occurs inside the pre-detonator with the help of Shchelkin spiral. For safety reasons, the chamber exit is connected to a vacuum tank via a 2 m length tube. The burned product is exhausted to the dump tank. In addition, the ambient pressure in the tank can be changed to take into account different altitudes.

The CDE operation sequence is shown in Fig. 7.4. First, the data acquisition system of the pressure transducer is switched on, ready for data collection. Next, the solenoid valves of oxygen and hydrogen supply for the pre-detonator are opened. Thereafter, oxygen and hydrogen flow into the combustion chamber. At this point, the pre-detonator is filled with the hydrogen/oxygen mixture and then the solenoid valves of hydrogen and oxygen supply for the pre-detonator are shut. After the time interval Δt_1, the reactive mixture in the pre-detonator is spark ignited. During the time interval Δt_2, the CDE is running continuously. Eventually, the supply of hydrogen and oxygen is shut down in order to protect the pressure transducer, stopping the running of the CDE.

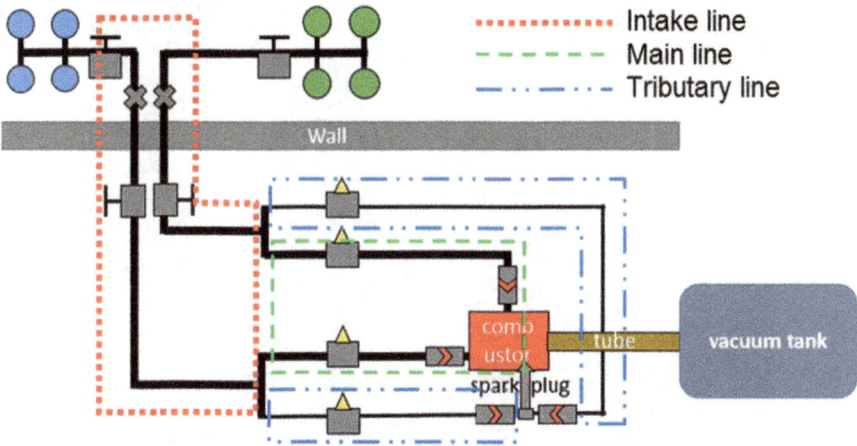

Fig. 7.1 Schematic of the experimental system

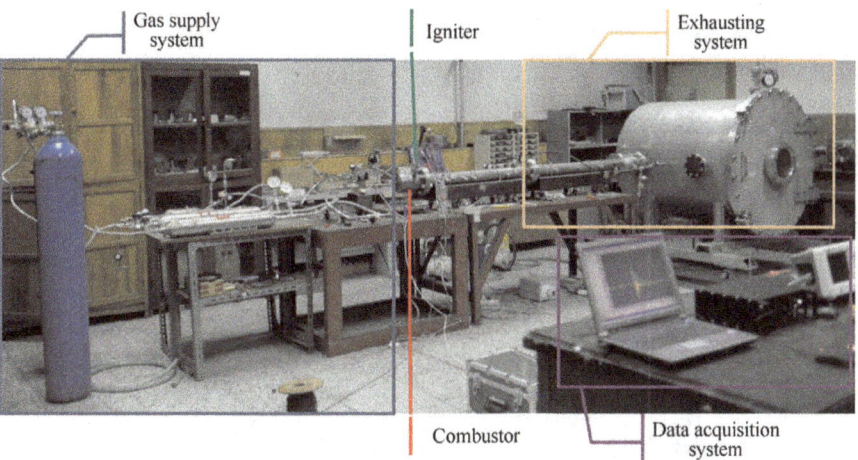

Fig. 7.2 Photograph of the experimental setup

Analysis of Stable Detonation Flow

Based on pressure-time profiles and high-speed shootings, propagation of the detonation wave is analyzed. The pressure record obtained in one of the experiments is shown in Fig. 7.5. Ignition occurred at $t = 0$ and it took around 250 ms before the stable detonation was generated. During the stable stage, the detonation wave propagated circumferentially in the annulus around the head wall. In this stage, the pressure varied within the range of 300 kPa. Since the back pressure was low (20 kPa) and the averaged value of the fresh reactant was of the same order of

Fig. 7.3 Combustor with pre-detonator

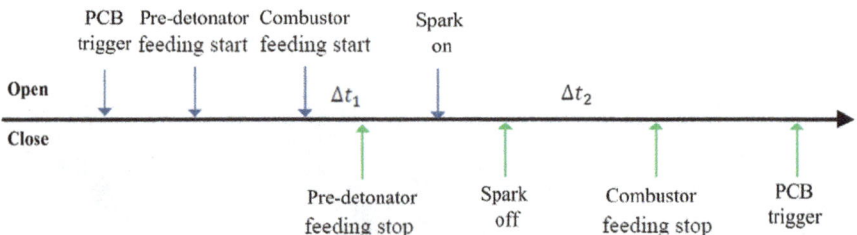

Fig. 7.4 Time sequence

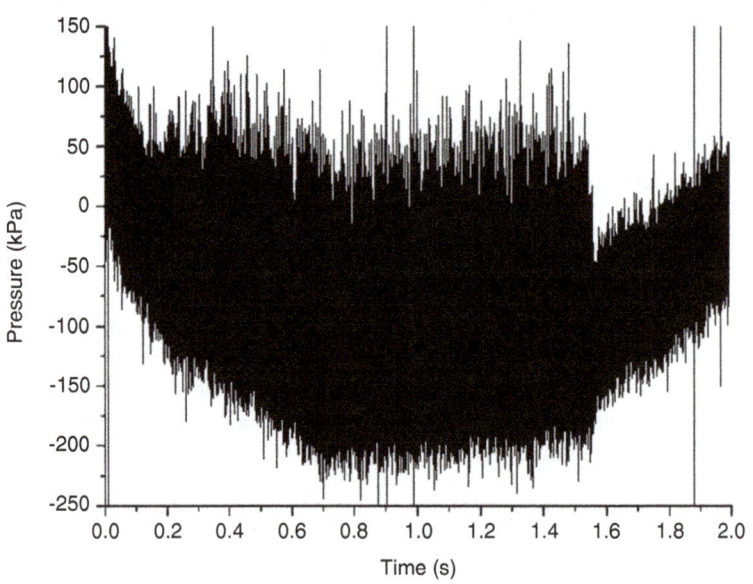

Fig. 7.5 A representative pressure record (Liu et al. 2015a)

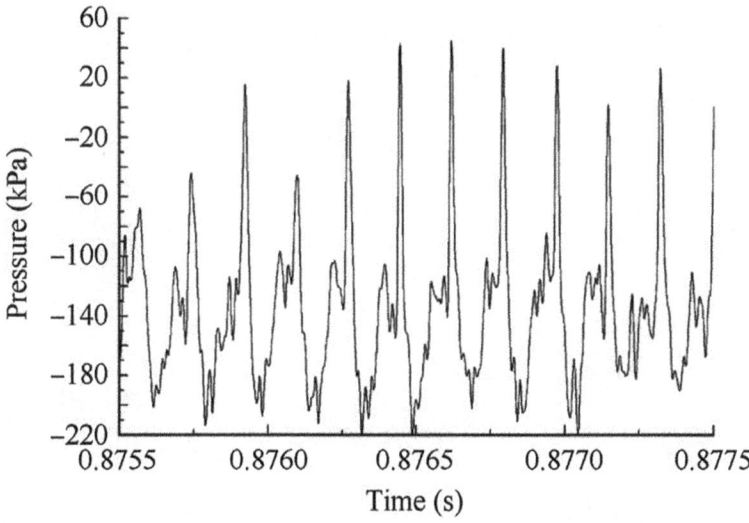

Fig. 7.6 Close-up of the pressure record in the stable stage (Liu et al. 2015a)

the magnitude, the pressure variation was within the reasonable range. The fuel supply was shut down at 1.5 s.

It was found that the time interval measured between two adjacent pressure peaks was 175 μs (Fig. 7.6). Also, it could be confirmed by the high-speed camera images (Fig. 7.7) that there was only one detonation wave. Therefore, it took 175 μs for the detonation wave to complete one cycle of rotating in the annulus. The calculated propagation velocity of the detonation wave was about 1417 m/s. Figure 7.8 shows the result of the FFT analysis of the pressure data of Fig. 7.6. The highest pressure amplitude occurred at about 6000 Hz. The basic frequency was not exactly 6000 Hz, but a little lower, around 5900 Hz. It implied that if there was only one detonation wave rotating in the annular chamber, it took the detonation wave about 170 μs ($\frac{1}{5900} \times 10^6$) to complete one cycle of rotating. This was in accordance with the previous analysis of the pressure-time profile.

Oscillation Phenomenon and Self-Adjusting Mechanism

Figure 7.9 shows a quasi-steady period of the pressure record shown in Fig. 7.5. It clearly shows that the values of the peak pressures are fluctuating rather than remaining constant. Wu et al. (2014a, b) also found the oscillation phenomenon of the detonation wave in the numerical simulation. We provided a possible underlying mechanism to explain this phenomenon – self-adjusting mechanism.

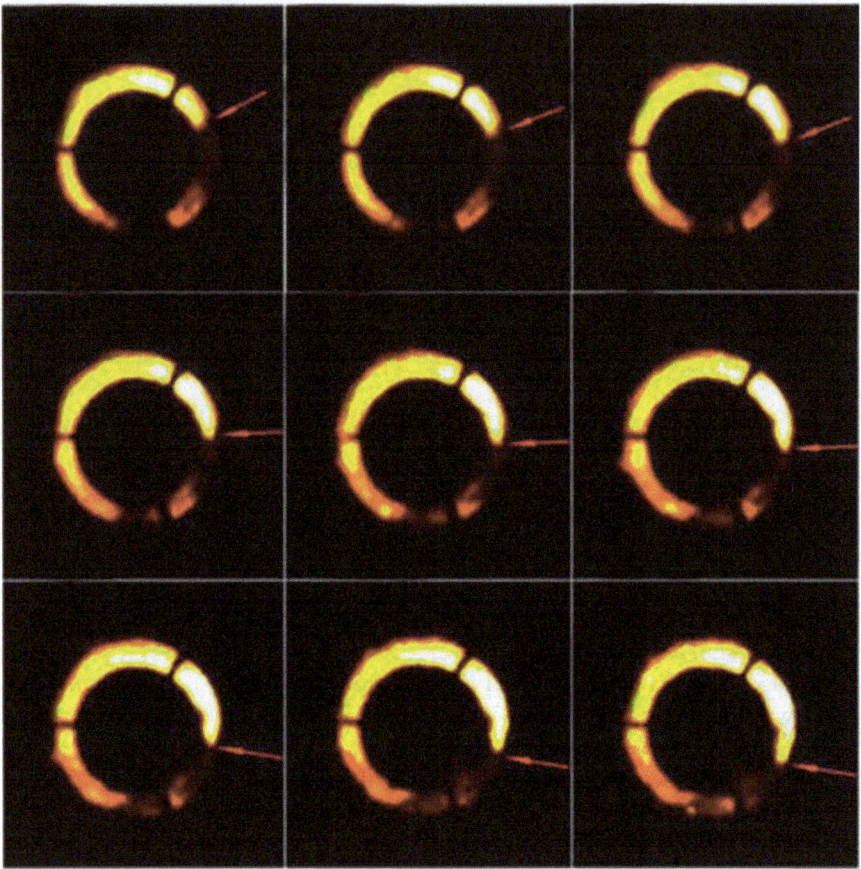

Fig. 7.7 Rotating detonation waves captured by high-speed camera (20,000 fps)

If there are more fresh reactants accumulating in front of the detonation wave, the detonation wave will become stronger. However, if the local pressure of the head-wall becomes higher than the fuel inlet pressure, the supply of fresh reactants will be blocked, which means the amount of reactants in front of the detonation wave will decrease, making the detonation weaker and the local pressure lower. Therefore, the fuel supply begins to recover. This adjusting process will continue until a balance is achieved. However, due to the unavoidable existence of disturbance, the balance is not guaranteed. Figure 7.10 demonstrates this self-adjusting mechanism. The adjusting process is likely to continue for more than one cycle, sometimes tens of cycles, represented by the continuous oscillation of pressure peaks. Wolanski (2011) described a similar mechanism, which was called a "galloping rotational detonation".

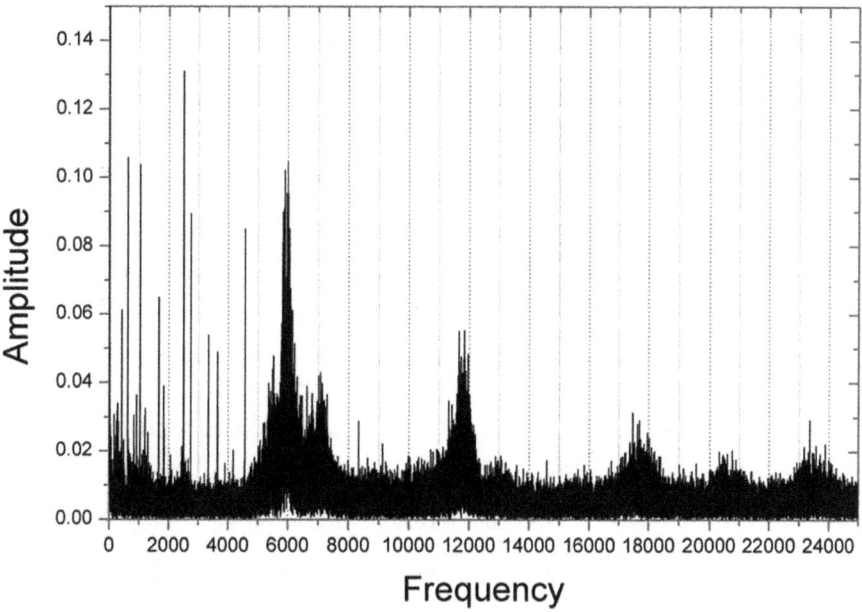

Fig. 7.8 Fourier transforms of pressure transducer data (Liu et al. 2015a)

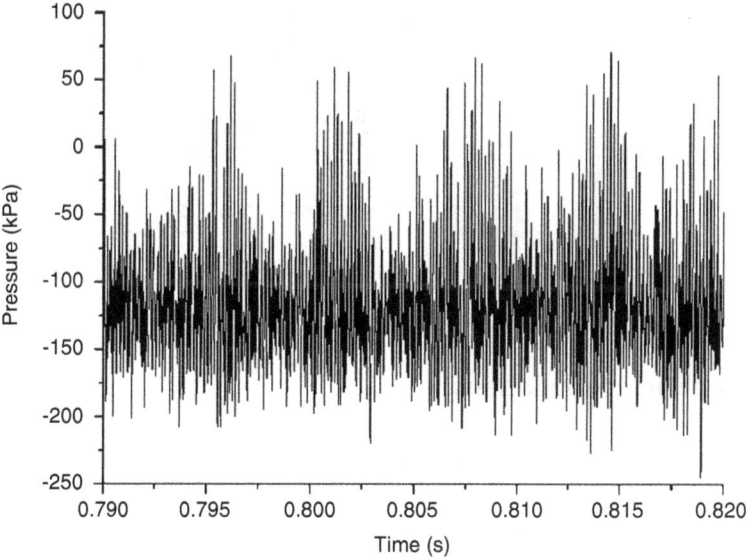

Fig. 7.9 A segment record of the pressure history (Liu et al. 2015a)

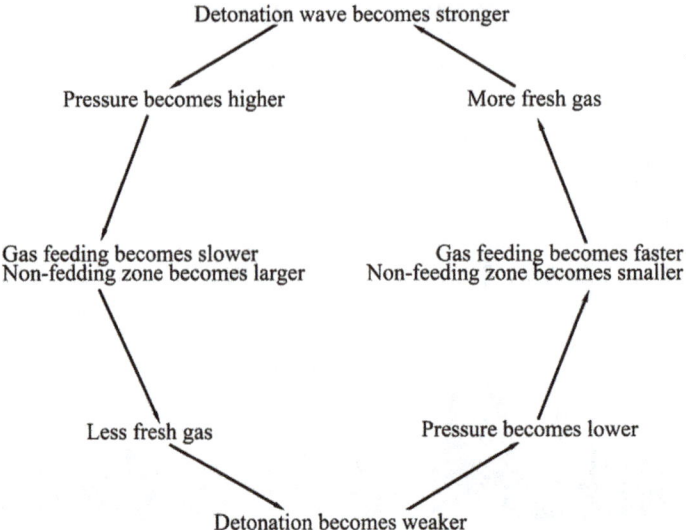

Fig. 7.10 The self-adjusting mechanism of continuously rotating detonation (Liu et al. 2015a)

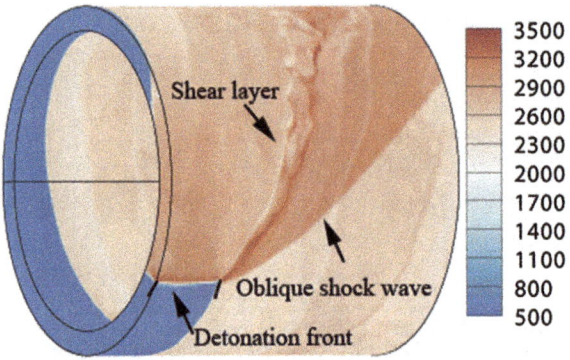

Fig. 7.11 Temperature distribution at stable stage

3 Numerical Simulations

Shao et al. (2010a, b) and Shao and Wang (2010) performed both two- and three-dimensional simulations of the CDE to validate its feasibility and to explore its physical mechanism. We obtained continuous and stable propagation of rotating detonation waves, and the flow field is shown in Fig. 7.11. In the study, we measured the propulsive performance of the CDE with different inlet stagnation pressures. Figure 7.12 shows that the propulsive performance varies when the injection stagnation pressure increases from 1.5 to 3 MPa. We also analyzed the injection mechanism of the CDE and verified its unique advantages: continuous detonations

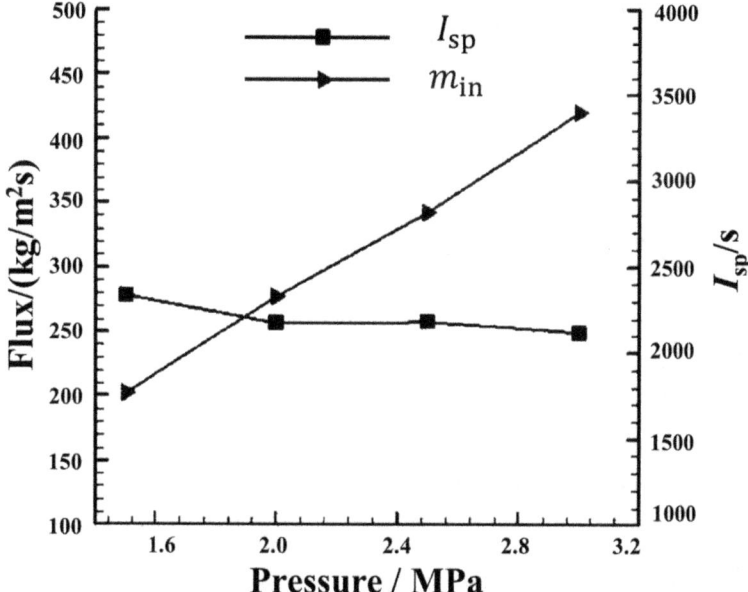

Fig. 7.12 Variations of specific impulse and mass flow rate

could be realized under different working conditions from subsonic to hypersonic inlet flows (Fig. 7.13).

For the first time, we carried out three-dimensional simulations of the CDE with four different types of nozzles, including the constant-area nozzle, Laval nozzle, diverging nozzle and converging nozzle, and analyzed their propulsive preformation in detail (Figs. 7.14 and 7.15). After calculating the thrust, gross specific impulse, net specific impulse and the average mass flow rate, we found that the Laval nozzle provided the best performance.

Liu et al. (2015b) first used GPU/CUDA for parallel computing to perform three-dimensional simulations of the CDE, shown in Fig. 7.16. It demonstrated that the computation was accelerated greatly. The computing speed was 18 times faster in the one-GPU-card case and 43 times in the two-GPU-card case, compared with that in the case without GPU/CUDA acceleration in parallel computing.

Furthermore, we captured the phenomenon of the spontaneous formation of multiple rotating detonation waves in the CDE (Yao et al. 2015), as shown in Fig. 7.17. We discussed the stability of rotating detonations, and suggested that stable propagation of detonation waves was not a necessity for running the CDE. In other words, the CDE could operate continuously even the detonation flow was unstable, though in this case it could not guarantee a continuous level of thrust.

We also proposed four new fuel injection layouts to improve the reliability of numerical set-up in CDE simulations (Liu et al. 2015c), including the radial strip injection, the oblique strip injection, the side-slit injection, and the mid-slit injection

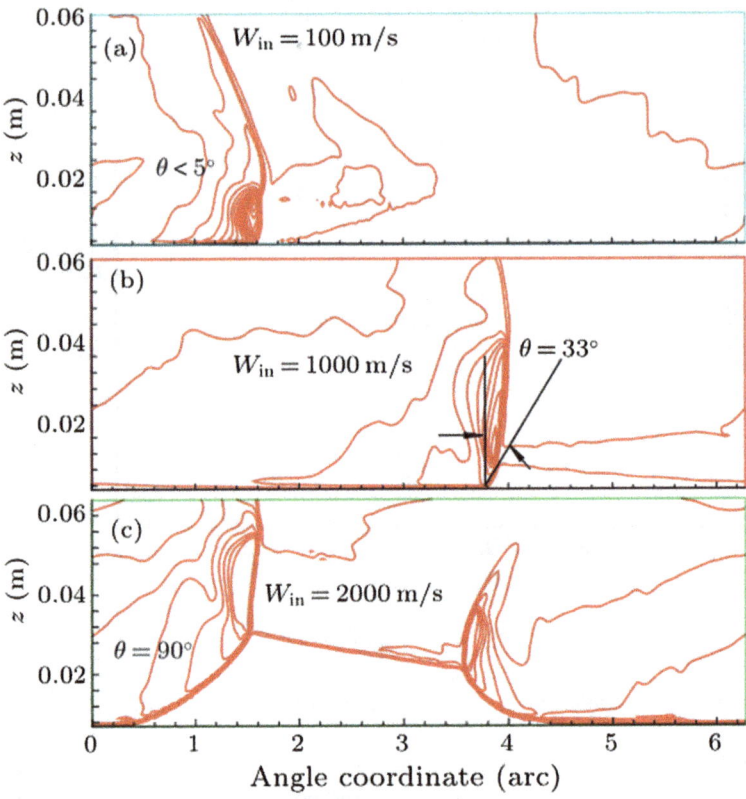

Fig. 7.13 Pressure counter on middle surface of the combustor with injection velocity W_{in} of 100, 1000 and 2000 m/s (Shao and Wang 2010)

(Fig. 7.18). We realized rotating detonations using these four new kinds of injection layouts (Fig. 7.19) and measured the corresponding propulsive performance.

Based on this, Yao et al. (2017a) proposed a more practical injection set-up in CDE simulations – injection via an array of orifices (Fig. 7.20). We performed two cases – one with two rows of orifices and the other with three, and achieved stable rotating detonations in both cases (Fig. 7.21). With this new injection model, we captured some important features, including the multi-wave rotating detonation phenomenon, transient process after initiation. These features have already been observed in experiments. But they are usually absent in previous CDE simulations since a very ideal injection approximation, the full area injection of mixture plenum on the head wall, is usually used for the sake of simplicity and computational efficiency.

Zhou and Wang (2012) for the first time investigated the paths of flow particles in an annular chamber of the CDE, based on two-dimensional numerical simulations. We found that the detonation wave, the deflagration wave, the oblique wave, and the contact surface had small influence on the paths of flow particles. The flow particles ejected rapidly almost along the axial direction. Also, we gave the *p-v* and

Fig. 7.14 Cross-sections of CDE configurations with different nozzles (Shao et al. 2010a)

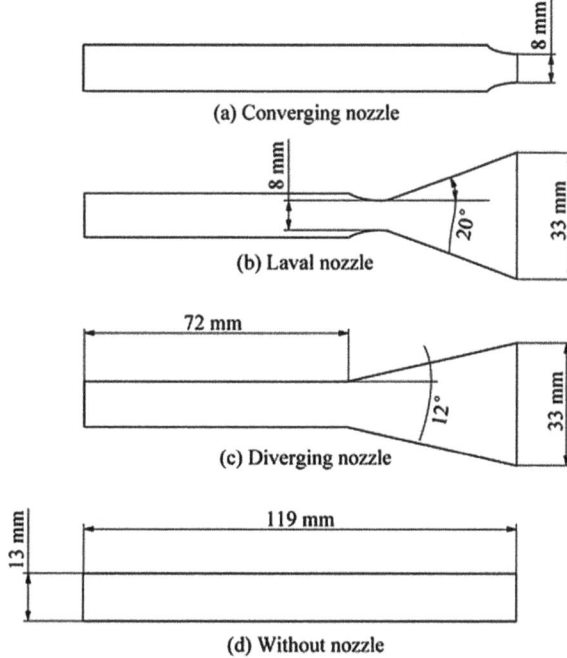

T-s diagrams which were qualitatively consistent with the ideal ZND model (Fig. 7.22). The thermal efficiency of a two-dimensional CDE model was calculated. The average thermal efficiency of detonation combustion in the two-dimensional CDE was 31% and the average net mechanical work was 1.3 MJ/kg. Zhou et al. (2014) also performed a follow-up study which used this particle path tracking method in three-dimensional CDE simulations.

Zhou and Wang (2013) investigated the wave reflections near the head-wall of the CDE (Fig. 7.23). Near the head-wall, shock waves reflected repeatedly between the inner and outer cylinders. Both regular and Mach reflections were found there in our study. It was also found that the length of the Mach stem increased with the increase of the chamber length. The increase of inner radius resulted in more shock waves between the inner and outer cylinders.

To overcome the overheat problem, Shao and Wang (2011) for the first time proposed a new model of the CDE, the CDE model with a cylindrical combustion chamber, which was also called the hollow CDE (Fig. 7.24).

The inner cylinder wall was removed to deal with the thermal load problem. The feasibility of this new model was verified by Tang et al. (2015) through numerical simulations. The flow fields inside the cylindrical chamber with different radii R_{outer} and chamber lengths L are shown in Fig. 7.25. We also found that in the annular model, the kinetic energy ratio in the axial direction was 89.9%, while that in the cylindrical model was 89.1% (Fig. 7.26). Therefore, little energy was lost in the

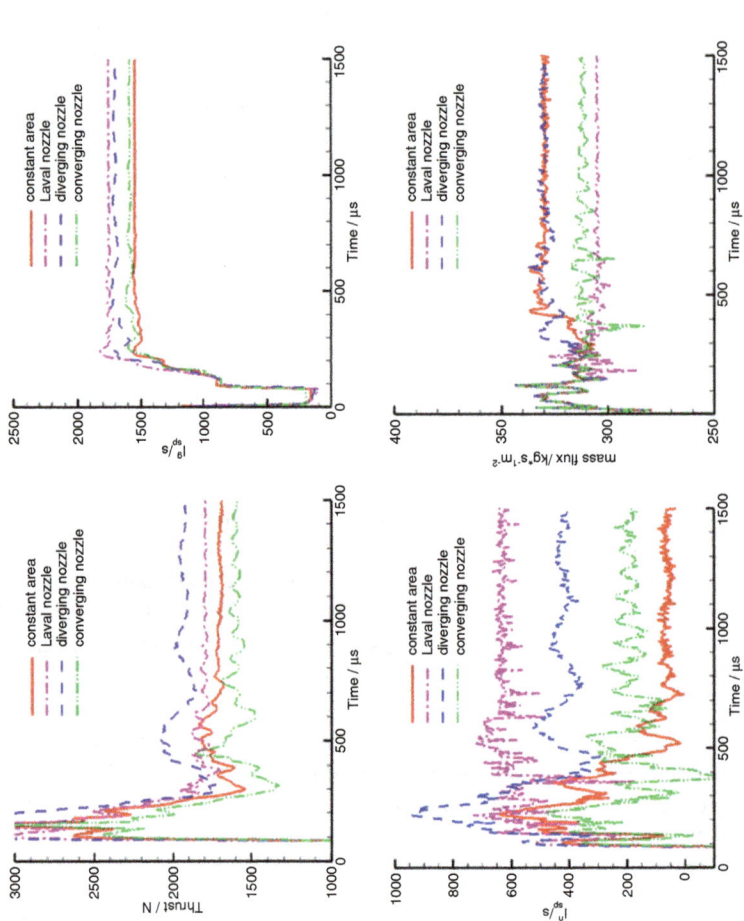

Fig. 7.15 Thrust history, gross specific impulse history, net specific impulse history and the average mass flow rate for four type of nozzles (Shao et al. 2010b)

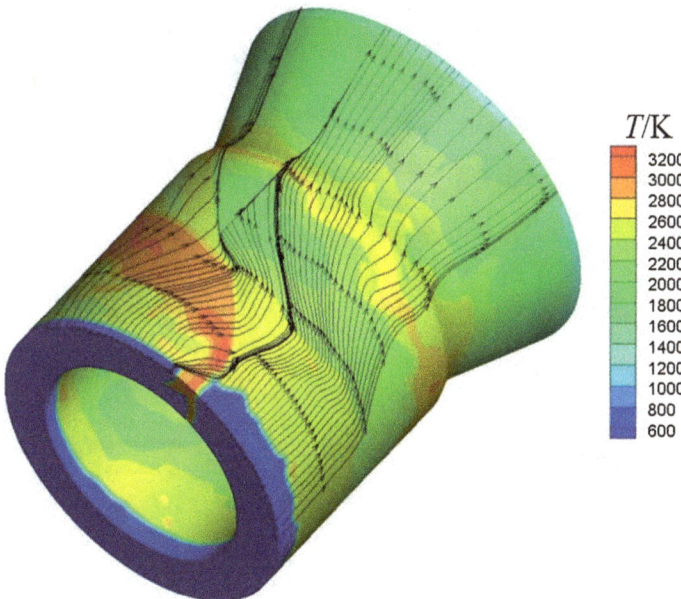

Fig. 7.16 The temperature contour and flow field on the outer wall of the combustion chamber at 1110 μs

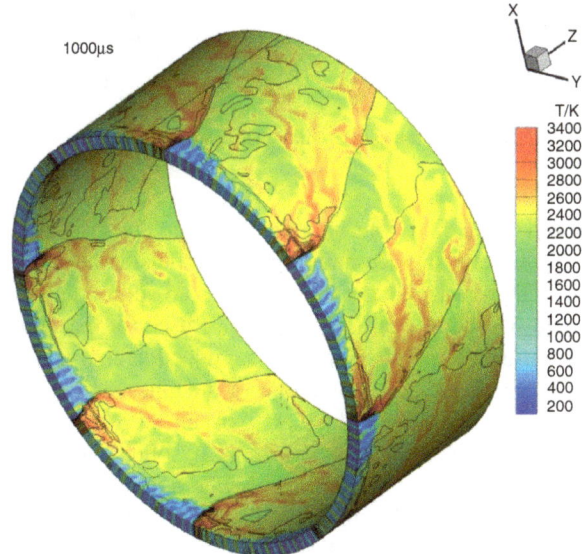

Fig. 7.17 Multi-wave rotating detonation flow field

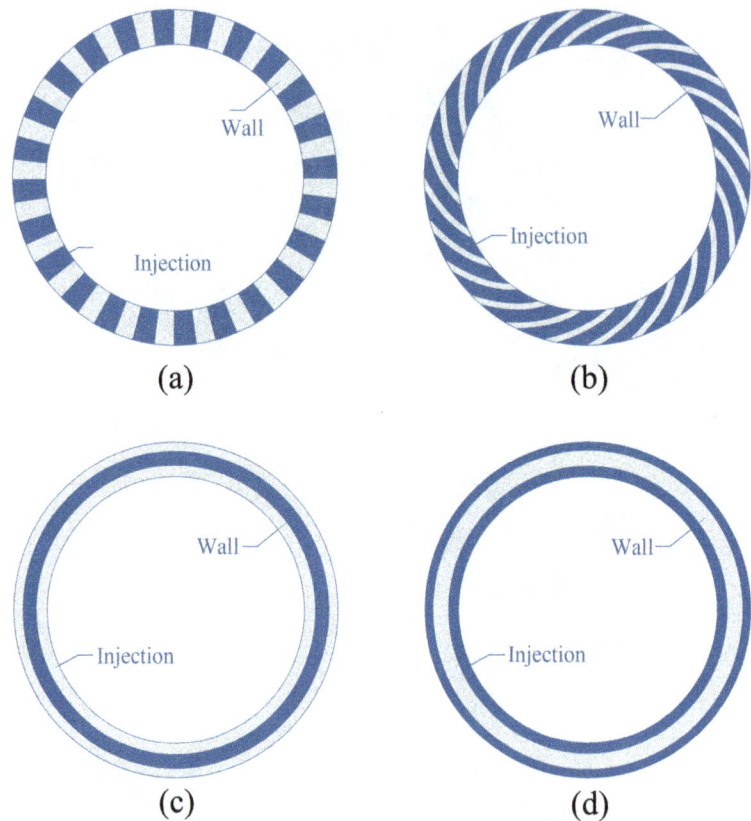

Fig. 7.18 Four fuel injection layouts: (**a**) radial strip, (**b**) oblique strip, (**c**) side-slit, and (**d**) mid-slit

circumferential and radial directions in both models. In this respect, this new model of the CDE is also a promising detonation-based propulsion device.

The removal of the inner cylinder also provides an additional benefit that theoretically the whole surface of the head wall can be used for reactant injection. Yao et al. (2017b) investigate the CDE with different injection area ratios, range from ψ = 55.6% to ψ = 88.9% (Fig. 7.27). The radius of the inner region is "R_{inner}" (R_{inner} is an auxiliary parameter that in fact does not exist in the hollow combustor). The inner region $r < R_{inner}$ on the head surface is a solid wall. The outer region $R_{inner} < r < R_{outer}$ is the reactant injection region where the fresh gas is continuously fed. The injection area ratio on the surface of the head wall played an important role in the flow field of the cylindrical CDE. If the chamber size is fixed, the number of detonation waves increased when the injection area ratio was higher (Fig. 7.28).

Wu et al. (2014b) discussed the effect of injection conditions on the stability of rotating detonation waves in simulations. It was found that as the injection stagnation pressure increased, the rotating detonation flow field became unstable and the

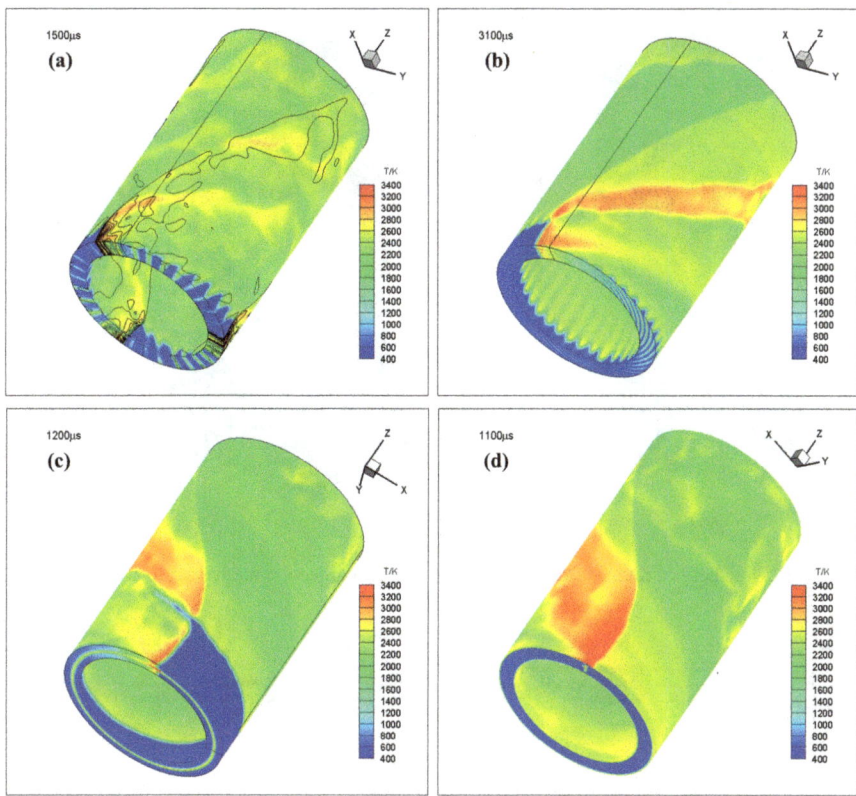

Fig. 7.19 Temperature contours of the flow field. (**a**) radial strip, (**b**) oblique strip, (**c**) side-slit, and (**d**) mid-slit

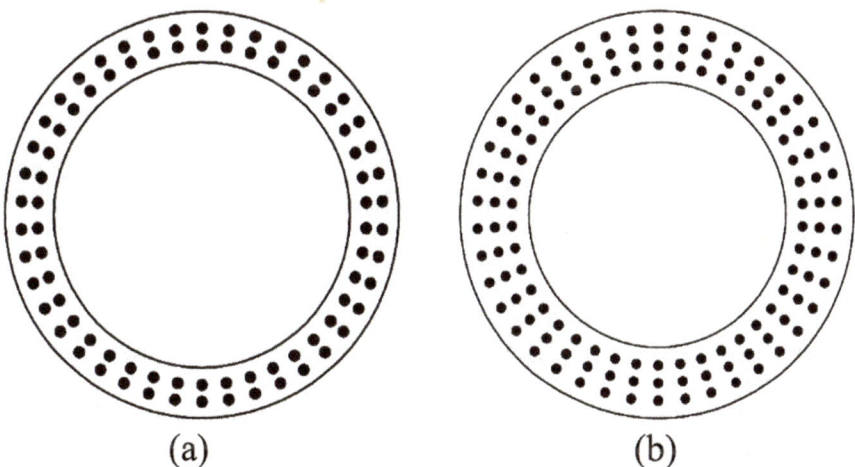

Fig. 7.20 Schematic of array injection method: (**a**) two-row and (**b**) three-row (Yao et al. 2017a)

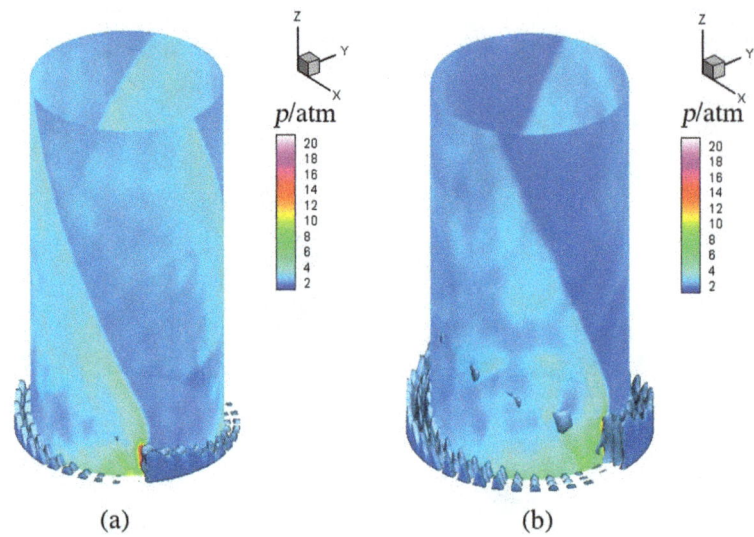

Fig. 7.21 Flow field of rotating detonations with the array injection method (Yao et al. 2017a)

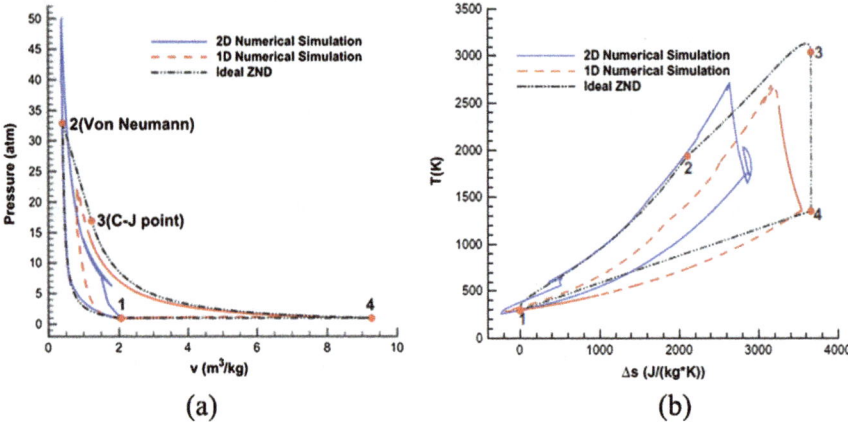

Fig. 7.22 Comparison of the p-v and T-s diagram from one-dimensional simulation, two-dimensional simulation, and the ideal ZND model. (**a**) p-v diagram; (**b**) T-s diagram (Zhou et al. 2014)

working condition of the CDE oscillated periodically and alternated between the strong and weak modes. Figure 7.29 shows that the detonation height and the average pressure of the reactant in front of the detonation wave oscillate periodically over time when the inlet stagnation pressure increased to 2 MPa. The propulsive performance was found to increase with higher inlet stagnation pressure, but it would oscillate when detonation waves became unstable.

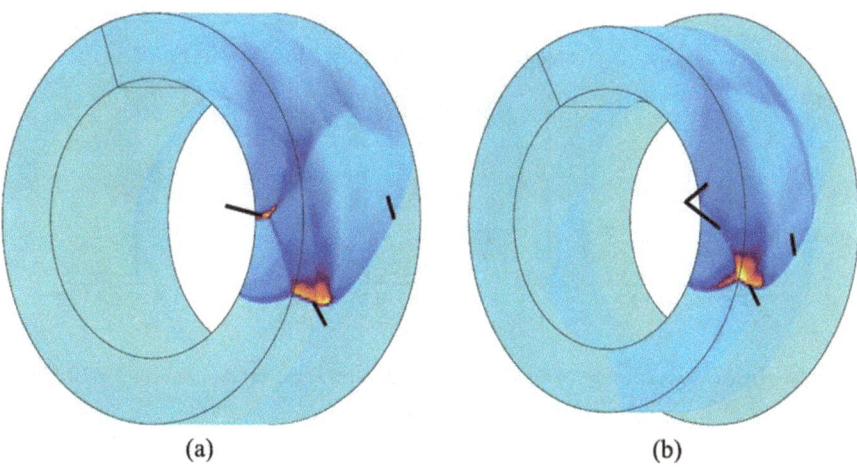

Fig. 7.23 Pressure contours of (**a**) the no-nozzle chamber and (**b**) the convergent-divergent nozzle chamber (Zhou and Wang 2013)

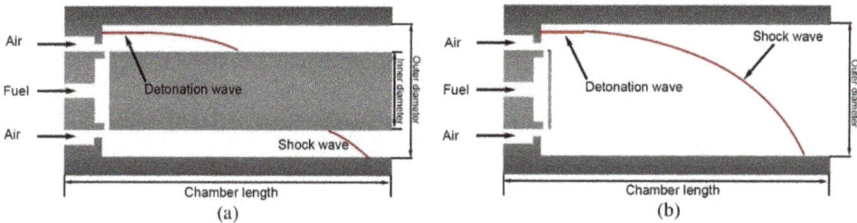

Fig. 7.24 Schematic of two models of the CDE. (**a**) Annular model and (**b**) Cylindrical model

In addition, the feedback on the operation of the CDE was investigated when the working conditions were changed (Wu et al. 2014a). When the inlet stagnation pressure was subject to sudden change, the time needed to complete the transition process was calculated. It was found that the sudden change of the inlet stagnation pressure had an immediate influence on the average axial flow velocity on the head surface of the CDE. After that, the velocity dropped and the average pressure increased gradually until the CDE reached a new stable state. When the inlet stagnation pressure increased from 2 to 2.5 MPa, the mass flux increased almost linearly to its peak value at the beginning of the transition process. The corresponding elapsed time was defined as Δt_1. After reaching the peak value, the mass flux dropped and reached a new stable state. The whole period of time of the transition process was defined as Δt_2. When the injection stagnation pressure decreased from 2 to 1.5 MPa, the transition process was almost the same. Figure 7.30 shows that the inlet stagnation pressure had little effect on Δt_1. But the total periods of time during the transition process were different.

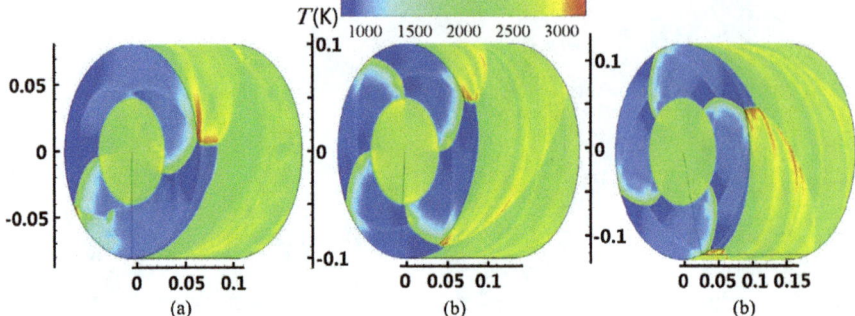

Fig. 7.25 Temperature contours of the flow field in the stable stage. (**a**) $R_{outer} = 8$ cm, $L = 10.67$ cm; (**b**) $R_{outer} = 10$ cm, $L = 13.33$ cm; (**c**) $R_{outer} = 12$ cm, $L = 16$ cm (Tang et al. 2015)

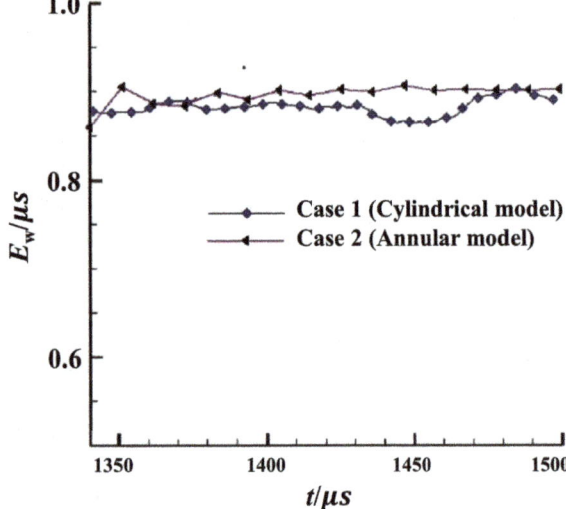

Fig. 7.26 Time history of kinetic energy ratio in axial direction (Tang et al. 2015)

4 Conclusions

The research team at Peking University has performed a series of numerical and experimental studies on the CDE over the past decade. We conducted the first successful hydrogen/oxygen rotating detonation experiment in China and proposed a new CDE combustor model, the cylindrical model. Our research covered many other important aspects of the CDE, especially our simulation work. We for the first time used the particle path tracking method in the research of the CDE and performed a three-dimensional simulation to reproduce the spontaneous formation of multiple rotating detonation waves. We also paid a lot of attention to the implementation of more practical injection set-up in simulations and the stability of the CDE.

Fig. 7.27 Schematic of the injection surface, $\psi = \left(R_{outer}^2 - R_{inner}^2\right)/R_{outer}^2$ (Yao et al. 2017b)

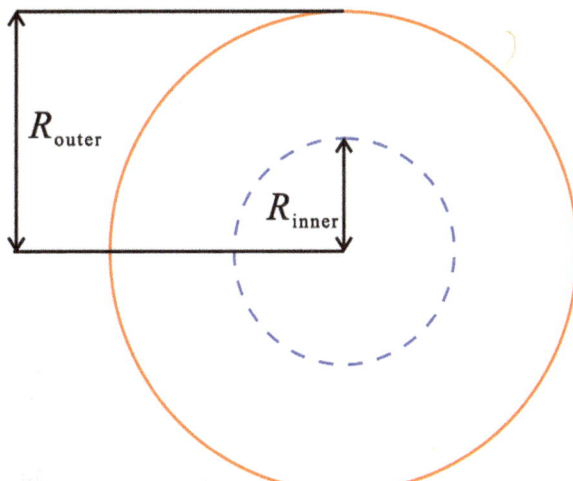

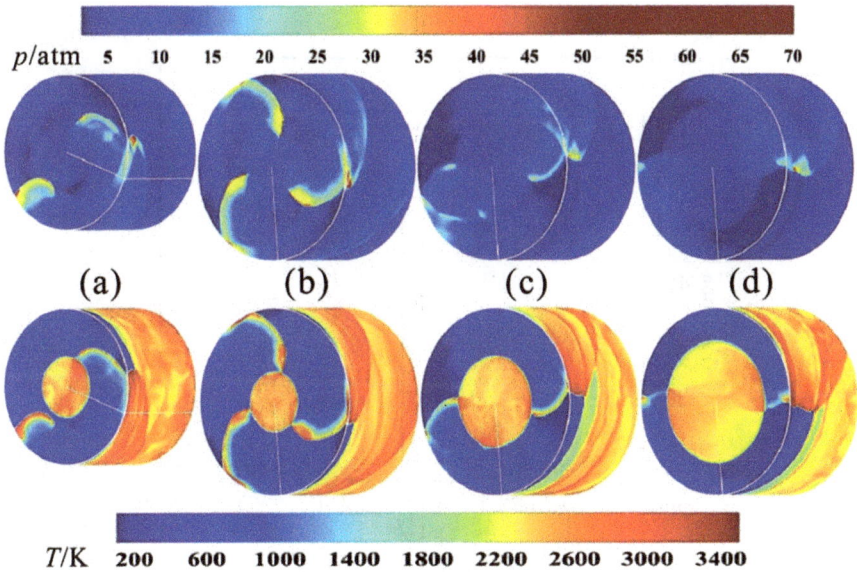

Fig. 7.28 Pressure and temperature contours of (**a**) case 1 (R_{outer} = 5 cm, ψ = 84%) at t = 2280 μs, (**b**) case 2 (R_{outer} = 6 cm, ψ = 88.9%) at t = 2050 μs, (**c**) case 3 (R_{outer} = 6 cm, ψ = 75%) at t = 1300 μs and (**d**) case 4 (R_{outer} = 6 cm, ψ = 55.6%) at t = 1700 μs (Yao et al. 2017b)

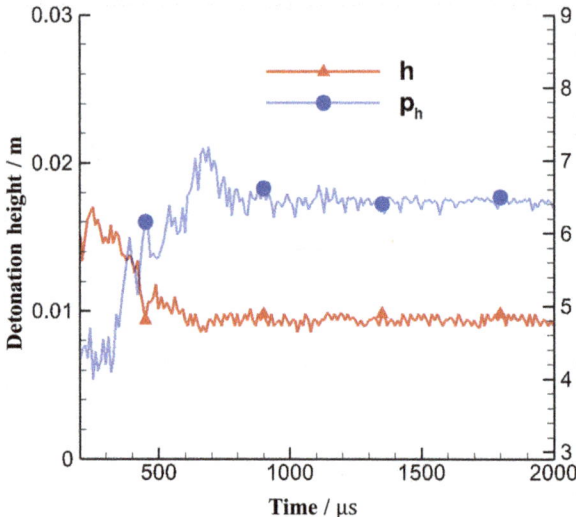

Fig. 7.29 Temporal evolution of the detonation height and the average pressure of the reactant ahead of the detonation. The injection stagnation pressure $p_0 = 2$ MPa

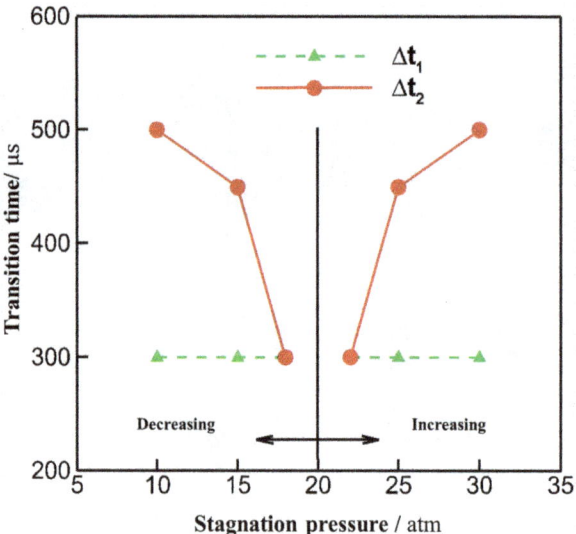

Fig. 7.30 Transition time versus stagnation pressure profile

References

Bykovskii, F. A., Zhdan, S. A., & Vedernikov, E. F. (2006). Continuous spin detonations. *Journal of Propulsion and Power, 22*(6), 1204–1216.

Liu, Y. S., Wang, J. P., Shi, T. Y., Wang, Y. H., Li, Y., & Li, Y. (2013). Experimental investigation on H2/O2 continuously rotating detonation engine. Paper presented at the 24th International Colloquium on the Dynamics of Explosions and Reactive Systems, Taipei, 28 July – 2 August 2013.

Liu, Y. S., Wang, Y. H., Li, Y., Li, Y. S., & Wang, J. P. (2015a). Spectral analysis and self-adjusting mechanism for oscillation phenomenon in hydrogen-oxygen continuously rotating detonation engine. *Chinese Journal of Aeronautics, 28*(3), 669–675.

Liu, M., Zhang, S., Wang, J. P., & Chen, Y. F. (2015b). Parallel three-dimensional numerical simulation of rotating detonation engine on graphics processing units. *Computers & Fluids, 110*, 36–42.

Liu, M., Zhou, R., & Wang, J. P. (2015c). Numerical investigation of different injection patterns in rotating detonation engines. *Combustion Science and Technology, 187*(3), 343–361.

Lu, F. K., & Braun, E. M. (2014). Rotating detonation wave propulsion: Experimental challenges, modeling, and engine concepts. *Journal of Propulsion and Power, 30*(5), 1125–1142.

Shao, Y. T., & Wang, J. P. (2010). Change in continuous detonation wave propagation mode from rotating detonation to standing detonation. *Chinese Physics Letters, 27*(3), 034705.

Shao, Y. T., & Wang, J. P. (2011). Three dimensional simulation of rotating detonation engine without inner wall. Paper presented at the 23rd International Colloquium on the Dynamics of Detonation and Reactive Systems, University of California, Irvine, U.S.A., 24–29 July 2011.

Shao, Y. T., Liu, M., & Wang, J. P. (2010a). Continuous detonation engine and effects of different types of nozzle on its propulsion performance. *Chinese Journal of Aeronautics, 23*(6), 647–652.

Shao, Y. T., Liu, M., & Wang, J. P. (2010b). Numerical investigation of rotating detonation engine propulsive performance. *Combustion Science and Technology, 182*(11–12), 1586–1597.

Tang, X. M., Wang, J. P., & Shao, Y. T. (2015). Three-dimensional numerical investigations of the rotating detonation engine with a hollow combustor. *Combustion and Flame, 162*(4), 997–1008.

Voitsekhovskii, B. V. (1959). Statsionarnaya dyetonatsiya. *Doklady Akademii Nauk SSSR, 129*(6), 1254–1256.

Wang, J. P., Shi, T. Y., Wang, Y. H., Liu, Y. S., & Li, Y. S. (2010). Experimental research on the rotating detonation engine. Paper presented at the 14th Shock Wave and Shock Tube Conference, Huangshan, Anhui, China, 14–16 July 2010.

Wang, Y. H., Wang, J. P., Shi, T. Y., & Liu, Y. S. (2012). Experimental research on transition regions in continuously rotating detonation waves. Paper presented at the 48th AIAA/ASME/SAE/ASEE Joint Propulsion Conference & Exhibit, AIAA Paper No. 2012-3946, Atlanta, Georgia, U.S.A., 30 July–1 August 2012.

Wang, Y. H., Wang, J. P., Li, Y. S., & Li, Y. (2014). Induction for multiple rotating detonation waves in the hydrogen-oxygen mixture with tangential flow. *International Journal of Hydrogen Energy, 39*(22), 11792–11797.

Wolanski, P. (2011). Rotating detonation wave stability. Paper presented at the 23rd International Colloquium on the Dynamics of Detonation and Reactive Systems, University of California, Irvine, U.S.A., 24–29 July, 2011.

Wu, D., Liu, Y., Liu, Y. S., & Wang, J. P. (2014a). Numerical investigations of the restabilization of hydrogen–air rotating detonation engines. *International Journal of Hydrogen Energy, 39*(28), 15803–15809.

Wu, D., Zhou, R., Liu, M., & Wang, J. P. (2014b). Numerical investigation of the stability of rotating detonation engines. *Combustion Science and Technology, 186*(10–11), 1699–1715.

Yao, S. B., Liu, M., & Wang, J. P. (2015). Numerical investigation of spontaneous formation of multiple detonation wave fronts in rotating detonation engine. *Combustion Science and Technology, 187*(12), 1867–1878.

Yao, S. B., Han, X. D., Liu, Y., & Wang, J. P. (2017a). Numerical study of rotating detonation engine with an array of injection holes. *Shock Waves, 27*(3), 467–476.

Yao, S. B., Tang, X. M., Luan, M. Y., & Wang, J. P. (2017b). Numerical study of hollow rotating detonation engine with different fuel injection area ratios. *Proceedings of the Combustion Institute, 36*(2), 2649–2655.

Zhou, R., & Wang, J. P. (2012). Numerical investigation of flow particle paths and thermodynamic performance of continuously rotating detonation engines. *Combustion and Flame, 159*(12), 3632–3645.

Zhou, R., & Wang, J. P. (2013). Numerical investigation of shock wave reflections near the head ends of rotating detonation engines. *Shock Waves, 23*(5), 461–472.

Zhou, R., Wu, D., Liu, Y., & Wang, J. P. (2014). Particle path tracking method in two-and three-dimensional continuously rotating detonation engines. *Chinese Physics B, 23*(12), 124704.

Chapter 8
Pulse Detonation Cycle at Kilohertz Frequency

Ken Matsuoka, Haruna Taki, Jiro Kasahara, Hiroaki Watanabe,
Akiko Matsuo, and Takuma Endo

Abstract To realize kilohertz and higher frequency of a pulse detonation cycle (PDC), enhancement of deflagration-to-detonation transition (DDT) is necessary. A novel semi-valveless PDC method, in which the inner diameter of the oxidizer feed line is equal to that of the combustor, can increase the pressure of detonable mixture by increasing total pressure of supplying oxidizer. In demonstration experiments, ethylene as fuel, pure oxygen as the oxidizer and the combustor having an inner diameter of 10 mm and length of 100 or 60 mm were used. A PDC was successfully operated at the frequency of up to 1916 Hz. Under the condition of 1010 Hz operation, the total pressure of supplying oxidizer were varied. As the results, it was found that the DDT distance and time decreased by approximately 50% when the total pressure of supplying oxidizer increased by 242%.

1 Introduction

Because the burning temperature in detonation is higher than that in conventional isobaric combustion, the theoretical thermal efficiency of a detonation engine is higher than that of a conventional internal combustion engine (ICE) (Endo et al. 2004b; Heiser and Pratt 2002; Wu et al. 2003). This is the primary reason why a detonation engine has been developed for decades worldwide as a pressure-gain ICE, particularly for propulsive applications. In addition, it is probably possible to

---

K. Matsuoka (✉) • H. Taki • J. Kasahara
Department of Aerospace Engineering, Nagoya University,
Furo-cho, Chikusa, Nagoya, Aichi 464-8603, Japan
e-mail: matsuoka@nuae.nagoya-u.ac.jp

H. Watanabe • A. Matsuo
Department of Mechanical Engineering, Keio University,
3-14-1 Hiyoshi, Kouhoku-ku, Yokohama, Kanagawa 223-8522, Japan

T. Endo
Department of Mechanical System Engineering, Hiroshima University,
1-4-1 Kagamiyama, Higashi-Hiroshima, Hiroshima 739-8527, Japan

realize more compact combustor because the detonation wave propagates at approximately 2000–3000 m/s.

In a conventional pulse detonation engine (PDE) (Kailasanath 2000, 2003; Matsuoka et al. 2016), the following processes are repeated in sequential order: (1) filling of fuel and oxidizer (detonable mixture), (2) ignition and deflagration-to-detonation transition (DDT), (3) propagation of a self-sustained detonation, (4) blowdown of the high-pressure hot burned gas, and (5) purging of the residual low-pressure hot burned gas. Of these processes, the duration of process (3) and (4) are governed by $\Delta t_3 = L_c/D_{CJ} = t_{CJ}$ and $\Delta t_4 =$ texhasut $\approx 7t_{CJ}$ (Endo et al. 2004a), and L_c and D_{CJ} are combustor length and Chapman − Jouguet detonation speed, respectively. In other words, if the detonation speed is constant at D_{CJ} and durations of process (1), (2) and (5) are negligibly small compared to that of processes (3) and (4), the gas-dynamic upper limit fupper of the operating frequency of the pulse detonation cycle (PDC) is given by fupper = $1/(\Delta t_3 + \Delta t_4) \propto 1/L_c$. To realize a high thrust density with a short combustor, it is necessary to realize a PDC at a high frequency, at the same order as the gas-dynamic upper limit $\alpha = f_{ope}/f_{upper} \approx 1$, using only the fuel and oxidizer, while keeping the system simple. Therefore, it is essential to shorten processes (1), (2), and (5) to achieve a higher operating frequency.

As a method for shortening processes (1) and (5), a valveless gas-feeding method has been proposed (Takahashi et al. 2012; Endo et al. 2016; Wang et al. 2014; Matsuoka et al. 2012, 2015, 2017a, b; Wu and Lu 2012). In the valveless-mode operation, the gas supply to a combustor is temporarily interrupted by the high-pressure burned gas generated in the combustor. A valveless PDC is suitable for high-frequency long-term operation because it is free from moving components and controlled only by the repeated ignitions. Recently, as shown in Fig. 8.1, Matsuoka et al. (2017b) proposed a semi-valveless PDC, in which the inner diameter id_o of the oxidizer feed line is equal to that of the combustor id_c.

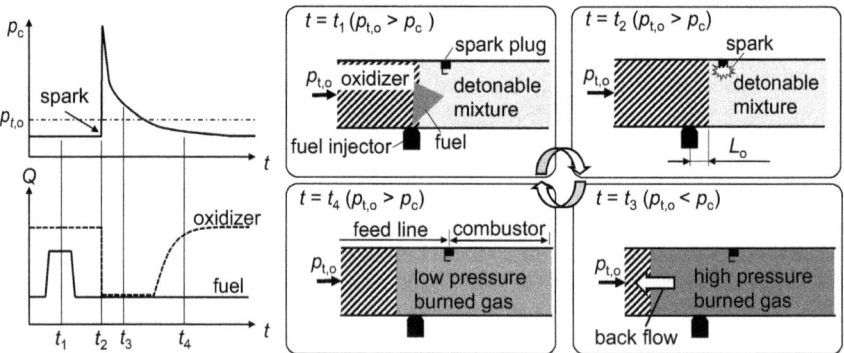

Fig. 8.1 Pulse detonation cycle using injector-feeding fuel and valveless-feeding gaseous oxidizer (Matsuoka et al. 2017b). Left: histories of pressure at spark plug (above) and flow rate of fuel and oxidizer (below). Right: schematic illustrations of inside of pulse detonation combustor at each specified time

8 Pulse Detonation Cycle at Kilohertz Frequency

The left figures in Fig. 8.1 show the typical pressure history at the spark plug of a combustor (upper) and the flow-rate history of the fuel and oxidizer corresponding to the pressure history (below).

The right figures show schematic illustrations of the material distribution in the combustor at each specified time shown in the left figures. The system consists of a spark plug for igniting a detonable mixture, valveless oxidizer feed line (on the left side of the spark plug), combustor (on the right side of the spark plug), and fuel injector installed at the side wall of the oxidizer feed line. The oxidizer at the total pressure $p_{t,o}$ is supplied from the left end. In contrast, the injector can inject fuel in a direction perpendicular to the axis of the combustor at any time because of the sufficiently high injection pressure p_{inj}. At $t = t_1$, the oxidizer is supplied to the combustor because the total pressure $p_{t,o}$ of the oxidizer is higher than the pressure p_c at the spark plug ($p_{t,o} > p_c$ at $t = t_1$). In addition, the fuel is simultaneously injected into the combustor in a direction perpendicular to its axis, and a detonable mixture is produced. At $t = t_2$, the injection of the fuel is stopped, and only the oxidizer is refilled between the fuel injector and spark plug (L_o in Fig. 8.1) before the ignition and subsequent initiation of a detonation ($p_{t,o} > p_c$ at $t = t_2$). At $t = t_3$, high pressure burned gas is generated after the initiation of the detonation by the spark plug ($p_{t,o} < p_c$ at $t = t_3$). The high-pressure burned gas in the combustor interrupts the oxidizer supply, and a portion of the burned gas flows backward to the oxidizer feed line. At $t = t_4$, once the gas pressure in the combustor drops below the oxidizer-supply pressure as a result of the expansion wave from the open end of the combustor, the oxidizer flows into the combustor again and purges the residual burned gas ($p_{t,o} > p_c$ at $t = t_4$). Matsuoka et al. (2017b) investigated the one-dimensional fluid motion in the semi-valveless PDC at an operating frequency of 500 Hz using 12 ion probes installed on the side wall of the oxidizer feed line and combustor. In addition, the experimental results were compared to those of a one-dimensional numerical calculation and found to be in good agreement. Up to now, Matsuoka (2016) demonstrated a maximum operating frequency of 800 Hz.

For shortening process (2), it is necessary to shorten the DDT time t_{DDT} (duration between spark time and time in which DDT occurs) sufficiently compared to that of a one PDC (i.e., $t_{DDT} << 1/f_{ope}$). In addition, with the shortening the combustor length L_c, it is more important to make the DDT distance x_{DDT} (distance between spark plug position and DDT point) sufficiently shorter than that of the combustor (i.e., $x_{DDT} << L_c$). There are many studies on shortening DDT process using a Shchelkin spiral (Shchelkin and Troshin 1965) and obstacles (Ciccarelli and Dorofeev 2008). However, there are some practical problems to install the obstacles in the combustor. For example, pressure loss and cooling of the obstacles themselves exist. Kuznetsov et al. (2005) investigated the influence of the initial pressure p_d of the detonable mixture on the DDT distance x_{DDT}. Stoichiometric hydrogen–oxygen mixtures were used in the single-shot tests with the initial pressure ranging from 0.2 to 8.0 bar. A dependence of the DDT distance x_{DDT} on the initial pressure p_d was found to be close to the inverse proportionality. However, the shortening of the DDT process by increasing the pressure of the mixture has not been demonstrated in the actual PDC operation.

In present study, to achieve higher operating frequency, two novel techniques, a piezoelectric fuel injector used in a commercial automotive engine and oxidizer feeding line for a choking flow, were introduced to a semi-valveless PDC proposed by Matsuoka et al. (2017b). In the combustion test, the PDC operations at 1010 Hz were carried out under the condition of the low and high pressure of the supplying oxidizer. As the results, it was found that the DDT process was shortened by increasing the pressure of the detonable mixture. In addition, with high-pressure condition, a PDC operation at 1916 Hz was confirmed.

2 Experimental Arrangement

Apparatus and Operation Sequence

Figure 8.2 shows the two combustors used in the experiments, which we call 'Type A' and 'Type B'. In the following, the x-coordinate is along the combustor axis in the direction toward the exit, with the origin corresponding to the spark plug position. The inner diameters of both the oxidizer feed line and combustor were constant at $id_o = id_c = 10$ mm. A spark plug SP (NGK Spark Plug Co., Ltd., CR8HSA) was installed at $x = 0$ mm. The combustor length, on the right side of the spark plug, was $L_c = 100$ mm and 60 mm for Type A and Type B, respectively. A pressure transducer PT2 (PCB Piezotronics, Inc, 113B24) was installed at $x = 40$ mm for Type A and $x = 20$ mm for Type B to measure the pressure p_c in the combustor. The cyclic combustion waves were detected by the ion probes installed on the side wall of the combustor (I1 − I4 for Type A, I1 and I2 for Type B).

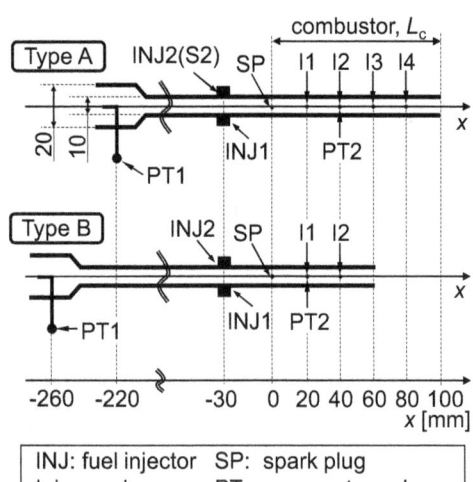

Fig. 8.2 Pulse detonation combustor used in experiments

8 Pulse Detonation Cycle at Kilohertz Frequency

The output voltage of the ion probe rapidly increased due to the passage of the combustion wave, and the cyclic flame propagation speeds V_f were measured using the rise time difference Δt and the interval between two ion probes, as 20 mm.

Gaseous oxygen as oxidizer was supplied into the combustor from the left end and the total pressure $p_{t,o}$ of the oxygen was measured by a low-speed pressure transducer PT1 (KELLER, 23 series) installed at $x = -220$ mm and -260 mm for Type A and Type B, respectively. In contrast, ethylene as fuel was injected into the combustor using a piezoelectric fuel injector INJ (BOSCH, HDEV4) located at $x = -30$ mm. Because both the temperature $T_{inj,f} = 292$ K and pressure $p_{inj,f} = 5.5$ MPa in the ethylene tank were higher than those of the critical state ($T_{cr,f} = 282.4$ K and $p_{cr,f} = 5.05$ MP), the injected ethylene was in the supercritical state. The single fuel injector INJ1 was used in the low-pressure condition (S1). In the high-pressure condition (S2 and S3), the two fuel injectors (INJ1 and INJ2) were used to maintain the equivalence ratio of the detonable mixture.

Figure 8.3 shows the input signal to the spark plug, input signal to the fuel injector (solid line), real spark and fuel injection (dashed line) and pressure history measured by the PT2 in the experiment S2. In contrast to the fuel injection and spark, oxidizer is steadily supplied to the combustor during the experiment. The time is defined so that the spark input time is set at $t = 0$. The delay times of the fuel injector and ignition device were investigated by visualization experiments of the intermittent fuel injection and spark. As per the results, shown in Fig. 8.3, the ignition delay Δt_{spark} from the on-signal time was $\Delta t_{spark} = 30 \pm 3$ μs. The standard deviations of 12 cycles were considered in the calculation of errors. The response delays (Δt_{start}, Δt_{stop}) of the injector from the on-signal time to the real injection start time and from the off-signal time to the actual injection stop time were $\Delta t_{start} = 36 \pm 10$ μs and $\Delta t_{stop} = 106 \pm 10$ μs, respectively. The frame rate of 100,000 frames/s of a high-speed camera was considered in the errors.

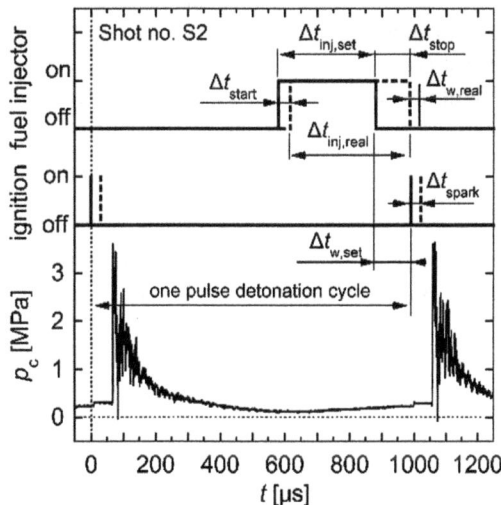

Fig. 8.3 Operational sequence diagram, in which the input signal to spark plug, fuel injector (solid line), and real open/closed state of fuel injector and spark (dashed line), and corresponding histories of pressure at pressure transducer PT2 at $x = 40$ mm (shot no. S2)

Table 8.1 Experimental conditions

Shot no.	Combustor type	Operating frequency f_{ope} [Hz]	Total pressure of supplying oxygen $p_{t,o}$ [MPa]	Ethylene injection pressure $p_{inj,f}$ [MPa]	Set ethylene injection duration $\Delta t_{inj,set}$ [μs]	Set duty ratio of fuel injector DR_{set} [%]
S1	Type A	1010	0.23 ± 0.01	5.5 ± 0.1	300	30
S2*			0.55 ± 0.01			
S3*	Type B	1916	0.57		180	34

Note: * means that two injectors were used. Ambient temperature: $T_a = 289 \pm 1$ K, ambient pressure: $p_a = 0.1 \pm 0.01$ MPa

Experimental Conditions

Table 8.1 shows the summary of the experimental conditions. We carried out combustion tests under three conditions (S1, S2 and S3). Type A of the combustor having a length of 100 mm was used in the condition of S1 and S2, and a PDC operation at an operating frequency of 1010 Hz was performed three times.

As the first sequence of a PDC operation, the steady flow of the oxidizer was produced. The total pressure of the oxygen immediately before starting the PDC operation was measured by the pressure transducer PT1 in Fig. 8.2, and the total pressure in the steady flow were $p_{t,o} = 0.23 \pm 0.01$ MPa and 0.55 ± 0.01 MPa for the low and high pressure condition, respectively. In a preliminary cold gas experiment with the oxygen, the mass flow rate of oxygen in the steady flow was $\dot{m}_o = 45.2 \pm 0.5$ g/s and 97.8 ± 3.9 g/s, respectively. The standard deviations of the three repeated measurements were taken into account in the errors.

The experimental mass flow rate $\dot{m}_o$ of oxygen was compared with a one-dimensional adiabatic flow with friction (Fanno flow) and an isentropic flow. The flow is assumed to be choked at the exit $x_2 = 100$ mm of the combustor (subscript 2). As shown in Fig. 8.2, the inlet of the combustor (subscript 1) having an inner diameter of $id_c = 10$ mm was $x_1 = -200$ mm. Assuming Fanno flow between the inlet and exit of the combustor, the following relationship holds (Matsuo 1994):

$$\chi(M_{o,1}) - \chi(M_{o,2}) = 4f \frac{x_2 - x_1}{id_c} \tag{8.1}$$

where f and M_o are friction factor and Mach number at each surface, respectively. $\chi(M_o)$ in Eq. (8.1) is given by:

$$\chi(M_o) = \frac{1 - M_o^2}{\gamma_o M_o^2} + \frac{\gamma_o + 1}{2\gamma_o} \ln \left[\frac{(\gamma_o + 1) M_o^2}{(\gamma_o - 1) M_o^2 + 2} \right] \tag{8.2}$$

8 Pulse Detonation Cycle at Kilohertz Frequency

In Eq. (8.2), $\chi(M_{o,2})$ becomes 0 because $M_{o,2}$ is assumed unity. If the friction factor is constant, the Mach number at the inlet $M_{o,1}$ can be calculated using Eq. (8.1). The mass flow rate $\dot{m}_{o,\text{Fanno}}$ at the inlet is obtained from the following equation:

$$\dot{m}_{o,\text{Fanno}} = \frac{A_c p_{t,o}}{\sqrt{R_o T_{t,o}}} \Gamma_o \sqrt{\gamma_o \left(\frac{2}{\gamma_o + 1}\right)^{\gamma_o + 1/\gamma_o - 1}} \tag{8.3}$$

where A_c, R_o, $T_{t,o}$, and γ_o are the cross-sectional area of the combustor, gas constant of the oxygen, total temperature of the oxygen and ratio of the specific heat of the oxygen, respectively. The total temperature $T_{t,o}$ of the oxygen was assumed to be equal to the temperature in the oxygen tank before starting the PDC operation ($T_{t,o} = 291 \pm 1$ K for S1, $T_{t,o} = 294 \pm 1$ K for S2). In Eq. (8.3), Γ_o is called gas dynamics function and is given as an expression of Mach number as follows:

$$\Gamma_o = M_o \left[\frac{2 + (\gamma_o - 1) M_o^2}{(\gamma_o + 1)}\right]^{-(\gamma_o + 1)/2(\gamma_o - 1)} \tag{8.4}$$

If $\Gamma_o = 1$, the mass flow rate $\dot{m}_{o,\text{isen}}$ of oxygen assuming an isentropic flow is obtained. The calculated Mach number at the inlet and mass flow rate in a Fanno flow was calculated at $M_{o,1} = 0.60$ and $\dot{m}_{o,\text{Fanno}} = 38.0$ g/s and 90.3 g/s for S1 and S2, respectively ($f = 0.004$, $A_c = 78.5$ mm², $R_o = 259.8$ J kg⁻¹ K⁻¹, and $\gamma_o = 1.4$). The experimental mass flow rate were 119% (S1) and 108% (S2) of $\dot{m}_{o,\text{Fanno}}$. On the other hand, the experimental mass flow rate were 100% (S1) and 91% (S2) of $\dot{m}_{o,\text{isen}}$. Consequently, the assumption of a Fanno flow underestimated the experimental mass flow rate. It was suggested that the influence of the friction on the total pressure loss was small because the total length of the oxidizer feed line and the combustor was short. Under the assumption of a isentropic acceleration from the stagnation state, the oxygen static pressure $p_{s,o}$ in the combustor is obtained from the following equation:

$$\frac{p_{t,o}}{p_{s,o}} = \left(1 + \frac{\gamma_o - 1}{2} M_o^2\right)^{\gamma_o/\gamma_o - 1} \tag{8.5}$$

The pressure ratio is calculated at $p_{t,o}/p_{s,o} = 1.89$ if the Mach number is unity and $\gamma_o = 1.4$. In other words, while the choking condition is maintained, the static pressure $p_{s,o}$ in the combustor can be increased with increasing the total pressure $p_{t,o}$ of supplying oxidizer. Using the total pressure measured in the combustion test (see Table 8.1), the static pressure in the low and high-pressure experiment was assumed to be $p_{s,o} = 122$ kPa and 291 kPa, respectively.

To calculate the mass flow rate $\dot{m}_f$ of the fuel during the injection, 5000-cycle fuel injection was carried out under the condition of S1 (same as S2), and $\dot{m}_f$ was obtained from the following equation:

$$\dot{m}_\mathrm{f} = \frac{\Delta m}{5000 \times \Delta t_\mathrm{inj,real}} \qquad (8.6)$$

where Δm is the tank mass difference (accuracy ±0.1 g) before and after the 5000-cycle injection to the atmospheric-pressure environment. As shown in Fig. 8.3, $\Delta t_\mathrm{inj,real}$ is the real injection duration $\Delta t_\mathrm{inj,real} = 370 \pm 10$ μs. The fuel mass flow rate of a single injector calculated with Eq. (8.6) was $\dot{m}_\mathrm{f} = 4.9 \pm 0.3$ g/s at the injection pressure of $p_\mathrm{inj,f} = 5.84 \pm 0.03$ MPa, and the injection pressure was identical to that of the combustion test within 7%. They were the averaged value of three-time measurement and the error was the standard deviation. Finally, the equivalence ratio of the detonable mixture was estimated to be $ER_\mathrm{d} = 0.38 \pm 0.03$ and 0.35 ± 0.04 for low and high-pressure condition, respectively.

Under the condition of S1 and S2, the input injection duration was set to $\Delta t_\mathrm{inj,set} = 300$ μs. Therefore, the set duty ratio (input injection duration/duration for one PDC) were $DR_\mathrm{set} = 30\%$. As shown in Fig. 8.3, the waiting duration for spark was empirically set to $\Delta t_\mathrm{w,set} = 110$ μs. Considering the real response delay, the real injection duration, duty ratio and waiting duration for spark were $\Delta t_\mathrm{inj,real} = 370 \pm 10$ μs, $DR_\mathrm{real} = 37 \pm 1\%$ and $\Delta t_\mathrm{w,real} = 34 \pm 13$ μs, respectively. If the delays of the fuel injection and spark were not changed under the high-frequency condition (S3), each values were estimated at $\Delta t_\mathrm{inj,real} = 250 \pm 10$ μs, $DR_\mathrm{real} = 48 \pm 2\%$ and $\Delta t_\mathrm{w,real} = 34 \pm 13$ μs, respectively. The sampling rate in all the experiments was 10 MHz.

3 Results and Discussion

Pressure History

Figure 8.4 (a, b) show the time history of the pressure measured by the pressure transducer (PT2) installed at $x = 40$ mm. It is known that the baselines of the pressure-transducer outputs during a PDC operation gradually shift downward, owing to the thermal effect. To cancel this characteristic, the baseline of the second and following cycle were corrected by the initial pressure p_d that was defined as the averaged pressure during 0 μs $< t < 50$ μs of the first PDC, and $t = 0$ means the time when the spark input signal switched on at the each PDC. Specifically, the initial pressure was $p_\mathrm{d} = 135 \pm 1$ kPa and 244 ± 3 kPa for low and high-pressure condition, respectively. The errors are the standard deviations of three experiments under each experimental condition.

The gray-color area in Fig. 8.4a shows the range of the Chapman − Jouguet detonation pressure p_CJ at the temperature range of $T_\mathrm{d} = 400$–500 K, estimated equivalence ratio $ER_\mathrm{d} = 0.38$, and pressure of detonable mixture $p_\mathrm{d} = 135$ kPa (Gordon and McBride 1996). The temperature range was roughly estimated by comparison between experiments and one-dimensional numerical calculation discussed in Sect.

8 Pulse Detonation Cycle at Kilohertz Frequency

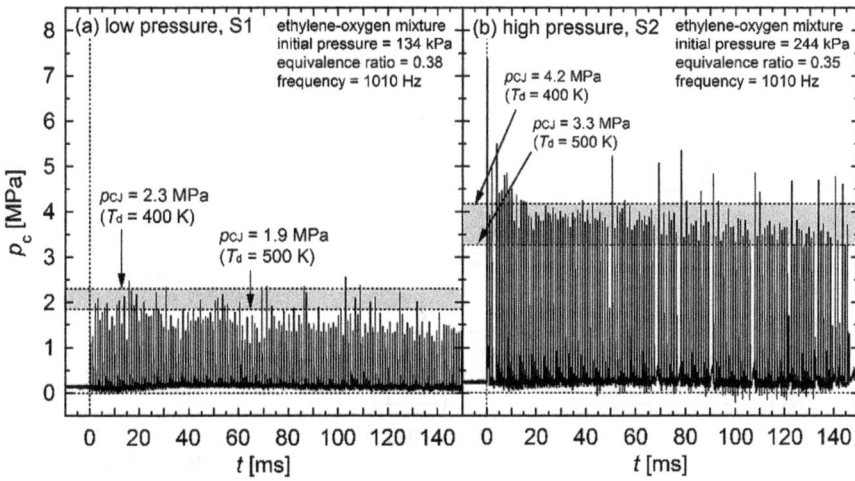

Fig. 8.4 Time history of pressure in combustor measured by pressure transducer located at $x = 40$ mm. (**a**): low-pressure condition S1, (**b**): high-pressure condition S2

3.3. Under the low-pressure condition S1, it was confirmed that the peak pressures in many cycles were lower than the estimated Chapman − Jouguet detonation pressure $p_{CJ} = 1.9$–2.3 MPa. As described in Sect. 3.2, it was probably due to the fact that the DDT occurred downstream of the pressure transducer PT2 installed at $x = 40$ mm. On the other hand, under the high-pressure condition S2 (Fig. 8.4b), the peak pressures in most cycles were in good agreement with $p_{CJ} = 3.3$–4.2 MPa ($T_d = 400$–500 K, $ER_d = 0.35$ and $p_d = 244$ kPa). It is considered that the DDT distance was shortened by increasing the pressure of the detonable mixture and the DDT occurred upstream of the PT2.

Flame Propagation Speed and Deflagration-to-Detonation Transition

Figure 8.5 shows the flame propagation speed measured with the ion probes installed on the side of the combustor. The plotted points indicate the center between the two ion probes. The error bar on the vertical axis shows the standard deviation of all cycles, and the error bar on the horizontal axis shows the distance between two ion probes.

The lower gray-color area in Fig. 8.5 shows the Chapman − Jouguet detonation speed D_{CJ} of the detonable mixture in static state under the temperature range of $T_d = 400$–500 K (Gordon and McBride 1996), and they were estimated to be $D_{CJ} = 1934$–1947 m/s for S1 and 1923–1934 m/s for S2. In the actual PDC operation, the detonable mixture flows at high speed u_d. Therefore, the apparent speed of

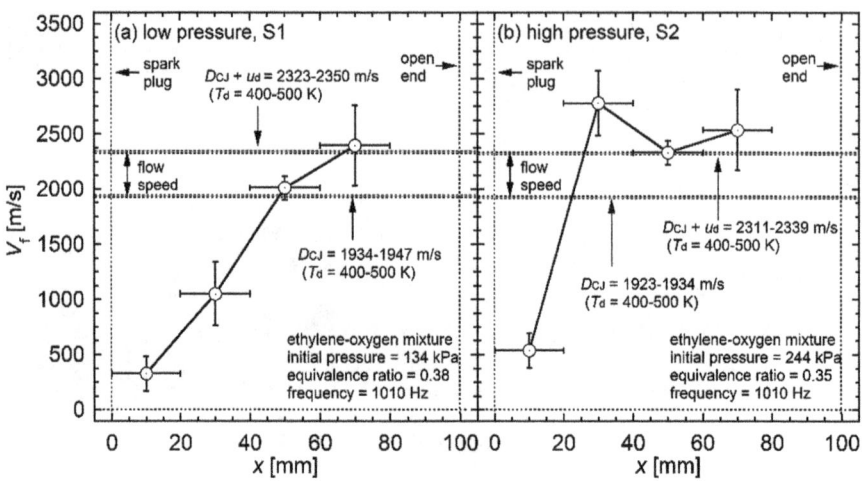

Fig. 8.5 Change in flame propagation speed in axial direction of combustor measured by ion probes. (**a**): low-pressure condition (S1), (**b**): high-pressure condition (S2)

the detonation wave was calculated by $D_{CJ} + u_d$. If the mixture flows at the sound speed, the speed is calculated from $u_d = \sqrt{\gamma_d R_d T_d}$. γ_d, R_d are the ratio of the specific heat and gas constant of the detonable mixture obtained by the estimated equivalence ratio ER_d, measured pressure p_d in the combustor, temperature T_d of the detonable mixture and NASA CEA (Gordon and McBride 1996). The apparent speed of the detonation waves $D_{CJ} + u_d$ = 2323–2350 m/s for S1 and 2311–2339 m/s for S2 are shown as the upper gray-color area in Fig. 8.5. If the mixing of fuel and oxidizer is not sufficient, the detonation speed and pressure decrease due to that the mixture was burned at lower equivalence ratio. Assuming that the equivalence ratio of S2 changed from ER_d = 0.35 to 0.20, the apparent speed of the detonation waves were estimated to be $D_{CJ} + u_d$ = 2101–2136 m/s. The experiment velocity in the region of 60 mm < x < 80 mm in Fig. 8.5b was 119–121% of the estimated velocity and the difference from the estimated value became larger than that at ER_d = 0.35. In addition, the Chapman – Jouguet detonation pressure p_{CJ} = 2.7–3.3 MPa at ER_d = 0.2 was much lower than the peak pressure confirmed in Fig. 8.4b. From the above, it was reasonable that the equivalence ratio was estimated by using experimental mass flow rate.

In the region of 60 mm $\leq x \leq$ 80 mm of Fig. 8.5a, the flame propagation speed increased to 102% of $D_{CJ} + u_d$ at T_d = 400 K. Therefore, DDT distance x_{DDT} was roughly estimated to be within 40 mm $\leq x_{DDT} \leq$ 60 mm. In addition, the region of DDT time t_{DDT} between the real spark time and the time when flames reached to the ion probes installed at x = 60 mm and 80 mm were 88 ± 45 < t_{DDT} < 98 ± 47 μs and the standard deviation of all cycles was taken into account. On the other hand, under the high-pressure condition S2, it was suggested that DDT occurred in the region of 0 mm $\leq x_{DDT} \leq$ 20 mm and 46 ± 8 < t_{DDT} < 54 ± 8 μs.

Under S1 and S2, the experimental conditions (spark system, equivalence ratio and combustor configuration) other than the oxidizer supply pressure are almost same. In fact, it was found that the DDT distance and time decreased by approximately 50% when the total pressure of the supplying oxidizer increased by 242%. Kuznetsov et al. (2005) proposed the relationship $x_{DDT} = a \cdot p_d^b = 0.7 p_d^{-1.17}$ for hydrogen-oxygen mixture, and the units of x_{DDT} and p_d in the equation were m and bar, respectively. By power approximation of the experiment S1 and S2, $a = 0.10 \pm 0.01$ and $b = -1.50 \pm 0.34$ were obtained. If the mixture pressure was increased up to $p_d = 10$ bar (1 MPa), the DDT distance was calculated at $x_{DDT} = 3$ mm.

Comparison with One-Dimensional Numerical Analysis

In the previous section, the equivalence ratio ER_d and pressure p_d of the detonable mixture at the spark time were estimated from the experimental results. However, it is difficult to estimate the temperature T_d of the mixture during PDC operation. Therefore, the measured pressure was compared with those of a numerical calculation using a one-dimensional compressible Navier–Stokes equation, in which a conservation equation for the chemical species was used. Yee's non-MUSCL (monotonic upstream schemes for conservation laws)-type second-order upwind scheme (Yee 1989) and second-order accurate central-difference scheme were employed for the convection and viscous terms, respectively. A chemical reaction model proposed by Singh and Jachimowski (1994) was used, in which 9 species and 10 reactions were considered in the ethylene–oxygen reaction. An MTS (multi-timescale) method (Gou et al. 2010) was used for the time integration of the source term. The gas was assumed to be a thermally perfect gas, and the thermodynamic properties were obtained from the NASA thermochemical polynomials.

As the initial condition, the steady flow of oxygen, in which the total pressure and temperature were equal to the experimental values (i.e., $p_{t,o} = 0.23$ MPa and $T_{t,o} = 291$ K for S1, $p_{t,o} = 0.55$ MPa and $T_{t,o} = 294$ K for S2), was prescribed in the combustor. Then, the detonable mixture was set in the combustor. In the experiments, the waiting duration for spark $\Delta t_{w,real}$ (see Fig. 8.3) exists and the oxygen is refilled during the duration (see Fig. 8.1). As the results of the steady calculation, the speed of oxygen was calculated at $u_o = 295 \pm 1$ m/s. Therefore, the oxygen refilled length during the waiting time was estimated to be $L_o = u_o \times \Delta t_{w,real} = 10$ mm. Consequently, the detonable mixture was set in the region of -20 mm $\leq x \leq L_c$. For the direct initiation of a detonation wave, a 2-mm-wide ignition source at 3000 K and 3 MPa was set at the spark position $x = 0$ mm.

The gray and black lines in Fig. 8.6 shows the enlarged pressure history of 100th cycle in Fig. 8.5 and calculation result varying the temperature of the detonable mixture, respectively. The vertical axis is represented by logarithm and the time τ of the horizontal axis indicated the time at which the pressure at the position of the pressure transducer PT2 ($x = 40$ mm) rapidly increased. Compared with the calculated pressure history, the low and high-frequency oscillations in the experiment

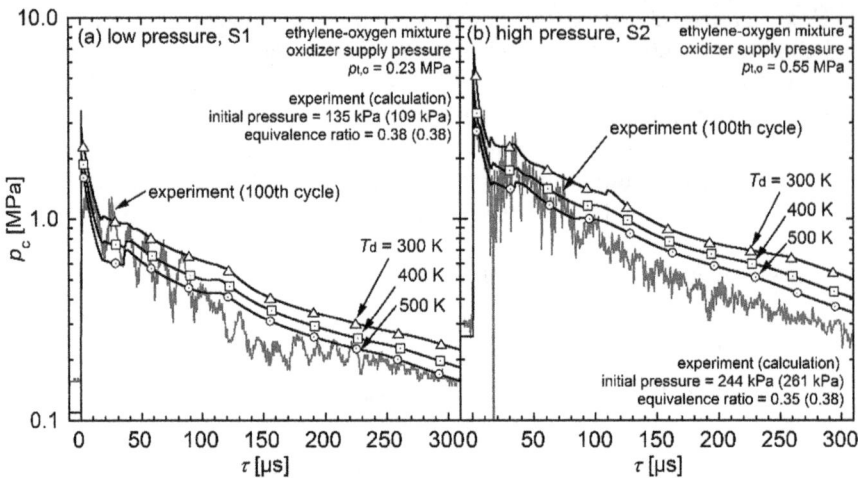

Fig. 8.6 Enlarged pressure history in 100th cycle of Fig. 8.5 (gray line) and calculation result of 1D numerical analysis (black lines with symbols). (**a**): low-pressure condition (S1), (**b**): high-pressure condition (S2)

pressure were confirmed. From the FFT analysis, the frequency of the low-frequency oscillation in S1 was $f_L = 64 \pm 1$ kHz. According to the visualization experiment of PDC using ethylene-oxygen mixture carried out by Matsuoka et al. (2012), the transverse shock waves to the center axis of the combustor were generated from DDT point and the velocity was approximately $V_s = 1538$ m/s. If the velocity was same as the experiment, the frequency was estimated at $f_s = V_s/2id_c = 77$ kHz and identical to f_L within $17 \pm 1\%$. It was suggested that the low-frequency pressure oscillation in Fig. 8.6 was due to the transverse shock generated by DDT point. In contrast, the high-frequency pressure oscillations in S1 was calculated at $f_H = 458 \pm 2$ kHz. The resonance frequency $f_r \geq 500$ kHz of the pressure transducer was closed to the frequency f_H and it was considered that the high-frequency pressure oscillation was due to the mechanical vibrations. Of course, the low and high-frequency pressure oscillations were not generated in the one-dimensional numerical calculation.

As shown in Fig. 8.6, the difference between the calculation and experiment increased after $\tau = 100$ μs. In the 100th cycle, the time-averaged pressure between 100 μs $\leq \tau \leq$ 300 μs was $81 \pm 5\%$ (S1) and $89 \pm 6\%$ (S2) of that at 50th cycle, and it is probably due to the influence of the heat on the pressure transducer. As shown in Fig. 8.6a, under the low-pressure condition S1, the peak pressure measured in the experiment was lower than the calculated value because the DDT occurred downstream of the pressure transducer PT2 (discussed in Sects. 3.1 and 3.2). However, in the decay region of the pressure, the experimental pressure history was in good agreement with the calculation, particular for the condition of $T_d = 400$–500 K. In contrast, under the high-pressure condition S2, the measured peak pressure and following decay region was in good agreement with the calculation.

According to the numerical calculation by Watanabe et al. (2017), if the composition of the detonable mixture was same and the oxidizer flow was choked, the exhaust duration of the burned gas depended on strongly the combustor length L_c. From the numerical analysis, the gas-dynamic upper limit f_{upper} of the operating frequency was evaluated as the inverse of the duration between the time at the direct initiation of detonation and the completion time of exhausting burned gas. As the results, in the range of T_d = 400–500 K, the upper limit was calculated at f_{upper} = 1389–1508 Hz for S1 and 1542–1694 Hz for S2, respectively. Therefore, the ratio of the operating frequency to the upper limit was $\alpha = f_{ope}/f_{upper}$ = 67–73% for S1 and 60–65% for S2, respectively. It was suggested that the burned gas was purged by oxygen for a long time ($t = t_4$ in Fig. 8.1). The frequency ratio $\alpha = f_{ope}/f_{upper}$ approaches unity when the duration of the purging by oxygen is eliminated and the burned gas is purged by low-temperature mixture.

Higher Frequency Operation of Pulse Detonation Cycle

As the third experiment S3, a PDC at higher operating frequency was performed one time under the same oxidizer supply condition as S2. As shown in Fig. 8.2, Type B of the combustor having a length of 60 mm was used because the durations of the propagation of a detonation and the following blowdown of the high-pressure hot burned gas strongly depend on the combustor length. Figure 8.7 shows pressure history measured by pressure transducer PT2 installed at x = 20 mm and 7-cycle pressure spike was confirmed. The initial pressure of the mixture was measured at p_d = 216 kPa. The gray-color area in Fig. 8.7 shows the Chapman – Jouguet detonation pressure p_{CJ} under the temperature range of T_d = 300–400 K, p_d = 216 kPa and

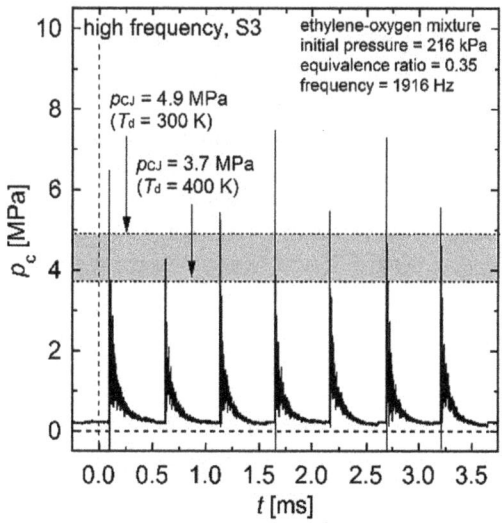

Fig. 8.7 Pressure history measured by pressure transducer installed at x = 20 mm under the operating frequency of 1916 Hz

$ER_d = 0.35$. Despite the same oxidizer supply condition as S2, the measured peak pressures were relatively higher than that in the condition of S2. When supercritical ethylene is injected to a low-pressure environment, low-temperature fuel is generated due to an irreversible isenthalpic change, which is called the Joule–Thomson effect (Matsuoka et al. 2017a). For example, assuming that the ambient pressure is $p_a = 0.1$ MPa, the supercritical ethylene is changed to a saturated vapor at 169 K. It was suggested that the low-temperature detonable mixture was filled in the combustor and the high Chapman – Jouguet detonation pressure was confirmed with a small number of cycles. The averaged flame propagation speed of 7 cycles with the standard deviations was $V_f = 2275 \pm 238$ m/s, and 100.1% and 98.6% of the estimated detonation speed at the temperature of $T_d = 300$ K and 400 K. Due to limitations of the ignition device, more cycles could not be confirmed.

By the similar method in Sect. 3.3, the upper limit of operating frequency in S3 was estimated using the same mixture condition as S2 ($L_c = 60$ mm, $p_{t,o} = 0.55$ MPa, $ER_d = 0.38$ and $T_d = 300$–400 K). As a result, the upper limit of the operating frequency was estimated to be $f_{upper} = 2058$–2322 Hz ($\alpha = f_{ope}/f_{upper} = 83$–93%).

4 Conclusions

The gas-dynamic upper limit of the operating frequency of a pulse detonation cycle (PDC) is dominated by the length of the combustor. To realize a high thrust density with a compact combustor, it is necessary to shorten the combustor length and increase the operating frequency. In order to solve this problem, it is required to sufficiently shorten the distance and time required for deflagration-to-detonation transition (DDT) compared with the combustor length and the duration for one PDC. However, from a practical point of view, enhancement of the DDT without obstacles is preferred. A novel PDC method, in which the inner diameter of the oxidizer feed line is equal to that of the combustor, can increase the static pressure in the combustor by increasing the total pressure because the inner flow is chocked. In the experiment, gaseous oxygen as oxidizer and supercritical ethylene as fuel were used. The PDC operation at 1010 Hz with the combustor having a length of 100 mm and an inner diameter of 10 mm was carried out under the two conditions of the total pressure of the supplying oxidizer at 0.23 MPa and 0.55 MPa. As the results, it was found that the distance and time for DDT process decreased by approximately 50% when the total pressure of supplying oxidizer increased by 242%. From the comparison with one-dimensional numerical analysis, the experimental frequency of 1010 Hz was range of 60–73% of the estimated gas-dynamic upper limit of the operating frequency. In addition, the experiment of a high-frequency PDC operation was carried out using the combustor of 60 mm in length, and a PDC at an operating frequency of 1916 Hz was confirmed and was 83–93% of the estimated upper limit of the operating frequency.

Using this DDT shortening method, a PDC at an operating frequency close to the gas-dynamic upper limit is realized with a compact combustor and high-pressure combustion. For a rocket system, the application to the small thrust system such as a reaction control system (RCS) and orbital maneuvering system (OMS) can be considered. For an air-breathing system, the effect on miniaturizing the combustor and increasing the thrust density becomes larger than a rocket system because the cell size of fuel-air mixture is larger than fuel-oxygen mixture. For example, the cell size of stoichiometric hydrogen-air mixture is approximately 10 mm at $p_d = 100$ kPa and $T_d = 293$ K. If the pressure of the mixture increases to $p_d = 500$ kPa, the sell size decreases to approximately 2–3 mm (Stamps and Tieszen 1991), and the DDT distance is also shortened. Moreover, for a thermal spray technique (Endo et al. 2016), spraying of powder with high mass flow rate is possible using compact combustor.

Acknowledgments This work was subsidized by a Grant-in-Aid for Scientific Research (B) (No. 26820371), the Toukai Foundation for Technology, the Paloma Environmental Technology Development Foundation, and Tatematsu Foundation.

References

Ciccarelli, G., & Dorofeev, S. (2008). Flame acceleration and transition to detonation in ducts. *Progress in Energy and Combustion Science, 34*(4), 499–550.
Endo, T., Yatsufusa, T., Taki, S., & Kasahara, J. (2004a). Thermodynamic analysis of the performance of a pulse detonation turbine engine. *Science and Technology of Energetic Materials, 65*(103), 103–110. (in Japanese).
Endo, T., Kasahara, J., Matsuo, A., Inaba, K., Sato, S., & Fujiwara, T. (2004b). Pressure history at the thrust wall of a simplified pulse detonation engine. *AIAA Journal, 42*(9), 1921–1930.
Endo, T., Obayashi, R., Tajiri, T., Kimura, K., Morohashi, Y., Johzaki, T., Matsuoka, K., Hanafusa, T., & Mizunari, S. (2016). Thermal spray using a high-frequency pulse detonation combustor operated in the liquid-purge mode. *Journal of Thermal Spray Technology, 25*(3), 494–508.
Gordon, S., & McBride, B. J. (1996). Computer program for calculation of complex chemical equilibrium compositions and applications. NASA Reference Publication 1311.
Gou, X., Sun, W., Chen, Z., & Ju, Y. (2010). A dynamic multi-timescale method for combustion modeling with detailed and reduced chemical kinetic mechanisms. *Combustion and Flame, 157*, 1111–1121.
Heiser, W. H., & Pratt, D. T. (2002). Thermodynamic cycle analysis of pulse detonation engines. *Journal of Propulsion and Power, 18*(1), 68–76.
Kailasanath, K. (2000). Review of propulsion applications of detonation wave. *AIAA Journal, 38*(9), 1698–1708.
Kailasanath, K. (2003). Recent developments in the research on pulse detonation engines. *AIAA Journal, 41*(2), 145–159.
Kuznetsov, M., Alekseev, V., Matsukov, I., & Dorofeev, S. (2005). DDT in a smooth tube filled with a hydrogen–oxygen mixture. *Shock Waves, 14*(3), 205–215.
Matsuo, K. (1994). *Compressible fluid dynamics — Theory and analysis in internal flow*. Tokyo: Rikogakusha Publ., Ltd.. (in Japanese).
Matsuoka, K. (2016). Experimental study on control technique of pulsed detonation. International workshop on detonation for propulsion 2016, Singapore, July 2016.
Matsuoka, K., Esumi, M., Ikeguchi, K., Kasahara, J., Matsuo, A., & Funaki, I. (2012). Optical and thrust measurement of a pulse detonation combustor with a coaxial rotary valve. *Combustion and Flame, 159*(3), 1321–1338.

Matsuoka, K., Mukai, T., & Endo, T. (2015). Development of a liquid-purge method for high-frequency operation of pulse detonation combustor. *Combustion Science and Technology, 187*(5), 747–764.

Matsuoka, K., Morozumi, T., Takagi, S., Kasahara, J., Matsuo, A., & Funaki, I. (2016). Flight validation of a rotary-valved four-cylinder pulse detonation rocket. *Journal of Propulsion and Power, 32*(2), 383–391.

Matsuoka, K., Muto, K., Kasahara, J., Watanabe, H., Matsuo, A., & Endo, T. (2017a). Development of high-frequency pulse detonation combustor without purging material. *Journal of Propulsion and Power, 33*. Special Section on Pressure Gain Combustion, 43–50.

Matsuoka, K., Muto, K., Kasahara, J., Watanabe, H., Matsuo, A., & Endo T. (2017b). Investigation of fluid motion in valveless pulse detonation combustor with high-frequency operation. *Proceedings of the Combustion Institute 36*(2):2641–2647.

Shchelkin, K. I., & Troshin, Y. K. (1965). *Gasdynamics of combustion*. Baltimore: Mono Book Corporation.

Singh, D. J., & Jachimowski, C. J. (1994). Quasiglobal reaction model for ethylene combustion. *AIAA Journal, 32*(1), 213–216.

Stamps, D. W., & Tieszen, S. R. (1991). The influence of initial pressure and temperature on hydrogen-air-diluent detonations. *Combustion and Flame, 83*(3), 353–364.

Takahashi, T., Mitsunobu, A., Ogawa, Y., Kato, S., Yokoyama, H., Susa, A., & Endo, T. (2012). Experiments on energy balance and thermal efficiency of pulse detonation turbine engine. *Science and Technology of Energetic Materials, 73*(6), 181–187.

Wang, K., Fan, W., Lu, W., Chen, F., Zhang, Q., & Yan, C. (2014). Study on a liquid-fueled and valveless pulse detonation rocket engine without the purge process. *Energy, 71*(15), 605–614.

Watanabe, H. Matsuo, A. Matsuoka, K., & Kasahara, J. (2017). Numerical investigation on burned gas backflow in liquid fuel purge method. 2016 AIAA Science and Technology Forum and Exposition, AIAA2017-1284, Jan. 9–13, 2017, Texas, USA.

Wu, M.-H., & Lu, T.-H. (2012). Development of a chemical microthruster based on pulsed detonation. *Journal of Micromechanics and Microengineering, 22*(10), Paper 105040.

Wu, Y., Ma, F., & Yang, V. (2003). System performance and thermodynamic cycle analysis of airbreathing pulse detonation engines. *Journal of Propulsion and Power, 19*(556), 556–567.

Yee, H. C. (1989). A class of high-resolution explicit and implicit shock-capturing methods. NASA Technical Memorandum 101088.

Chapter 9
On the Investigation of Detonation Re-initiation Mechanisms and the Influences of the Geometry Confinements and Mixture Properties

Lei Li, Jiun-Ming Li, Chiang Juay Teo, Po-Hsiung Chang, Van Bo Nguyen, and Boo Cheong Khoo

Abstract The topic of detonation re-initiation is studied through both experimental measurements and numerical simulations using a bifurcation channel and the detonation research facilities in Temasek Laboratories. The main objective is to understand the re-initiation mechanisms through shock reflections, and investigate the performance of detonation re-initiation at different test conditions. Stable and unstable detonation waves are both taken into consideration. It is found that the re-initiation through shock reflection is mainly achieved through the interactions of the multiple transverse waves. The details of the generation and evolution of the transverse waves are also clarified. The influence of the geometry confinement to detonation re-initiation is investigated. It is found that the length of the bifurcation channel can affect the re-initiation results by limiting the shock reflection times, which is discovered to be the main reason leading to the discrepancies between the previous similar studies. The width of the bifurcation channel is also critical as it can directly affect the induction length during detonation diffraction which determines the shock reflection strength. The differences of re-initiation using various mixture properties are also addressed, and a sudden transitional behavior of detonation re-initiation is found between stable and unstable detonation waves. Regarding the reason why a certain number of shock reflections are required before successful re-initiation, it can be explained using the relative relation between the shock reflection strength and the corresponding marginal solution curve of a quasi-steady detonation.

L. Li (✉)
Temasek Laboratories, National University of Singapore,
5A Engineering Drive 1, 02-02, Singapore 117411, Singapore
e-mail: lilei@u.nus.edu

J.-M. Li • P.-H. Chang • V. Bo Nguyen • B.C. Khoo
Temasek Laboratories, National University of Singapore, Singapore, Singapore

C.J. Teo
Mechanical Engineering Department, National University of Singapore, Singapore, Singapore

1 Introduction

Detonation phenomenon is associated with a supersonic exothermic combustion front and a preceding shock wave. Due to its fast propagation speed and the relatively high combustion temperature, the combustion efficiency of detonation can be higher than deflagration form of combustion. So many attempts have been made to implement detonation into engine applications. Considering the various applications of detonation phenomenon, understanding of detonation re-initiation can be utilized to meet certain industrial needs. In fuel pipelines and mine tunnels, detonation wave is anticipated to be suppressed by certain geometric designs due to safety concerns, so that disastrous outcome could be prevented. However, in engine applications, e.g. Pulse Detonation Engine (PDE) and Rotating Detonation Engine (RDE), the engine has to be designed to easily achieve detonation transition inside a compact engine design. So such knowledge of detonation re-initiation could serve as a guideline for the engine tube design.

Detonation re-initiation is essentially a detonation transition process. There are extensive studies which focused on deflagration to detonation transition (DDT), shock to detonation transition, detonation propagation and detonation direct initiation in straight tubes and channel (Laderman et al. 1963; Oppenheim et al. 1962; Roy et al. 2004). Other geometries including obstructed tubes and bent tubes are also of great interests regarding their potential applications in industrial safety and propulsion (Frolov et al. 2007a, b; Oran and Gamezo 2007; Ciccarelli and Dorofeev 2008; Gamezo et al. 2007; Lee 1977; Silvestrini et al. 2008; Boeck et al. 2017). Through testing on various parameters including fuel type, mixture composition, geometry confinement and configuration of obstacles, it was concluded that the interactions among shock waves, flame, flow turbulence and boundary layer is important in assisting detonation transition. Other factors including thermal expansion of combustion products, flame-vortex interaction and instability are also shown to be responsible for initiating DDT with the assistance from obstacles.

One of the most extensively studied re-initiation topics is detonation spontaneous re-initiation in an unconfined space. Through many studies using experiment, numerical simulation and analytical method (Mitrofanov and Soloukhin 1964; Edwards et al. 1979; Edwards and Thomas 1981; Knystautas et al. 1982), the well-known conclusion of 13 times cell width has been proved to be the minimum tube geometry for successful spontaneous detonation re-initiation if a detonation wave diffracts into an unconfined space. If rectangular channel cross section was employed, a slightly different conclusion regarding the critical channel geometry size has also been summarized. In addition, it was also discovered that spontaneous re-initiation depends on the nature of mixture properties, especially the degree of regularity of a detonation wave. When unstable detonation waves were used, the critical geometry may deviate from the one that was summarized for stable detonation.

Except for spontaneous re-initiation, detonation re-initiation can also be achieved in confined geometries after interactions with the wall. Detonation re-initiations in

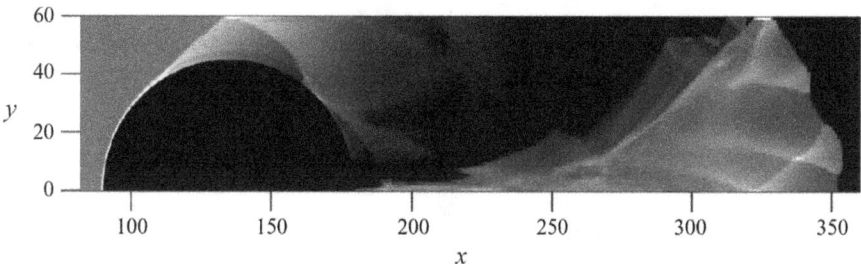

Fig. 9.1 Maximum exothermicity field of detonation re-initiation process by Radulescu and Maxwell (2011)

a straight tube or channel with sudden geometry expansion or gradual conical expansion have been studied due to their potential applications in propulsion engines and industrial processes. Brophy et al. (2003) have conducted detonation re-initiation experiments in a geometry where a detonation wave moves from a small channel to a larger channel. They concluded that the shock reflection on the wall of the larger channel is responsible for the successful detonation re-initiation. The Mach stem generated after reflection can release sufficient heat which can eventually lead to the successful transmission of the detonation wave. Thomas and Williams (2011) studied interactions of detonation waves with wedges and expansion nozzle geometries. It was found that when the cross section changes in a gradual way (less than 45 degrees), it becomes increasingly easier for detonation to propagate through. In order to generate a quenched detonation wave for re-initiation study, other geometry confinements were also employed.

In addition to the numerical simulations and experimental observations of the phenomenon, some researchers have been focusing on the re-initiation mechanisms. Bhattacharjee (2013) and Bhattacharjee et al. (2013) investigated the sequence of detonation formation when a decoupled shock-flame complex undergoes single Mach reflection (as shown in Fig. 9.1). Five mechanisms were reported to be responsible for re-initiation according to some previous studies. It was also found that among all the possible mechanisms, the jet formation after Mach reflection, transverse detonation and local instabilities can assist detonation re-initiation. Lv and Ihme (2015) summarized detonation re-initiation mechanisms into three categories, namely by regular reflection, Mach reflection and hydrodynamic instability. It was found that for more reactive Ar-diluted mixture, ignition first appeared after regular reflection which is then followed by spontaneous re-initiation through SWACER mechanism. By replacing Ar with N_2 dilution, re-initiation was found through Mach reflection. Other mechanisms, e.g. flame-acoustic and wave-wave interactions were also observed to contribute to final re-initiation in the N_2 diluted mixture. Radulescu and Maxwell (2011) studied detonation re-initiation through multiple reflections. A mechanism which is associated with the leading front slow acceleration and the local small scale re-acceleration through multiple reflections was identified. It was also found that the relative sensitivity of the mixture to the geometry dimensions directly determines the times of reflections that are required

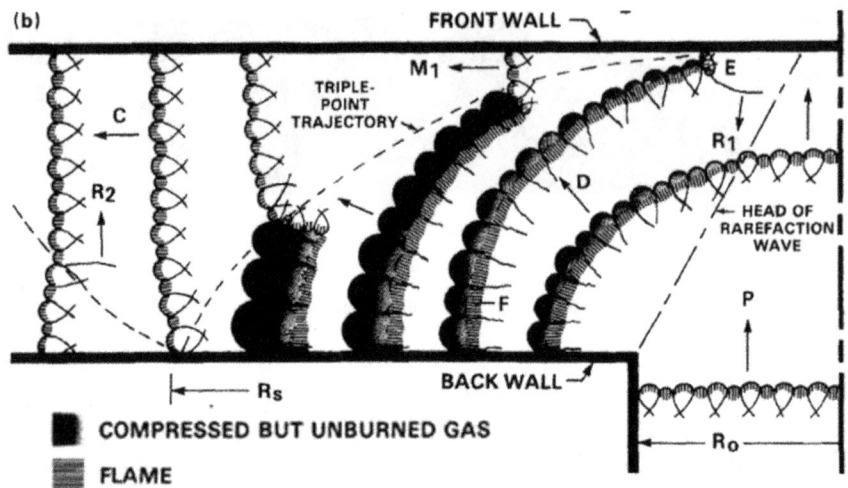

Fig. 9.2 Sketch of detonation reflected re-initiation process by Murray and Lee (1983)

for successful re-initiation. However, in their study, the experiment was not conducted in the same geometry of that applied for numerical simulations.

Except for the above straight geometry confinements, other geometries with various deviation angles have also been investigated for detonation re-initiation. Wang et al. (2008) observed the detonation re-initiation process in a channel branch which is 90-degree deviated from the inlet channel. The detonation wave was observed to be decoupled after diffraction and re-initiated from the opposite wall of the vertical channel through a shock reflection. Numerical simulations were also employed to compare with experimental results and look for the re-initiation mechanisms. It was found that an embedded jet which can be formed from the interactions between the unreacted recirculation region and small vortices contributed to detonation re-initiation. Even though this study revealed re-initiation mechanisms, it only targeted re-initiation through single reflection.

There are three other similar studies which are worth noting regarding their summaries on the critical conditions for successful re-initiation. Murray and Lee (1983), Polley et al. (2013) and Wang et al. (2015) investigated the transmission of a planar detonation wave into a diverging cylindrical detonation using soot foil technique. It was depicted in Murray and Lee's study (as shown in Fig. 9.2) that the first reflection on the opposite wall is the main triggering source leading to detonation re-initiation. Polley et al. (2013) (as shown in Fig. 9.3) revealed a new re-initiation process, namely multiple reflected re-initiation. It points out that the re-established detonation after the first reflection may still fail temporarily, and the detonation can be re-initiated after another reflection process on the back wall. Wang et al. revealed through experiments that the limiting channel width for successful detonation re-initiation is approximately one cell width. All the three studies summarized critical conditions for successful detonation re-initiation on the basis of the vertical channel

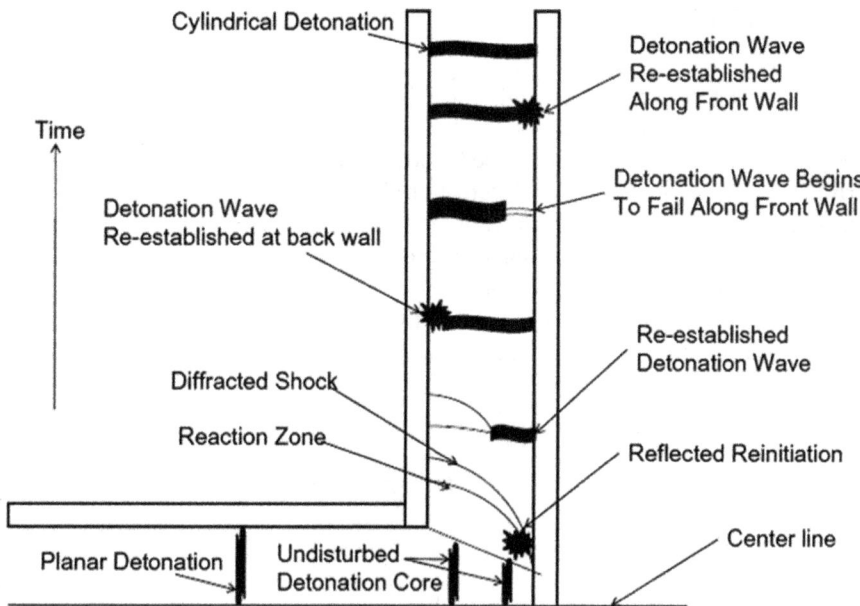

Fig. 9.3 Sketch of detonation reflected re-initiation through multiple shock reflections by Polley et al. (2013)

width and the cell size. But it should be noted that there are big discrepancies among three critical conditions. As such, the re-initiation topic is worth working on to further address the disagreement.

According to the literature review, it can be concluded that, except for the spontaneous re-initiation in which no interaction between waves and wall exists, most re-initiation cases involve some forms of shock reflection process. Through reflection, a quenched detonation wave is subject to instantaneous energy generation which can re-accelerate the wave. Thus the reflection process is the key to re-initiation. The present research is to systematically study the mechanism of re-initiation through shock reflections. By focusing on detonation re-initiation through multiple shock reflections, it is intriguing to know what could really happen during the early reflections that is unable to re-initiate detonation, and what can be achieved through the subsequent reflections that could finally contribute to the successful re-initiation, thus allow us to locate the important mechanisms. Another topic to address is the differences of detonation re-initiation phenomenon between stable and unstable detonations. Even though clear differences between the propagation mechanisms of the two types of detonation waves have been identified which could probably affect the re-initiation process, the topic has yet to be discussed in detonation re-initiation. Furthermore, by conducting experiments in various geometries using different mixtures, hopefully, we could also find the causes to the discrepancies of the critical conditions for successful re-initiation from different studies.

2 Experimental Setup

Detonation experiments were mainly carried out using a valveless detonation driver operated in a single-shot fashion when the experiment initial pressure is 1 bar. The description of the driver section can be found in details in (Li et al. 2013, 2017). The driver section is connected with a bifurcation chamber as shown in Fig. 9.4 (Li et al. 2017). A detonation wave at the inlet of the bifurcation channel can diffract around the corner, and then become a quenched detonation wave which can be used for the re-initiation study through shock reflections on the wall. The highlighted green area has quartz windows placed at both sides of the chamber so that the inside space can be optical accessible for direct visualization of detonation transmission. The optical

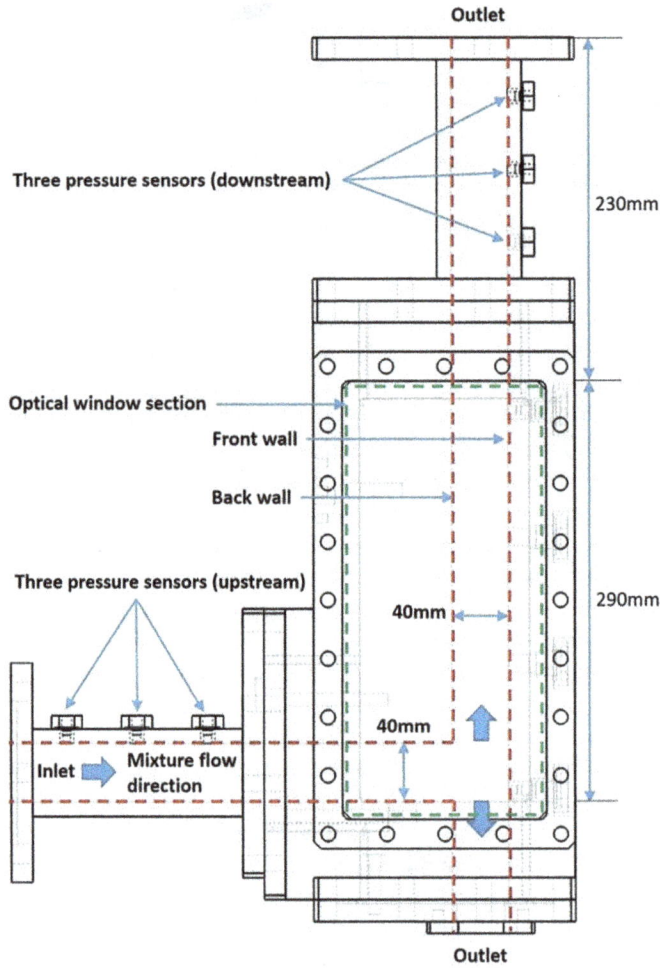

Fig. 9.4 Schematic of the bifurcation channel

windows allow an optical accessible area of 270× 120 mm. The mixture is ethylene (C_2H_4) + air (initial mixture condition at 303 K, 1 bar) at the stoichiometric ratio. Multiple diagnostic methods including Schlieren photography, planar laser induced fluorescence, open shutter photography and soot foil technique were implemented to reveal the phenomenon from different perspectives.

Experiments were also conducted at a different initial pressure other than 1 bar. In such cases, the detonation channel was designed with one driver section and one driven section. The driven section is the same with the bifurcation chamber from the previous description, and the driver section can be filled with ethylene and oxygen to facilitate the generation of a detonation wave in a short distance (less than 0.5 m). To separate different mixtures between driver section and driven section, the moisture-resistant polyester (PET) diaphragm is installed in between. The top and the bottom outlet on the driven section are covered with quartz and PET, respectively to allow laser access and ensure gas tightness. An electric spark plug which is placed at the upstream end of the driver section with a total energy of 120–150 mJ is used to ignite the mixture.

3 Detonation Re-initiation Mechanisms

Unstable Detonation

Detonation experiments were carried out using a valveless detonation driver operated in a single-shot fashion. The description of the driver section can be found in details in (Li et al. 2013, 2017). The driver section is connected with a bifurcation chamber as shown in Fig. 9.4 (Li et al. 2017). The mixture is ethylene (C_2H_4) + air (initial mixture condition at 303 K, 1 bar) at the stoichiometric ratio. Multiple diagnostic methods including Schlieren photography, planar laser induced fluorescence, open shutter photography and soot foil technique were implemented to reveal the phenomenon from different perspectives.

As Schlieren photography is not capable of determining whether detonation transition is successful or not, especially during the transition process, open shutter photography is applied as a supplement to provide cell structure information from images to tell the difference between detonation wave and deflagration wave. Figure 9.5 shows open shutter photography images before and after the bifurcation channel. As the channel is too long to fit in the frame without sacrificing resolution, images were taken from three different locations.

The imaging frequency is 50.4 kHz and the exposure time for each individual frame is 18.84 μs. By overlaying the multiple frames of images all together, the complete re-initiation process can be shown. Based on the open shutter photography images results, it can be found that re-initiation is successful, but only after the third reflection on the front wall. The disappearing cell structures after the first and second reflection indicate that detonation quenched at both locations. From Fig. 9.5,

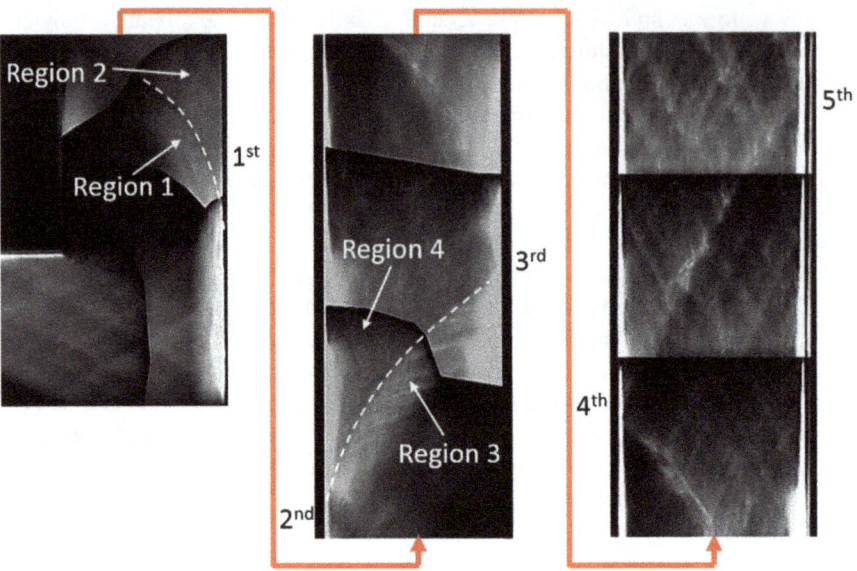

Fig. 9.5 Sequential records of detonation using open shutter photography

two distinct regions can be found after the first reflection. Region 1 is filled with very fine cell features. It has a banded structure which arises due to the transversely propagating detonation wave. Region 2 also shows traces of fine cellular structures. There is a distinct dividing line separating the two regions. According to the corresponding Schlieren images of the whole process in Fig. 9.6, this dividing line is actually the trajectory of the main triple point. We can also see from this figure that the width of the banded region is actually determined by the decoupled induction length between the diffracted shock and the flame. The successful detonation re-initiation started only from half of the channel cross section, as indicated by the cell structures which only existed at the right side of the channel cross section. According to the Schlieren images, the dividing line between the cell region and no cell region was the trajectory of the main triple point. When the main triple point reflected on the back wall, the cells started to fill in the whole cross section, and the detonation was then fully re-initiated after the fourth reflection.

In order to further investigate the detonation re-initiation mechanism at the third reflection location, Schlieren measurements were also conducted further downstream to record the wave dynamics. Figure 9.7 shows the wave propagation before and after the third reflection. It was found that when the second reflection occurs, multiple small transverse waves at the wave front could be clearly identified which are critical to detonation re-initiation. In the frames after the main transverse wave hit the front wall, the trailing small transverse waves all followed the same process of head on reflection with the front wall. As a result, a region in which waves went across each other as shown in Fig. 9.7(6) could be generated. Due to these interactions, we could expect the formation of high pressure and temperature regions in

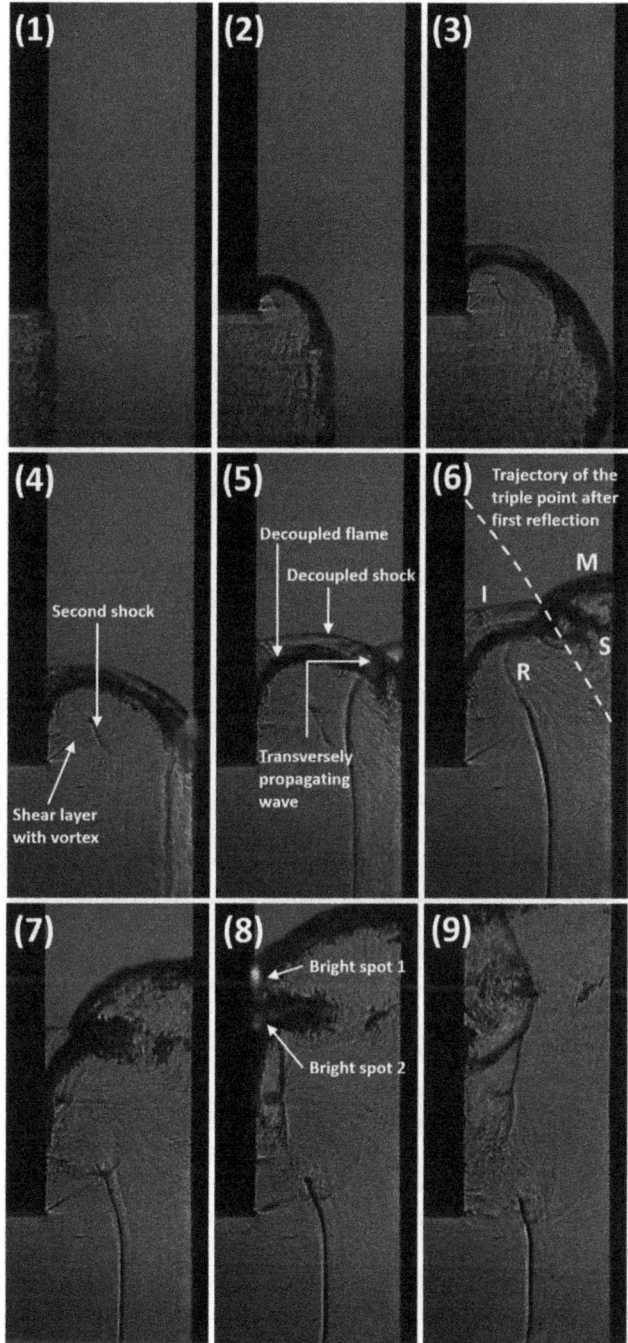

Fig. 9.6 Schlieren photography of detonation multiple reflections at the bifurcation region. The time difference between two consecutive images Δt is approximately 9.5 μs (M: Mach stem, R: Reflected wave, S: Slip line, I: Incident wave)

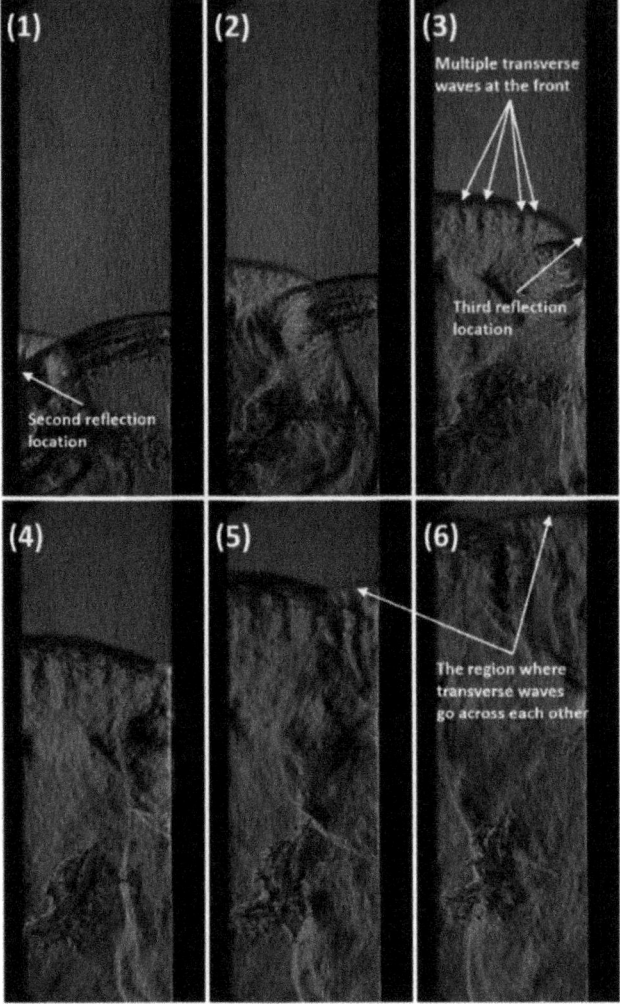

Fig. 9.7 Schlieren images of detonation re-initiation at the third reflection

which chemical reactions can be enhanced. So the interactions of the transverse waves can provide punctuated impetus which served as extra energy to prevent the further decay of detonation, thus to achieve re-initiation at this region instead of the other reflection locations. A drawing to depict the detailed re-initiation process is shown in Fig. 9.8. It should be noted that the small transverse waves' interactions found in Fig. 9.7 resemble the transverse waves interactions that one could find in a typical self-sustained detonation wave. As transverse waves were found playing an important role for detonation propagation, especially for irregular detonation waves (Radulescu 2003), it further addresses the importance of the multiple small transverse waves to the re-initiation process in the present geometry confinement.

Fig. 9.8 Schematic of transverse waves interaction at the second reflection

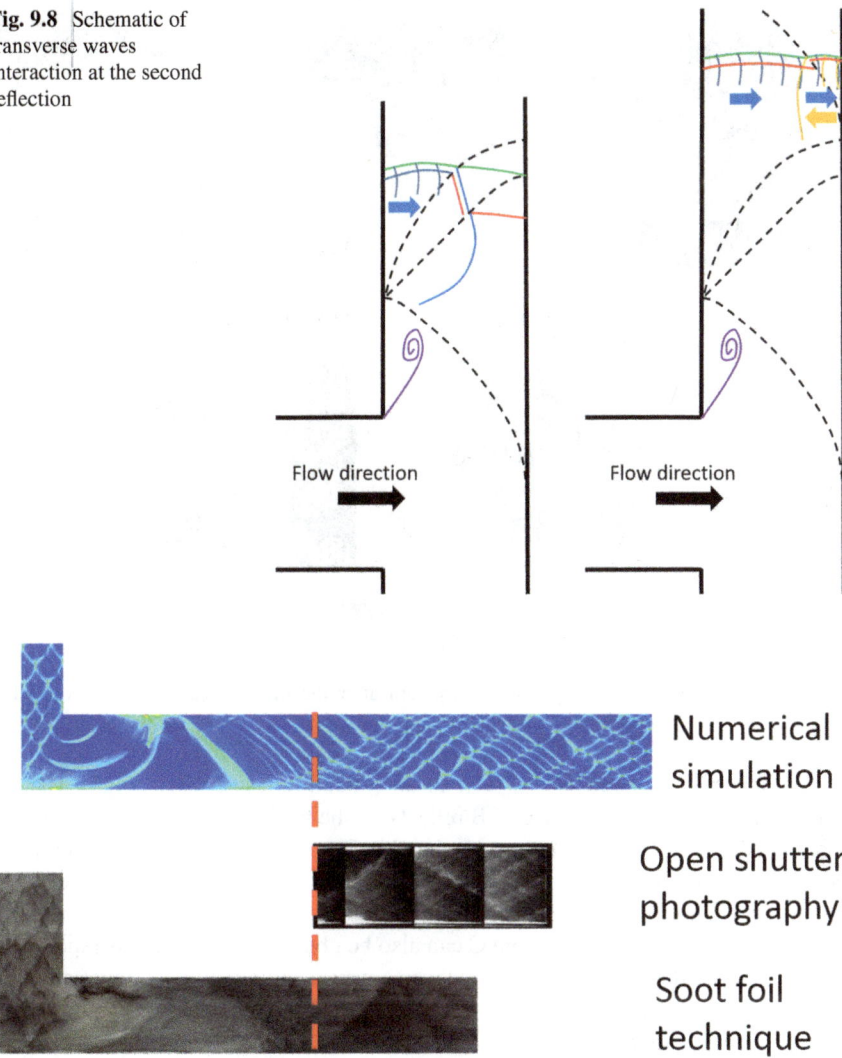

Fig. 9.9 Comparison between open shutter photography, soot foil technique and CFD time history of pressure maxima at the second shock reflection

As experimental measurements are limited by the spatial and temporal resolution, numerical simulations have also been performed to provide insights into the re-initiation mechanisms (Li et al. 2017). Figure 9.9 shows the time history of the pressure peak value at the downstream window area. The simulation reconstructed the same re-initiation pattern with that observed in experiments, which shows successful re-initiation after three shock reflections. More details regarding the generations of the multiple transverse waves were also revealed (Li et al. 2017). As shown in Fig. 9.10, there are triple points A and B (also local hot spots) propagating towards

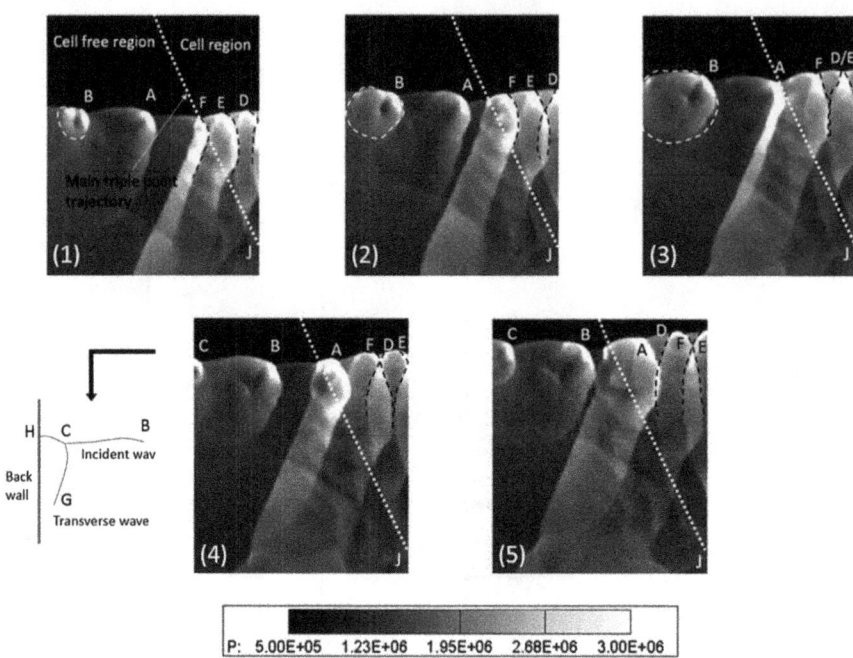

Fig. 9.10 Sequential pressure profile downstream after the third shock reflection from CFD simulations

the right with their behaviors similar to an expanding blast wave (Fig. 9.10a–c). When the expanding blast wave of B reflects on the back wall, a small Mach reflection configuration can be found near point C in Fig. 9.10d. The latest triple point C, which has a relatively high temperature and pressure, reflects off the back wall together with a new transverse wave tagging along. Later on, a similar expanding blast wave released outwards from C can also be observed. This process repeats for certain times and then the multiple transverse waves and the oblique lines as shown in Fig. 9.9 are formed.

The multiplication mechanism shows the generation of these transverse waves from the previous wave. However, the question regarding the origin of the first transverse wave is still unknown. An experiment using simultaneous Schlieren and PLIF measurement, which has been proven to be a promising technique (Boeck et al. 2017; Pintgen 2004), was conducted to look for the origin.

Figure 9.11 qualitatively shows shock and flame structures after the first and the second reflections. It was discovered that a transversely propagating detonation wave existed between the leading Mach stem and the diffracted flame after the first reflection, as indicated by the bright line in Fig. 9.11(1). However, according to the subsequent frame, it quickly quenched and faded away in Fig. 9.11(2). As a result, the mixture behind the transverse shock was not ignited immediately after the wave. When this transverse wave further moved across from the front wall to the back

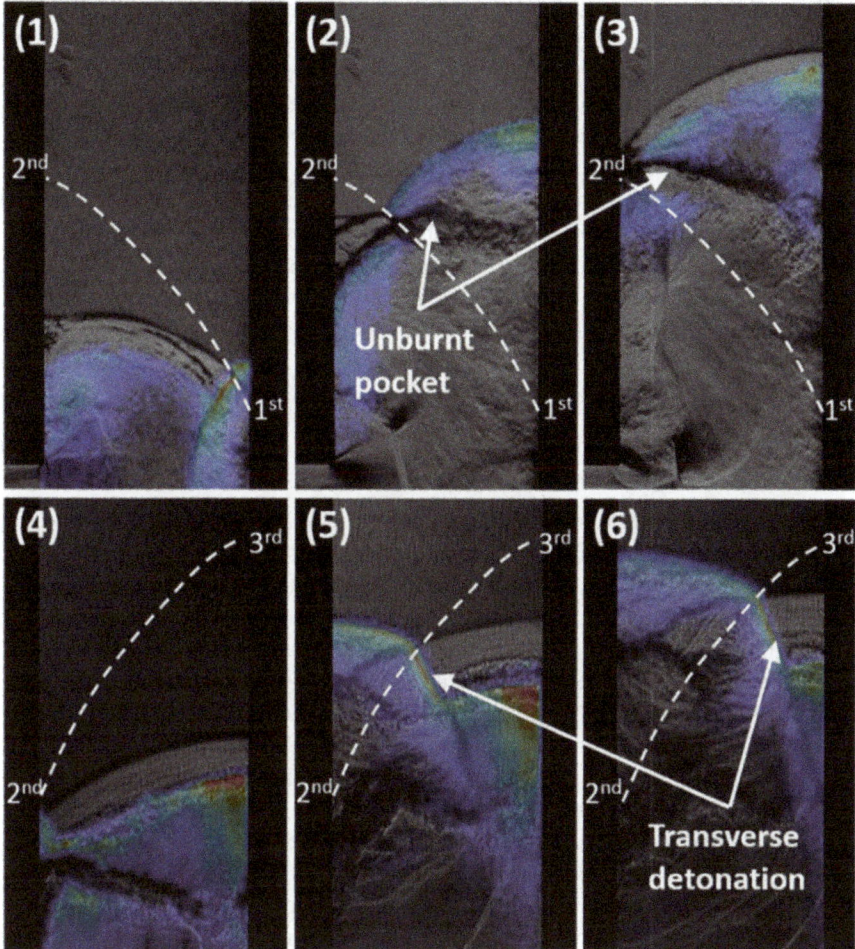

Fig. 9.11 Superposition of Schlieren and PLIF results before and after the second reflection

wall, a whole band of an unburnt area was left behind and appeared as the unburnt pocket as shown in Fig. 9.11(3, 4). Figure 9.11(4–6) illustrate the shock and flame after the second reflection. According to Fig. 9.11(4), a circular local explosion wave was formed after the reflection. This explosion wave could also be clearly observed in Fig. 9.6(9) which immediately penetrated the unburnt pocket, culminating in ignition. As the explosion wave propagated, it was found to recompress the shocked mixture in front of the decoupled flame and ignite it to form a sharp transversely propagating detonation (or super-detonation). Being quite different from the first reflection, the transversely propagating detonation wave did not decay at this time, indicating that the local explosion successfully caused a transverse shock to undergo transition into a transverse detonation.

Regarding the origin of the local explosion, two bright spots as identified in Fig. 9.6(8) were observed to be responsible. Based on the bright spots location, the top one was found due to the reflection of the decoupled leading shock wave which recompressed the mixture in the induction zone and led to an instantaneous explosion core. The bottom one was at the head of the unburnt pocket, which was caused by the shock reflection that propagated into the unburnt pocket and finally led to the bright ignition spot. So to summarize the origins of the local explosion, the reflections of both the leading shock wave and the transverse shock were observed to play an important role. As the strong local explosion started, small disturbances were able to initiate and grow at the wave front which subsequently became the early stage of the first few transverse waves as shown in Fig. 9.7.

Stable Detonation

In order to investigate re-initiation phenomenon of stable detonation and find the differences from that of unstable detonation, mixtures with a high percentage of Argon dilution have been tested. In order to compare the stable detonation results with the unstable cases, the mixture which can generate approximately the same cell width (20 mm for ethylene (C_2H_4) air mixture at 1 bar and 300 K temperature) was tested. According to the cell width database (Kaneshige and Shepherd 1997), we tested $C_2H_4 + 3O_2$ with 88% dilution of Argon gas at the initial condition of 0.3 bar and 300 K which has a reduced activation energy of 5.9.

Figure 9.12 shows the high speed re-initiation images of a stable detonation wave. Generally, the pictures of stable detonation are much cleaner than those of unstable detonation, especially at the region behind the leading shock wave where very little density variations exist within. According to the figure, it can be observed that the re-initiation failed at the present testing condition. In the first frame, we can see a detonation wave which was moving in the horizontal channel (6% lower than the corresponding CJ velocity according to CEA program (Gordon and McBride 1994; McBride and Gordon 1996)). When the detonation wave diffracted at the bifurcation region as shown in frame 2–3, a clear induction length can be induced immediately. As the wave reflected on the front wall, a typical Mach reflection configuration with a Mach stem, an incident wave, a reflected wave and a shear layer can be observed. Even though the Mach stem in frame 4 was observed to propagate at 1355 m/s, it immediately quenched out to a shock wave and a trailing combustion front in frame 5. Before the third reflection took place on the front wall again, the leading wave velocity can drop to 800 m/s. Immediately after the third shock reflection, the Mach stem quickly evolved into a shock-flame complex with a clear induction length in between in frame 8. The final evolution of the third reflection was out of the camera view area, but according to the pressure signal of the sensors further downstream, the velocity of the transmitted wave was approximately 1000 m/s, indicating that the re-initiation was unsuccessful.

9 On the Investigation of Detonation Re-initiation Mechanisms and the Influences... 213

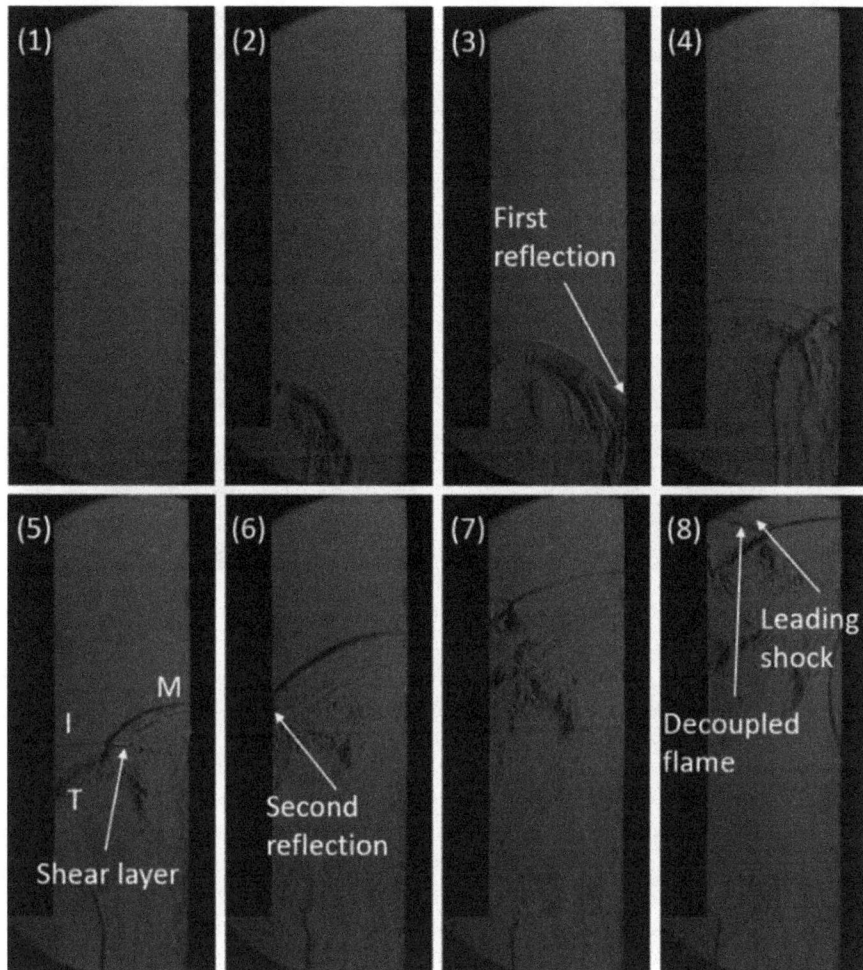

Fig. 9.12 High speed Schlieren imaging of unsuccessful detonation re-initiation at bifurcation region (SS58)

Generally, the Schlieren results of stable detonation at the first two reflections show similar features with those for unstable detonation, despite some minor differences mentioned above. Figure 9.13 shows the simultaneous measurement of the shock wave and the combustion front after the first reflection. From the figure, we can clearly see that a big induction length was developed after the shock diffraction. Near the back wall, the induction length was measured to be almost 20 mm. Due to such a long diffraction period, the flame region behind the diffracted shock wave was completely quenched as no OH radical has been detected near the back wall through PLIF. Behind the incident wave, combustion process only exists at the curved reflected wave front which is induced by the vortex near the bifurcation

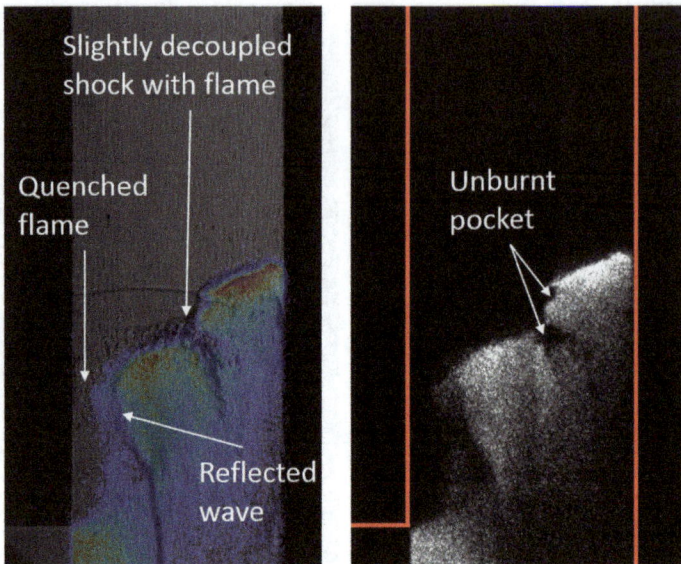

Fig. 9.13 Simultaneous measurement of firing shot SS58 using Schlieren and PLIF after the first reflection

corner. This is probably because of the shock compression effect from the reflected wave which increased the local pressure and temperature and thus ignited the pre-compressed mixture behind the incident wave. A small and clear distance between the transverse shock and the trailing flame can be identified, indicating that a transverse detonation was not generated. Owing to the weak transverse wave, an unburnt pocket behind the transverse wave was developed and started to accumulate. Compared with the simultaneous measurement in Fig. 9.11 of the unstable detonation wave, we can see that both cases show unburnt pockets and weak transverse shock waves. However, for the stable detonation, the combustion tends to quench out faster and more complete than that for the unstable detonation.

4 Critical Conditions for Detonation Re-initiation

Unstable Detonation

Since previous results showed that the cell width is also an important factor to influence the re-initiation, experiments with mixtures which have other cell widths have also been conducted in the present study. In order to better compare with the previous baseline experiment, we change the detonation cell width by altering the initial pressure but at the same time keep the same mixture mole fraction and channel width (40 mm). Various tests were conducted with the initial pressure from 0.8 to

9 On the Investigation of Detonation Re-initiation Mechanisms and the Influences... 215

Table 9.1 Detonation re-initiation results when initial pressure changes

Firing shot number	Initial pressure (bar)	Equivalence ratio	Nitrogen dilution ratio	Results
SS74	0.792	1.02	72.9%	Unsuccessful
SS75	0.904	1.02	72.7%	Unsuccessful
SS76	0.954	1.01	72.8%	Unsuccessful
SS77	0.981	1.01	72.8%	Successful but after more than three shock reflections
SS73	1	1.00	72.8%	Successful
SS79	1.043	1.01	72.8%	Successful
SS78	1.053	0.99	72.9%	Successful but after more than three shock reflections

1.05 bar as shown in Table 9.1. Even though pressure only varies in a relatively small range, multiple re-initiation patterns have been observed. It was found that when initial pressure is smaller than 0.95 bar, the re-initiation will be unsuccessful. However, if the pressure is larger than 0.95 bar but smaller than 1.05 bar, detonation re-initiation is usually achieved through more than three reflections. The location of detonation transition was also found a bit random in this case. When further increase the initial pressure, the re-initiation pattern became relatively predictable with less than three times of reflections required. In addition, the re-initiation usually occurs with the aid from the multiple transverse waves' interaction, which is the same with that as illustrated in Fig. 9.8.

Since previous results showed that the cell width is also an important factor to influence the re-initiation, experiments with mixtures which have other cell widths (by changing the initial pressure) have also been conducted in the present study. In addition to the high speed images, velocity information of the leading waves after each shock reflection has also been extracted for all the firing shots in Table 9.1 to identify their differences. Figures 9.14 and 9.15 show typical wave propagation for unsuccessful and successful detonation re-initiation cases, respectively. Main differences could be identified from the second reflection which shows that an over-driven detonation wave can always be initiated at the second reflection for successful detonation re-initiation. Since shock reflection strength directly depends on the wave speed and the angle (in this case, angles for all the cases are the same 90 degrees), it could be concluded that 1000 m/s is the critical velocity. When the leading waves before the second shock reflection reach more than 1000 m/s in firing shot SS73 and SS79, the re-initiations were found all successful through three shock reflections. In addition to the critical wave velocity, another velocity feature could also be discovered. It was found that for successful detonation re-initiation, even though the wave after the second reflection would decelerate, its propagation velocity fluctuates in a big range. This is probably due to the existence of the multiple transverse waves generated from the second shock reflection as observed in Fig. 9.8, which could interact with each other and act as the pulsed impetus to push the leading wave ahead over and over again.

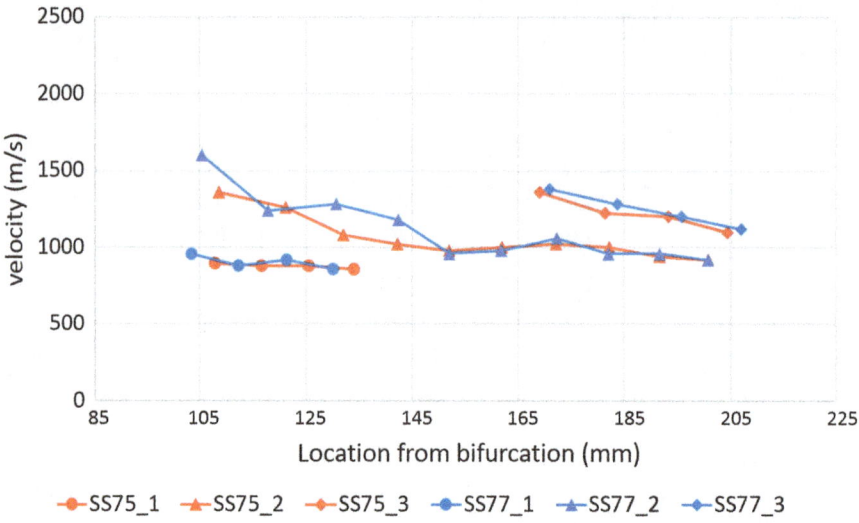

Fig. 9.14 Wave velocity change in unsuccessful detonation re-initiation cases (firing shots SS75 and SS77, subscripts indicate waves after the specific times of shock reflection)

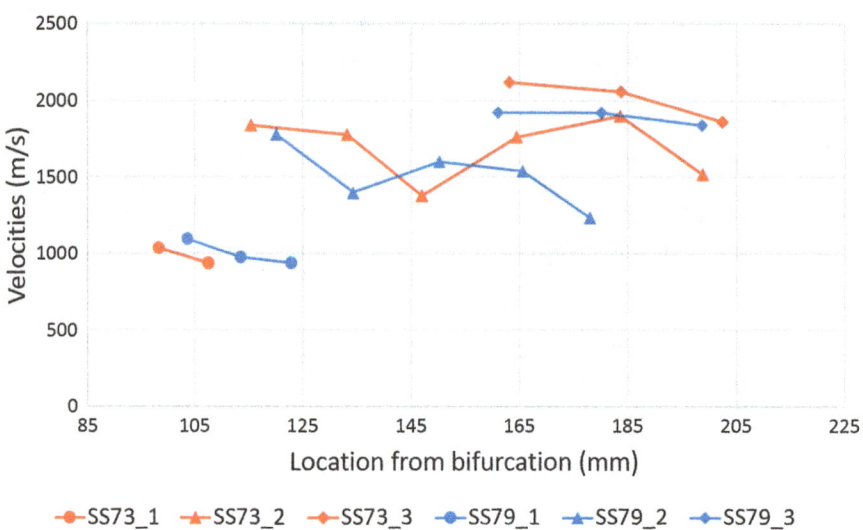

Fig. 9.15 Wave velocity change in successful detonation re-initiation cases (firing shots SS73 and SS79)

5 Stable Detonation

Similar experiments were conducted in the pressure range from 0.3 to 0.7 bar, which corresponds to the cell width from 20 mm to approximately 7 mm for stable detonation, and velocity information was extracted as well. It was found in Table 9.2 that only if the initial pressure is higher than 0.51 bar could the re-initiation be successful. To further dig into the details, Figs. 9.16 and 9.17 shows the velocity of Mach stems after the first, the second and the third reflection in successful and

Table 9.2 Testing conditions of Argon diluted mixtures

Firing number	Initial pressure (bar)	Equivalence ratio	Ar dilution ratio	Re-initiation result
SS66	0.302	1.08	87.9%	Unsuccessful
SS71	0.456	1.02	87.9%	Unsuccessful
SS65	0.501	1.12	87.7%	Unsuccessful
SS67	0.504	0.97	88.2%	Unsuccessful
SS81	0.513	1.02	88.0%	Successful
SS80	0.532	1.05	87.9%	Successful
SS70	0.558	0.99	88.1%	Successful
SS69	0.608	0.99	88.0%	Successful
SS68	0.700	0.98	88.0%	Successful

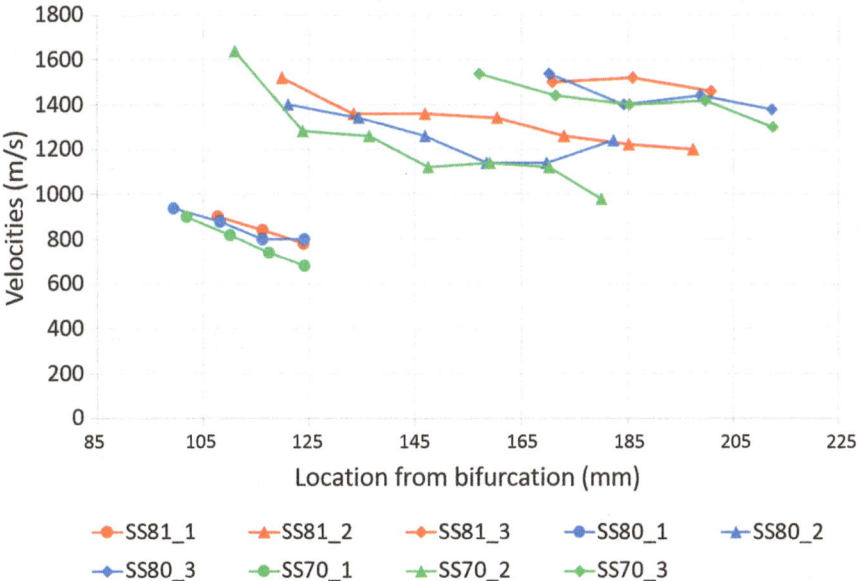

Fig. 9.16 The velocity of the front leading waves after three shock reflections in three successful detonation re-initiation cases. The subscript denotes the wave after certain times of shock reflections

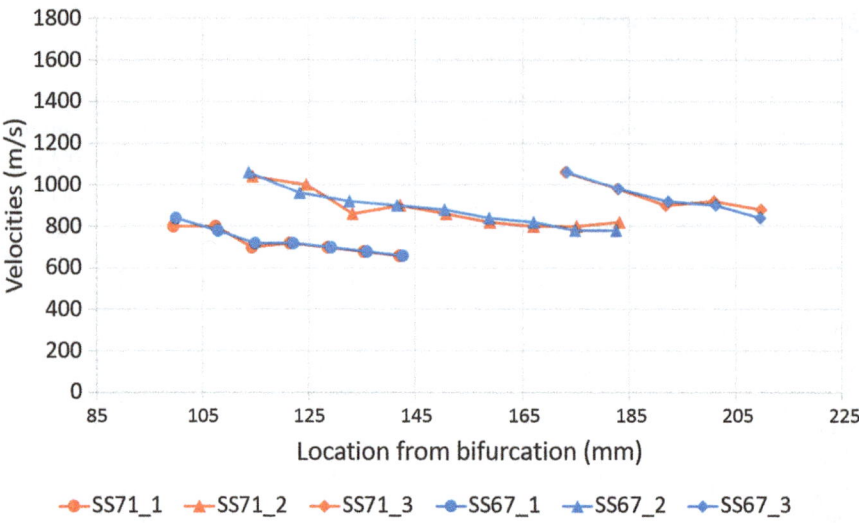

Fig. 9.17 The velocity of the front leading waves after three shock reflections in two unsuccessful detonation re-initiation cases. The subscript denotes the wave after certain times of shock reflections

the Mach stem velocities after the first reflection are all slower than 900 m/s. Even though this velocity is only slightly smaller than that of the successful cases, the second shock reflection failed to induce a reactive Mach stem. The reflection was shown to be only able to result in a non-reactive Mach stem with the velocity of approximately 1000 m/s, which is much smaller than the corresponding CJ velocity. During subsequent propagation, the Mach stem after each reflection was found to repeat the same velocity variation, changing from 1100 to 800 m/s until the next shock reflection occurred. Due to such a slow leading shock wave velocity, the induction time and length would increase significantly. Therefore, the flame and the leading shock wave were detached with each other all the time which eventually made it impossible to re-initiate a detonation wave.

Discrepancies Between the Present Study and the Previous Studies About the Critical Conditions

Except for the cell width which can significantly change the re-initiation results, channel width was also observed to be an important factor (Murray and Lee 1983; Polley et al. 2013; Wang et al. 2015). Figure 9.18 shows soot foil results of detonation propagation inside three different vertical channel geometries (30 mm, 20 mm and 10 mm). Channel with even wider width was not tested as it would lead to detonation spontaneous re-initiation without shock reflection. It was found that when channel width changes, completely different re-initiation process could take place.

Fig. 9.18 Soot foil results of detonation re-initiation in other channel geometries

For 10 mm channel width, all detonation shots quenched immediately after an incoming detonation wave propagated into the vertical channel. When the channel width increases to 20 mm, the situation changes. For all the seven detonation shots ran in this geometry, two were found with successful re-initiation and all the other five were not. Two cases with 20 mm channel width in Fig. 9.18 shows typical successful and unsuccessful re-initiation processes. In the unsuccessful case, the soot foil result looks very similar with that in 10 mm channel except that a noticeable trajectory from the second reflection exists. When the second reflection occurs, tiny cells appeared from the reflection location. The cells were found disappeared immediately after shock reflection and the main triple point trajectory also faded out on the soot foil. For the successful re-initiation case, the first two reflections were also observed not sufficient for detonation re-initiation. No evident cell structures exist until the third reflection occurs on the front wall. Very fine cell structures together with the triple point trajectory of the main reflected wave suddenly become evident only from the third reflection, indicating that an overdriven detonation was formed from the third reflection which could instantaneously lead to the final re-initiation. When further increase the channel width to 30 mm, successful detonation re-initiation can be found in all the detonation firing shots. According to the cell structures, the re-initiation pattern was found similar to that in 40 mm channel as shown in Fig. 9.6. The high resemblance of the re-initiation patterns between 40 and 30 mm channel width indicates that the same re-initiation mechanisms can probably apply to this 30 mm channel case.

It should be noted that, even though successful detonation re-initiation can be obtained from the channel geometries which are wider than 20 mm, this result is not consistent with either of the three critical conditions which were summarized by Murray and Lee (1983); Polley et al. (2013) and Wang et al. (2015). Figure 9.19 shows the summarized criteria. In order to find the reason leading to the discrepancies, we analyze the differences applied in testing conditions and methodologies. One of the major differences is the length of the vertical channel width. The length of the present vertical channel is longer than 600 mm, which is almost two times longer than those utilized by Murray and Lee (1983) and Polley et al. (2013). Since re-initiation through multiple reflections require a long channel (at least 450 mm bifurcation channel in two directions according to Fig. 9.6) to achieve, it could be concluded that the discrepancies may probably due to the insufficient length of the vertical channel from the previous studies which is not capable for sufficient times of reflection to take place. To verify the speculation, we also re-evaluated the re-initiation results at the same vertical channel location (228 mm from the channel bifurcation) with that used by Polley et al. (2013). Owing to the limited channel length which could only allow less than two times of reflections to take place, the successful re-initiation cases as shown in Figs. 9.6 and 9.18 would now be unsuccessful. So it could be concluded that even though discrepancies exist between the present study and the previous studies, the summarized results are all correct with respect to the certain geometries confinements.

6 Other Features of Detonation Re-initiation

Influence of the Channel Width to Re-initiation

Another important question to address is to explain why a wider vertical channel width is more favorable for successful detonation re-initiation. According to the previous imaging results and the literature (Murray and Lee 1983; Polley et al. 2013; Wang et al. 2015), it can be discovered that longer channel width will lead to a longer diffraction period before the first reflection which could make the resultant diffracted wave weaker than that in a shorter channel width. So the result in Fig. 9.19 seems counter intuitive. In order to answer this question, some preliminary detonation transition results have been obtained using deflagration waves as inlet waves. It was found that the leading shock wave can directly initiate a strong local explosion before the trailing flame reached the front wall. This is because the reflected shock wave from the front wall can recompress the mixture in the induction length which induced an extremely high temperature zone for local explosion. In this case, deflagration waves are more likely to cause detonation initiation. So it is reasonable to speculate that the induction length generated before the first shock reflection is the deciding factor for detonation re-initiation, rather than the leading shock strength.

In order to further prove the speculation, another two special channels were designed for experiments as shown in Fig. 9.20. Each of them was designed with an extrusion into the vertical channel. One has an extrusion of 15 mm into the vertical

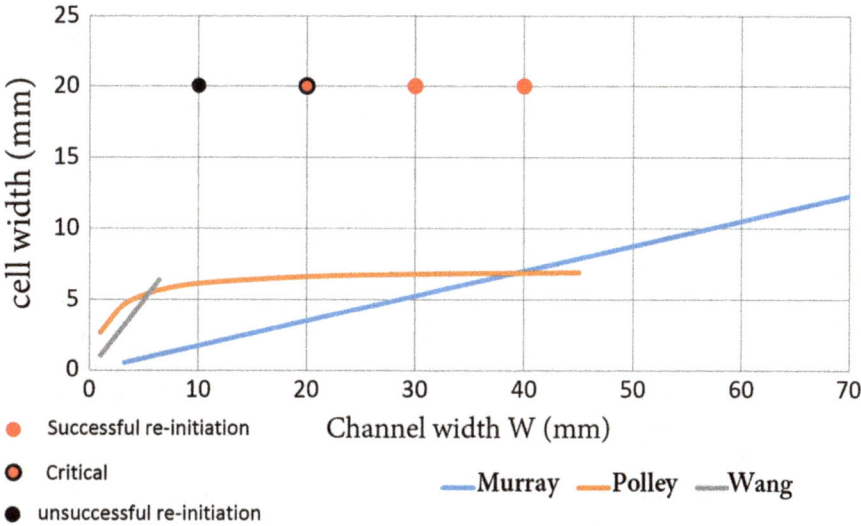

Fig. 9.19 Comparisons between the present study and the previous studies

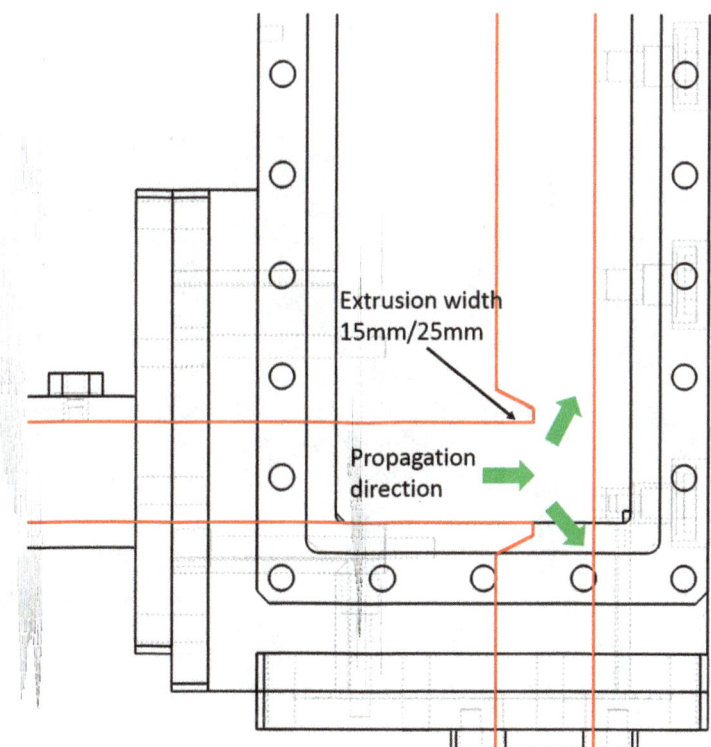

Fig. 9.20 New geometry confinement with extrusion into the vertical channel (15 mm/25 mm)

channel which decreases the local channel width to 25 mm, and the other one has an extrusion of 25 mm into the vertical channel which decreases the local channel width to 15 mm. The reason for such a design is to decrease the detonation diffraction period before the first reflection so that to decrease the time for the development of induction length. At the same time, such a geometry design can keep the same 40 mm channel width at downstream which has been proven to be capable for successful detonation re-initiation through all the subsequent reflections. Thus the only parameter that changes is the induction length, which would allow us to evaluate its influence to the re-initiation phenomenon.

Figures 9.21 and 9.22 shows detonation re-initiation results in the new channels with different extrusion geometries. It can be discovered that when the extrusion narrows the initial channel width to 15 mm (Fig. 9.21), the re-initiation is unsuccessful even though the downstream channel width keeps 40 mm. When the extrusion narrows the channel width to 25 mm which is larger than the critical geometry identified in Fig. 9.22, the re-initiation was found to be successful again. So according to the results, we can conclude that the changes made to the first reflection (induction length) can determine the re-initiation process, and the induction length is the main reason why longer channel width is more favorable for detonation re-initiation. So to summarize the results, we can make a conclusion that the reason why longer channel width is preferable for detonation re-initiation is due to the reactive mixture in the induction length. If the channel width is longer than a critical value, the distance would allow sufficient time for the diffracted detonation to

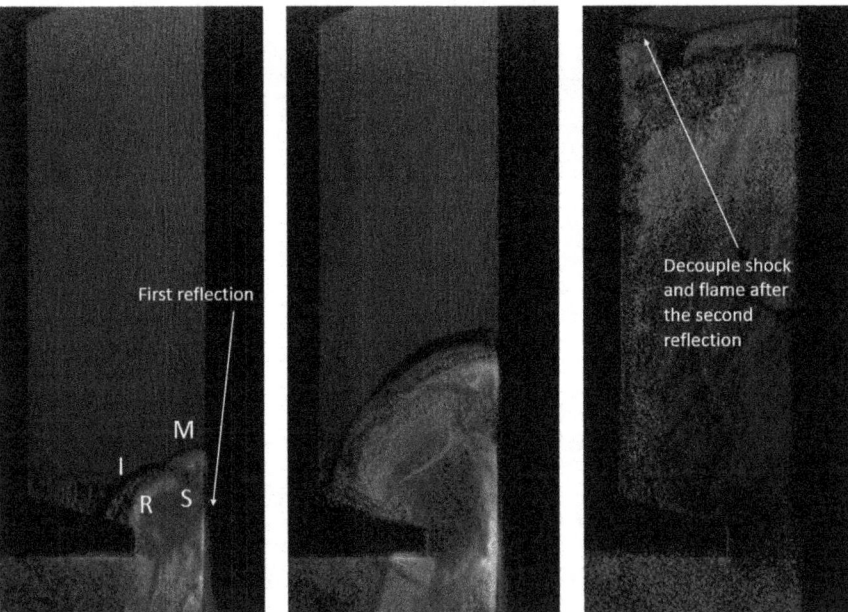

Fig. 9.21 Simultaneous measurement of 25 mm extrusion into the vertical channel in time sequence

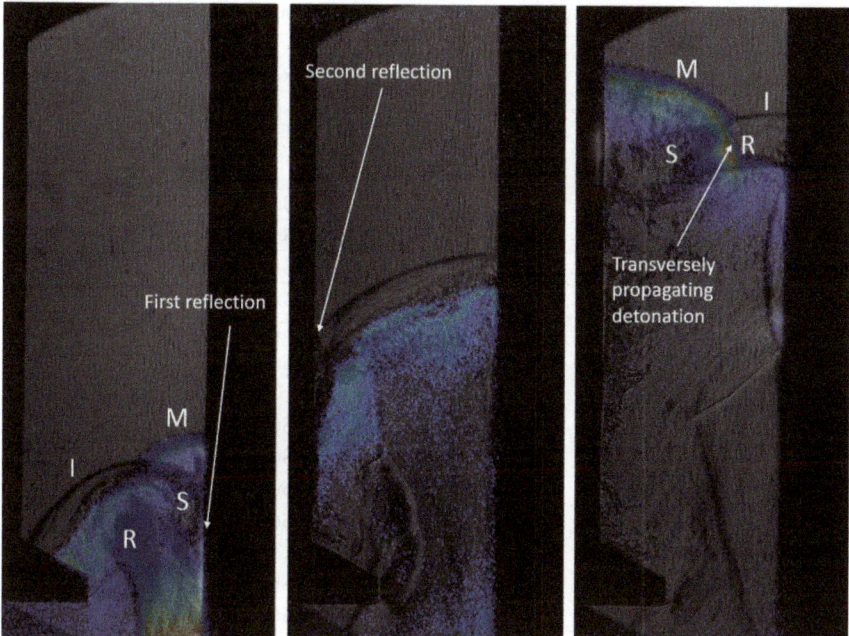

Fig. 9.22 Simultaneous measurement of 15 mm extrusion into the vertical channel in time sequence

develop a long induction length which can subsequently contribute to a strong shock reflection, and thus result in detonation transition at downstream.

Transitional Behavior of Re-initiation from Stable Detonation to Unstable Detonation

The results obtained experimentally using stable and unstable detonation for re-initiation show different critical initial pressure and cell size, indicating that single general criteria regarding the channel geometry and the mixture property may not be able to apply to two distinct detonation types. So there may exist some re-initiation mechanism differences when we gradually change the detonation waves from unstable to stable, and the change may take place either in a gradual way or a sudden way. In order to investigate the transitional behavior, some experiments were conducted at various Ar and N_2 dilution ratios. The testing conditions were selected based on the linear interpolation of all the parameters including fuel volume ratio, dilution type and dilution ratio of the two extreme cases (stable and unstable detonation). Other parameters, e.g. equivalence ratio, ignition energy were kept the same. The test conditions are shown in Table 9.3.

Table 9.3 Detonation shots for experiments of re-initiation transitional behavior

Firing shot	O$_2$/fuel	Ar/fuel	N$_2$/fuel	Initial pressure	Results	Ar/Ar + N$_2$
SS73	2.99	0	11.26	1	Successful	0
SS74	2.95	0	11.11	0.792	Unsuccessful	0
SS75	2.93	0	11.06	0.904	Unsuccessful	0
SS76	2.95	0	11.12	0.954	Successful	0
SS77	2.96	0	11.14	0.981	Successful	0
SS78	3.03	0	11.43	1.052	Successful	0
SS79	2.97	0	11.18	1.043	Successful	0
SS86	3.06	8.38	8.25	0.8	Successful	0.50
SS87	3.09	8.38	8.43	0.603	Successful	0.49
SS88	3.09	8.27	8.36	0.446	Unsuccessful	0.49
SS89	2.90	7.92	7.72	0.517	Successful	0.50
SS90	2.73	7.42	7.18	0.491	Successful	0.50
SS91	3.14	16.31	5.29	0.457	Successful	0.75
SS92	2.88	14.93	4.87	0.358	Successful	0.75
SS93	2.85	14.96	4.90	0.308	Unsuccessful	0.75
SS96	2.80	20.14	2.77	0.403	Successful	0.87
SS97	2.78	19.98	2.71	0.344	Successful	0.88
SS98	2.90	20.97	3.03	0.27	Unsuccessful	0.87
SS99	2.89	20.89	2.89	0.316	Successful	0.87
SS100	2.96	3.295	9.59	0.648	Successful	0.25
SS101	2.93	3.27	9.46	0.546	Unsuccessful	0.25
SS102	2.99	3.30	9.54	0.598	Successful	0.25
SS103	2.93	25.88	1.23	0.399	Successful	0.95
SS104	3.11	27.40	1.30	0.354	Successful	0.95
SS105	3.10	27.07	1.35	0.303	Unsuccessful	0.95
SS64	2.88	28.89	0	0.301	Unsuccessful	1
SS65	2.67	26.13	0	0.501	Unsuccessful	1
SS66	2.77	27.47	0	0.302	Unsuccessful	1
SS67	3.08	30.60	0	0.504	Unsuccessful	1
SS68	3.06	29.86	0	0.700	Successful	1
SS69	3.02	29.64	0	0.608	Successful	1
SS70	3.02	29.78	0	0.557	Successful	1
SS71	2.94	28.74	0	0.455	Unsuccessful	1
SS80	2.84	28.02	0	0.532	Successful	1
SS81	2.94	29.04	0	0.513	Successful	1

By gradually changing the initial pressure of the various test conditions, successful and unsuccessful detonation re-initiation cases were observed. The experimental results are plotted in the graph as shown in Fig. 9.23. The vertical axis is the detonation initial pressure and the horizontal axis is the Ar volume ratio as in all the dilution gases (Ar + N$_2$). We can see that when the dilution gas has less than 75% Argon, the critical initial pressure for a successful detonation re-initiation decreases with

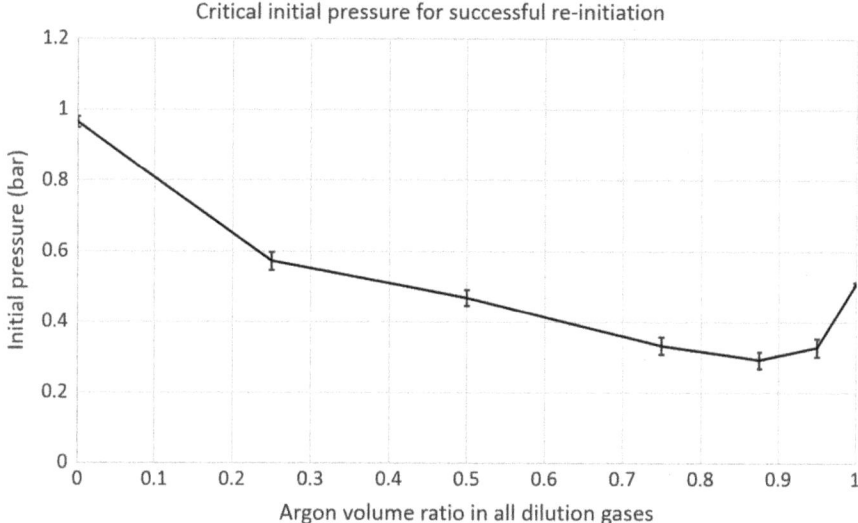

Fig. 9.23 Critical initial pressure against different Argon volume dilution ratio for successful detonation re-initiation

increasing Argon ratio. However, when Argon ratio further increases from 75% to 100%, the critical initial pressure starts to increase. This result suggests that the critical initial pressure is not decreasing all the way from pure N_2 dilution to pure Argon dilution, and there probably exists a sudden change of detonation re-initiation process when Argon consists 75% of all the dilution gases.

In order to find out the specialty of 75% Argon volume ratio as dilution, another graph was plotted in Fig. 9.24 showing Argon dilution ratio against the critical reduced activation energy for successful detonation re-initiation. Here the chemical kinetics mechanism employed to obtain both shocked state temperature and induction time for the calculation of reduced activation energy is from Wang and Frenklach (1997), which has been previously validated by Schultz and Shepherd (2000) with reasonable accuracy. The constant volume calculation adopted Shepherd's algorithm (Browne and Shepherd 2004). We can see that the critical reduced activation energy decreases from 13.5 to 6 when the argon dilution ratio changes from 1% to 75%, and then keeps almost constant when further increases Argon ratio to 100%. According to the calculation of neutral stability curve, we know that detonation waves become stable when reduced activation energy is approaching 6 (For Mach number > 3). So we could also conclude that the critical initial pressure for successful re-initiation decreases with increasing Argon dilution when detonation waves are in stable range, and would increase with increasing Argon dilution when detonation waves are unstable. We also analyzed the result based on the volume percentage of Argon dilution gas in the overall mixture. We noticed that 75% of Argon in N_2 and Argon dilution can be converted to approximately 63% of Argon gas in the whole mixture. This critical number of 63% is

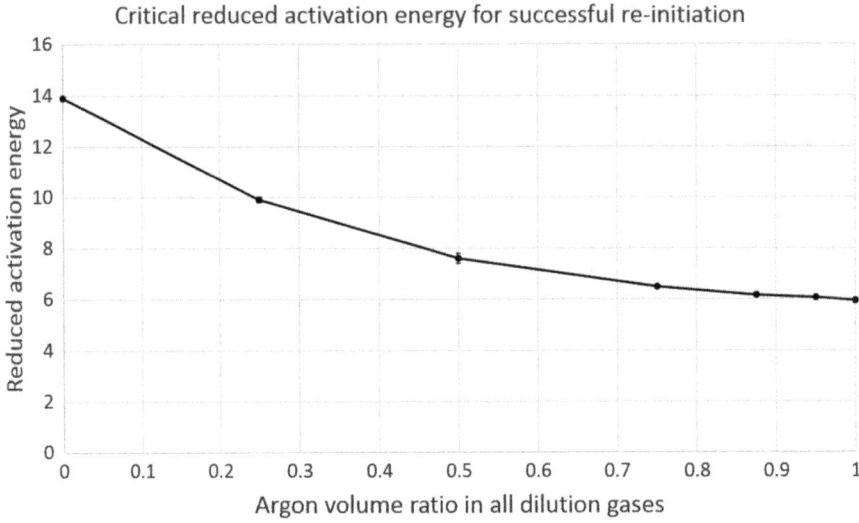

Fig. 9.24 Critical reduced activation energy against different Argon volume dilution ratio for successful detonation re-initiation

consistent with the critical value that has been summarized as the transition point between regular and irregular structures of detonation waves in the previous studies (Vandermeiren and Van Tiggelen 1984). It also matches very well with the discontinuous bifurcation point of detonation failure mechanisms change which was proposed by Radulescu (2003). Since detonation re-initiation in the present study is associated with detonation failure, the explanation to the critical value which was proposed by Radulescu could also shed some light on what we've discovered in the re-initiation phenomenon. It was discovered that the critical value of 60% of Argon dilution is due to the sudden detonation dynamics change. According to the calculation of detonation dynamics length scale values (cell size/reaction time and cell size/reaction time) at various Argon dilution ratio, it was discovered that both values change drastically at 60% of Argon. It appears that when the Argon dilution is below 60%, the length of induction region is proportional to cell width, which suggests that cell size in this dilution ratio range is better represented by the induction length. While for Argon dilution ratio of more than 60%, cell width/reaction length becomes constant, indicating that the exothermic region determines the cell spacing. Such changes of detonation dynamic length scale at 60% Ar dilution ratio could influence both detonation propagation and failure mechanisms, thus it is very likely to be responsible for the transitional behavior of detonation re-initiation in the present study.

Number of Shock Reflections Before Detonation Re-initiation

Even though the velocity change during the whole process could provide us with some insight about how re-initiation was formed. It remains unsolved how re-initiation could be achieved even if the waves after each reflection continue decaying to sub-CJ, and why the final re-initiation takes place certain times of shock reflections instead of at an earlier or later reflection.

Previous attempts have been made to solve the problem. Eckett (2001) investigated the detonation reaction zone structure equation and looked for all the factors that could affect the propagation of a detonation wave. By analyzing the temperature reaction zone structure equation for the one-step reaction model, it was found that the propagation of a detonation wave is essentially controlled by reaction heat release, the wave curvature and the wave unsteadiness. In 1996, He (1996) proposed an analytical solution to explain the influence of the leading wave's curvature to direct initiation of a self-sustained detonation wave. Through numerical simulations, it was concluded that a detonation wave can only exist if its radius is larger than the critical value Rc and the velocity larger than the critical value Dc. For a detonation wave which has a smaller curvature than Rc, the generalized CJ solution doesn't exist. So the curved detonation wave will always be weakened by the rarefaction wave, thus the wave cannot sustain. By solving the governing equations, a non-linear curve can be derived as shown below to be the marginal solution curve for quasi-steady detonation:

$$\left(2\theta \frac{D_{CJ} - D}{D_{CJ}}\right) \exp\left(-2\beta \frac{D_{CJ} - D}{D_{CJ}}\right) = \frac{8}{1-\gamma^{-2}} \left(\theta \frac{L_{CJ}}{R_s}\right) \tag{9.1}$$

The critical velocity and radius point on the non-linear curve exist on the left of the curve which is exhibited by the above non-linear relation as:

$$\frac{R_c}{L_{CJ}} = \frac{8e}{1-\gamma^{-2}} \theta \tag{9.2}$$

$$\frac{D_{CJ} - D_c}{D_{CJ}} = \frac{1}{2\theta} \tag{9.3}$$

Here using the Matlab environment Cantera, the induction length can be directly calculated based on Shepherd's algorithm (Browne and Shepherd 2004). Considering the importance of wave curvature, the leading wave diameter after each reflection was also measured from the images. We assume that the leading wave after each reflection has a circular shape. For each image, random three points can be selected on the leading wave front and then used to determine a circle.

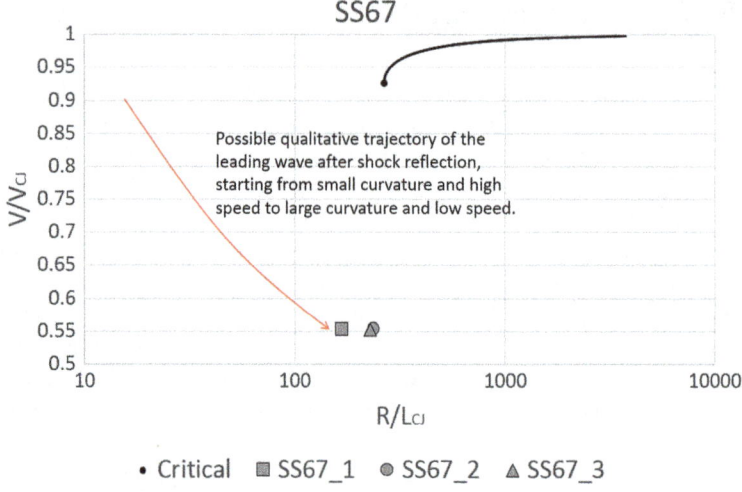

Fig. 9.25 Leading wave curvature and velocity of firing shot SS67

Starting from stable detonation waves, Fig. 9.25 plots the wave velocity information against the extracted curvature radius from firing shot SS67. Each individual point in the figure represents the wave curvature and velocity information at one frame after each shock reflection. Due to the large uncertainties of curvature measurement when the leading wave is small, only radiuses of big front waves were measured, so only one frame after each shock reflection was chosen. The marginal solution curve for quasi-steady detonation and the corresponding critical curvature/velocity point have also been drawn on the graph to show its relative position with the individual points. Due to the relatively low activation energy of the mixtures, the lower branch of the marginal quasi-steady state solution does not correspond to stable detonation regime (He 1996). So here only upper branch was plotted. Point SS67_1 represents the leading wave of one image frame after the first reflection. It can be found that when the leading wave was at 90 mm from the horizontal channel center axis, or alternatively at the location immediately before the second reflection occurs, it was propagating at $0.55V_{CJ}$ with a curvature radius of $168L_{CJ}$. Based on Eqs. (9.2 and 9.3), the critical velocity and curvature radius are calculated to be $0.926V_{CJ}$ and $266L_{CJ}$, respectively. So this point locates much below the critical point of the marginal solution curve. Individual points SS67_2 and SS67_3 represent the leading wave after the second and the third shock reflection, respectively. According to the location of the two points, they were both found far below the critical point. If we assume that the leading wave curvature radius immediately after the each reflection is small, and it gradually expanded outwards with increasing radius, then the arrow trajectory which represents the wave evolution on the graph will be from the region of small radius and high velocity to the region of large radius and low velocity (and the trajectory goes through the point). As shown in Fig. 9.25, the trajectory, in this case, could never intersect with the marginal solution curve. Since

SS67 experimental results indicate unsuccessful re-initiation after each shock reflection, thus we can see that generally, the analysis based on the marginal solution curve complies well with the experimental findings.

Then we turn to the successful re-initiation case, e.g. firing shot SS81 as shown in Fig. 9.26. The wave after the first reflection for each firing shot shows a similar point far below the critical, indicating that it is impossible to re-initiate a detonation wave through single reflection. Regarding the second shock reflection, the corresponding point of the wave is very close to the critical point. It is thus difficult to judge if the wave propagation trajectory could intersect with the marginal solution curve or not. Some estimations of the trajectory can be conducted. First, we directly adopted the same way that has been applied by He and Clavin (1994) using blast wave propagation. Here the wave after each shock reflection is approximated as a blast wave originating from a point source. According to Taylor (1950) and Sedov (1946), the leading wave velocity and the curvature radius of such a cylindrical blast wave can be represented using a hyperbola function. So based on one existing point, the trajectory can be drawn as shown in Fig. 9.26 which intersects with the marginal solution curve after the second shock reflection. In order to see if the prediction of the trajectory curve is accurate, more direct measurements of the images on the curvature radius and the velocity were conducted and the other trajectory is then predicted. Due to the chemical reactions behind the leading shock which can result in relatively slow velocity decrease during propagation, it was found that the real trajectory is quite different from the prediction based on blast wave. One major difference from the previous prediction is that the real trajectory based on velocity and radius measurement does not intersect with the marginal solution curve. This result complies with the experimental images which indicate that successful re-initiation

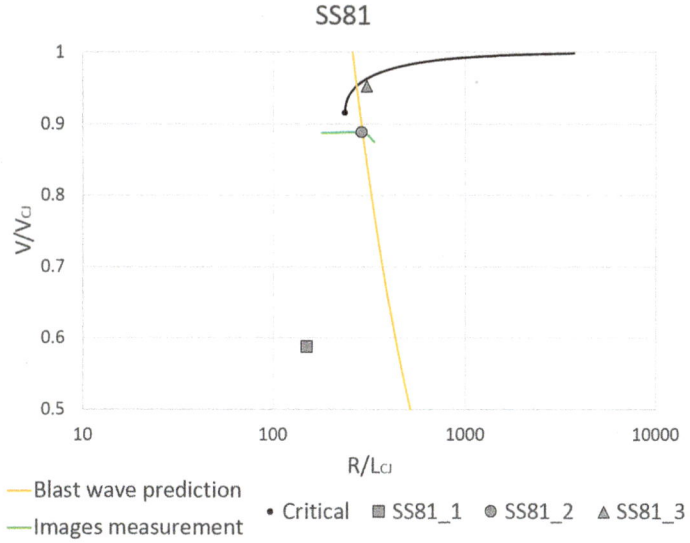

Fig. 9.26 Leading wave curvature and velocity of firing shot SS81

under this test condition is only successful after the third reflection. So we could see that the prediction using blast wave may not reflect the real wave propagation due to the trailing combustion reactions.

Since predicting detonation re-initiation can be taken as the same process of evaluating if the waves generated after each shock reflection can be self-sustained separately, the quasi-steady solution need to be applied to each of the waves from shock reflections, and see if the trajectory of each wave during development and further propagation can intersect with the marginal solution curve. Even though the re-initiation process in the present study is different from the direct initiation of detonation in (He 1996), we can still simplify it as a detonation initiation process from a diffracted detonation wave with multiple sudden energy inputs through shock reflections. So a similar and qualitative graph in Fig. 9.27 with respect to the wave curvature radius and leading shock wave velocity based on firing shot SS81 can be plotted to illustrate the process. It shows that the marginal solution curve can be used to predict the re-initiation process.

Another example of detonation shot SS68 is shown in Fig. 9.28. The test condition is similar to that of SS81 except for the higher initial pressure of 0.7 bar. Due to such a high initial pressure, the cell width would become smaller which is more favorable for detonation re-initiation, thus fewer times of shock reflection would be needed for successful detonation re-initiation. According to Fig. 9.28, the analysis based on blast wave prediction and real image measurements was conducted. We can see that both trajectories which represent the second shock reflection intersect with the marginal solution curve. So it suggests that only two times of shock

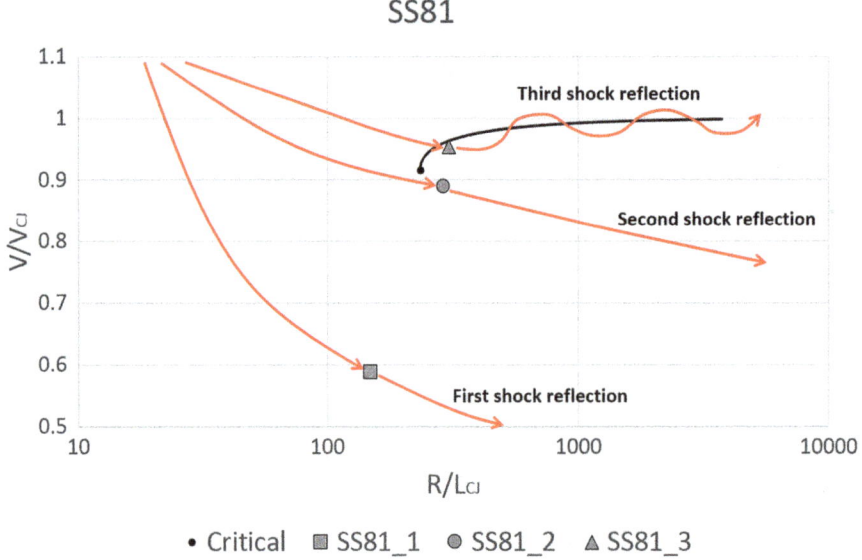

Fig. 9.27 Illustration of successful re-initiation based on marginal solution curve

9 On the Investigation of Detonation Re-initiation Mechanisms and the Influences...

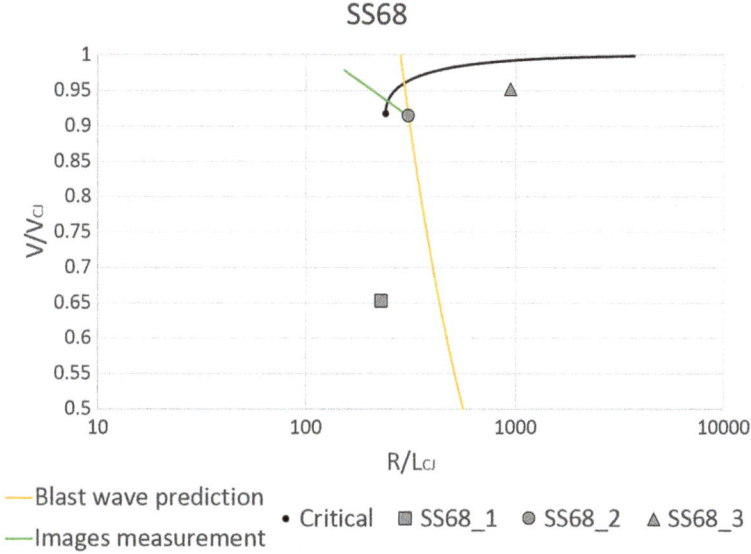

Fig. 9.28 Leading wave curvature and velocity of firing shot SS68

reflection would be required to achieve successful detonation re-initiation at 0.7 bar initial pressure. It also complies well with the experimental results which shows a much smaller gap between the leading shock wave and the trailing flame after the second shock reflection.

So from the firing shot near the critical condition, we see that the one-dimensional quasi-steady model for detonation can perfectly explain the re-initiation phenomenon of stable detonation waves after each shock reflection. As for unstable detonation waves, the same experimental measurement regarding the curvature and radius can also be conducted. One successful re-initiation case of SS73 is shown in Fig. 9.29. The curvature radius was measured to be far smaller than the critical radius $1056L_{CJ}$. From the second reflection, the local velocity after the second reflection reaches the critical velocity of $0.96V_{CJ}$. After the second shock reflection on the wall, the curvature radius was very difficult to measure because the whole Mach stem is no longer an arc shape. Due to the existence of the multiple transverse waves which could greatly result in big local velocity fluctuations, the curvature of the Mach stem is not a constant throughout the curve. So it is difficult to use the quasi-steady solution to correlate with the experimental results. Neither can we determine the validity of this method to unstable detonation waves. One point to note for this method is that it is based on one-dimensional instability theory, which doesn't take multi-dimensional effect into account. For highly unstable detonation, the multi-dimensional disturbance may become more important which was not accounted for in the method. Besides, the intrinsic unsteadiness effect from the unstable detonation front can induce additional failure to the detonation wave which dominates the quenching mechanism (He 1996; He and Calvin 1994).

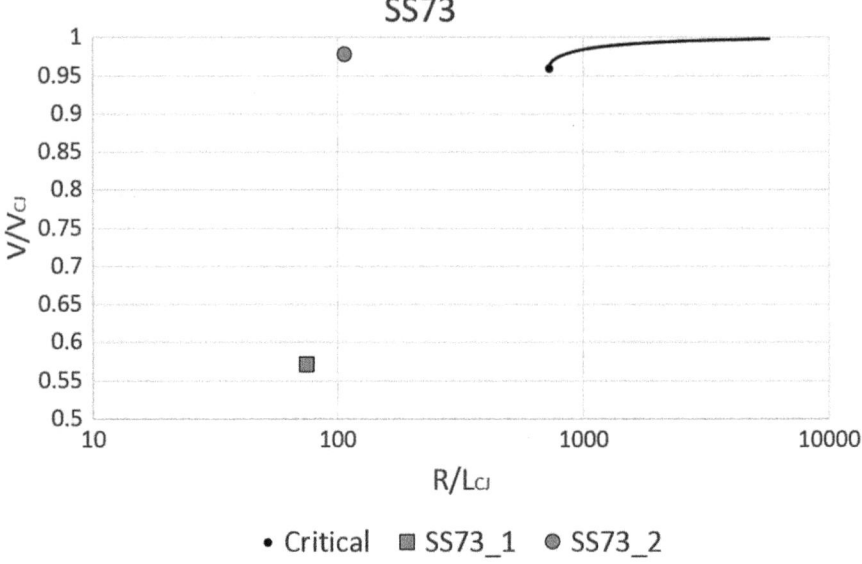

Fig. 9.29 Leading wave curvature and velocity of firing shot SS73

7 Conclusions

Detonation re-initiation through multiple shock reflections has been investigated in the present study as it has wide applications in industrial explosion hazard preventions and propulsion applications. The present study is conducted to explain the discrepancies between the previous similar studies, reveal the corresponding re-initiation mechanisms and find the influence of geometry confinement and mixture properties to the re-initiation phenomenon. The research methodology is mainly based on optical measurement techniques, including Schlieren photography, PLIF, open shutter photography and other non-optical measurement technique to provide a clearer visual of the detailed re-initiation processes from several different perspectives. Numerical simulations are also applied to reveal more details.

Due to various propagation mechanisms, the corresponding re-initiation mechanisms for the stable and the unstable detonation waves can also expect to be different. According to the optical measurement of the multiple reflected re-initiation phenomenon using unstable detonation waves, it was found that the second last reflection is the most critical for re-initiation instead of the last reflection. This is because, during the second last reflection, a train of multiple transverse waves can be generated. They were observed to reflect back from the wall, interact with one another and form the cellular shaped structures which resemble the transverse waves in a typical detonation wave. Besides, a transverse wave multiplication mechanism was discovered to play an important role to assist re-initiation. Velocity information extracted from high speed images revealed a critical velocity of 1000 m/s of the wave before the second shock reflection for successful re-initiation. This critical velocity is acting as a threshold below which the multiple reflections cannot progressively accelerate the wave and achieve detonation.

Re-initiation of stable detonation was observed to be similar with that of unstable detonation. A critical velocity threshold of stable detonation waves was identified to be 900 m/s for successful detonation re-initiation. The successful re-initiation was also found characterized by a transverse detonation after the second shock reflection. Some differences can be observed based on the experimental results. It was discovered that the transverse waves' interactions were not evident in this case. No velocity fluctuations exist during the propagation of the waves.

Except for the extreme cases of stable and unstable detonation waves, more experiments were conducted in between to investigate the transitional behavior of detonation re-initiation. It can be concluded that there exists a critical condition in which a sudden change of critical initial pressure occurs at Argon dilution ratio of 75%. This value is found consistent with other previous studies, which can be explained by the sudden change of the detonation dynamics between stable and unstable detonation at this value of dilution. To reveal the reason why re-initiation is always achieved through certain times of shock reflections, a marginal solution curve which is based on quasi-steady solution of detonation was found able to give a reasonable explanation. Through analysis, it shows that re-initiation can be simplified as several detonation initiation problems after all shock reflections. It has been proven that the last shock reflection for successful re-initiation always corresponds to the trajectory which crosses the marginal solution curve for the first time.

Regarding the influence of the geometry confinement dimensions to detonation re-initiation, it is discovered that a wider channel width is preferable for detonation re-initiation. Even though the wider gap between the channel walls would allow a longer time for the leading wave to quench, it is, in fact, the highly reactive mixture in the induction zone which could facilitate the shock reflection by inducing a strong reactive Mach stem and thus finally induce detonation transition. The length of the bifurcation channel can also affect the re-initiation phenomenon. As certain times of shock reflections are required for successful re-initiation, a bifurcation channel with insufficient length would affect the re-initiation by limiting the overall reflection times, thus a case which is supposed to be a successful re-initiation case would be interpreted as unsuccessful. This is found to be the main reason leading to the discrepancies of the previous similar studies. If taking this factor into consideration, the previous summaries of the critical conditions for successful re-initiation could now comply well with each other.

References

Bhattacharjee, R. R. (2013). Experimental investigation of detonation re-initiation mechanisms following a Mach reflection of a quenched detonation. *Dissertation for the fulfilment of Master degree*, Ottawa University, Ottawa.

Bhattacharjee, R. R., Lau-Chapdelaine, S. S. M., Maines, G., Maley, L., & Radulescu, M. I. (2013). Detonation re-initiation mechanism following the Mach reflection of a quenched detonation. *Proceedings of the Combustion Institute 34*: 1893–1901.

Boeck, L. R., Kellenberger, M., Rainsford, G., & Ciccarelli, G. (2017). Simultaneous OH-PLIF and schlieren imaging of flame acceleration in an obstacle-laden channel. *Proceedings of the Combustion Institute 36*: 2807–2814.

Brophy, C. M., Werner, L. T. S., & Sinibaldi, J. O. (2003). Performance characterization of a valveless pulse detonation engine. AIAA Paper 2003–1344.

Browne, S., & Shepherd, J. (2004). Numerical solution methods for shock and detonation jump conditions. GALCIT technical report FM2006.006.

Ciccarelli, G., & Dorofeev, S. (2008). Flame acceleration and transition to detonation in ducts. *Progress in Energy and Combustion Science, 34*(4), 499–550.

Eckett, C. A. (2001). Numerical and analytical studies of the dynamics of gaseous detonations. *Dissertation for the fulfilment of degree of Doctor of Philosophy*, California Institute of Technology, Pasadena.

Edwards, D. H., Nettleton, M. A., & Thomas, G. O. (1979). The diffraction of a planar detonation wave at an abrupt area change. *Journal of Fluid Mechanics, 95*, 79–96.

Edwards, D. H., & Thomas, G. O. (1981). Diffraction of a planar detonation in various fuel-oxygen mixtures at an area change. *Progress in Astronautics and Aeronautics, 75*, 341–357.

Frolov, S. M., Aksenov, V. S., & Shamshin, I. O. (2007a). Shock wave and detonation propagation through U-bend tubes. *Proceedings of the Combustion Institute 31*(2): 2421–2428.

Frolov, S. M., Aksenov, V. S., & Shamshin, I. O. (2007b). Reactive shock and detonation propagation in U-bend tubes. *Journal of Loss Prevention in Process Industry, 20*(4–6), 501–508.

Gamezo, V. N., Ogawa, T., & Oran, E. S. (2007). Numerical simulations of flame propagation and DDT in obstructed channels filled with hydrogen–air mixture. *Proceedings of the Combustion Institute 31*: 2463–2471.

Gordon and McBride (1994): National Aeronautics and Space Administration, Lewis Research Center, Cleveland, Ohio.

He, L. (1996). Theoretical determination of the critical conditions for the direct initiation of detonations in hydrogen-oxygen mixtures. *Combustion and Flame, 104*, 401–418.

He, L., & Clavin, P. (1994). On the direct initiation of gaseous detonations by an energy source. *Journal of Fluid Mechanics, 277*, 227–248.

Kaneshige, M., & Shepherd, J. (1997). Detonation database. GALCIT Technical Report FM97-8, http://www.galcit.caltech.edu/detn_db/html/

Knystautas, R., Lee, J. H. S., & Guirao, C. M. (1982). The critical tube diameter for detonation failure in hydrocarbon–air mixtures. *Combustion and Flame, 48*, 63–83.

Laderman, A. J., Urtiew, P. A., & Oppenheim, A. K. (1963). On the generation of a shock wave by flame in an explosive gas, Ninth Symposium (International) on Combustion, The Combustion Institute, Pittsburgh 265–274.

Lee, J. H. S. (1977). Initiation of gaseous detonation. *Annual Review of Physical Chemistry, 28*, 75–104.

Li et al. (2013): 49th AIAA/ASME/SAE/ASEE Joint Propulsion Conference San Jose, CA

Li, L., Li, J., Nguyen, V. B., Teo, C. J., Chang, P. H., & Khoo, B. C. (2017). A study of detonation re-initiation through multiple reflections in a 90-degree bifurcation channel. *Combustion and Flame, 180*, 207–216.

Lv, Y., & Ihme, M. (2015). Computational analysis of re-ignition and re-initiation mechanisms of quenched detonation waves behind a backward facing step. *Proceedings of the Combustion Institute 35*(2): 1963–1972.

MacBride and Gorden (1996): National Aeronautics and Space Administration, Lewis Research Center, Cleveland, Ohio

Mitrofanov, V. V., & Soloukhin, R. I. (1964). The diffraction of multifront detonation waves. *Soviet Physics Doklady, 9*(12), 1055.

Murray, S. B., & Lee, J. H. S. (1983). On the transformation of planar detonations to cylindrical detonation. *Combustion and Flame, 52*, 269–289.

Oppenheim, A. K., Laderman, A. J., & Urtiew, P. A. (1962). The onset of retonation. *Combustion and Flame, 6*, 193–197.

Oran, E. S., & Gamezo, V. N. (2007). Origins of the deflagration-to-detonation transition in gas-phase combustion. *Combustion and Flame, 148*(1–2), 4–47.

Pintgen, F. (2004). Detonation diffraction in mixtures with various degrees of instability. *Dissertation for fulfilment of Degree of Doctor of Philosophy*, California Institute of Technology, Pasadena.

Polley, N. L., Egbert, M. Q., & Petersen, E. L. (2013). Methods for re-initiation and critical conditions for a planar detonation transforming to a cylindrical detonation within a confined volume. *Combustion and Flame, 160*, 212–221.

Radulescu, M. I. (2003). The propagation and failure mechanism of gaseous detonations: experiments in porous-walled tubes. *Dissertation for the fulfilment of degree of Doctor of Philosophy*, Mcgill University, Montreal.

Radulescu, M. I., & Maxwell, B. M. (2011). The mechanism of detonation attenuation by a porous medium and its subsequent re-initiation. *Journal of Fluid Mechanics, 667*, 96–134.

Roy, G. D., Frolov, S. M., Borisov, A. A., & Netzer, D. W. (2004). Pulse detonation propulsion: Challenges, current status, and future perspective. *Progress in Energy and Combustion Science, 30*, 545–672.

Schultz, E., & Shepherd, J. (2000). Validation of detailed reaction mechanisms for detonation simulation, Explosion Dynamics Laboratory Report FM99-5, California Institute of Technology, Pasadena.

Sedov, L. I. (1946). Propagation of strong blast waves. *Journal of Applied Mathematics and Mechanics, 10*, 241–250.

Silvestrini, M., Genova, B., Parisi, G., & Leon Trujillo, F. J. (2008). Flame acceleration and DDT run-up distance for smooth and obstacles filled tubes. *Journal of Loss Prevention in Process Industry, 21*(5), 555–562.

Taylor, G. I. (1950). The dynamics of the combustion products behind plane and spherical detonation front in explosives. *Proceedings of the Royal Society of London A 200*: 235–247.

Thomas, G. O., & Williams, R. L. (2011). Detonation interaction with wedges and bends. *Shock Waves, 11*(6), 481–492.

Vandermeiren, M., & Van Tiggelen, P. J. (1984). Cellular structure in detonation of acetylene-oxygen mixture. *Progress in Astronautics and Aeronautics, 94*, 104–117.

Wang, C. J., Xu, S. L., & Guo, C. M. (2008). Study on gaseous detonation propagation in a bifurcated tube. *Journal of Fluid Mechanics, 599*, 81–110.

Wang, H., & Frenklach, M. (1997). Detailed kinetic modeling study of aromatics formation in laminar premixed acetylene and ethylene flames. *Combustion and Flame, 110*, 173–221.

Wang, J., Lee, J. H. S., & Ng, H. S. (2015). Velocity deficits in thin channels for a cylindrically expanding detonation. 25th International Colloquium on the Dynamics of Explosions and Reactive Systems, Leeds.

2016 International Workshop on Detonation for Propulsion: Panel Discussion

The details of the Panel Discussion after the 3 days of workshop are described below. Chiang Juay Teo (National University of Singapore, Singapore) chaired this panel and the panel comprised Kailas Kailasanath (Naval Research Laboratory, USA), Piotr Wolanski (Institute of Aviation, Poland), Matthew Fotia (Innovative Scientific Solutions Inc., USA), Christopher Stevens (Innovative Scientific Solutions Inc., USA), Kenneth Yu (University of Maryland, USA), Jiro Kasahara (Nagoya University, Japan), Christopher Brophy (Naval Postgraduate School, USA), Jian-Ping Wang (Peking University, China), Boo Cheong Khoo (National University of Singapore, Singapore), Jeong-Yeol Choi (Pusan National University, Korea), Pierre Vidal (Institute Pprime. France), Nobuyuki Tsuboi (Kyushu Institute of Technology), Akiko Matsuo (Keio University, Japan), Bing Wang (Tsinghua University, China), Ken Matsuoka (Nagoya University, Japan), Po-Hsiung Chang (National University of Singapore, Singapore) and Jiun-Ming Li (National University of Singapore, Singapore).

Piotr Wolanski: In the PDE, we are basically moving to direct propulsion, like What Jiro Kasahara is doing on the real application for attitude control system, a very excellent example of application of the PDE. Here one has to understand the benefit of low pressure supply and relatively high specific impulse in PDE. I also noted the direct propulsion of PDE still needs a lot of further research. On the other hand, on the RDE, we have more or less basically understand the nature of the rotating detonation. The rotating detonation is actually getting more attraction from the researchers whatever for application in direct propulsion, rocket and ramjet or otherwise. Now, we are moving also to more practical things. Of course, the Workshop has seen some presentations, like Jimmy (Jiun-Ming Li) on the application of jet fuel which is very essential and basic because we can apply hydrogen and oxygen in rocket for the momentum generation but not in the air-breathing propulsion. For the air-breathing propulsion, we do need to develop very efficient system for mixing. Still, we also have to look at other important part as shown in Kailas's presentation, with the calculation by the researchers from American, is very encouraging and supporting about the efficiency. And in this regard lies the importance of heat exchanger in long duration propulsion. I surmise that there are many more questions to be addressed. That would need everyone contributions. Thank you.

Chiang Juay Teo: I think one of the main challenges is to use liquid fuel for RDE. Thus far, we haven't seen the presentations of liquid fuel, like jet fuel, for RDE. Most of the interest centered on gaseous fuel. So, does anyone have any comments for challenges and application?

Kailas Kailasanath: In the discussion of the research community for past years, it used to that we have few works carried out in the pressure, velocity of injection system or efficiency of injection system. I was kind of surprised that there is no discussion about the efficiency of injection system. Is it because it is a solved problem? Seems the Workshop won't talk about that.

Piotr Wolanski: You (Kailas) mentioned a very important problem we got. During the last workshop in Beijing, we got some research work on pressure gain and application efficiency for hydrogen and air. At least, our research provides not only practical, but also fundamental evidences. So, there are more and more evidences to support our research. I know it is not easy. But of course, we have to optimize the injection system and try to minimize the pressure loss in the system arising from the injection. Kailas is right, the pressure gain is one of the most important issue to be addressed.

Matthew Fotia: Getting a rotation detonation engine to work, just to detonate doesn't seem to be the only problem. Now, it is taking about the rotating engine and the point to build a device, inclusive of incorporated components. That is the thing what temperature it goes to the materials. How could you be close the propulsion cycle to delineate whatever combustion system you have in combustors, and what application you aim (major applications in the gas turbine and ramjet) and then you can tailor the geometry based the application. We can see many inlet systems but they are application dependent. Each one has a solid different set of heat drivers and you can tailor the combustion technology and get it close the available engineering systems.

Christopher Stevens: It is very well to see different injection systems in different applications. It looks different when you use air as your oxidizer. You get the air from the compressor in the case of gas turbine or get the air from ram intake in the case of ramjet. The homogeneity of course is important for all the gaseous mixtures. As for liquid fuel, the homogeneity must be at best situation I guess.

Jiro Kasahara: The RDE is a huge system in totality. For the next step of research, I think the injection system is the most important part to get more pressure gain. I feel we should concentrate at the injection modeling, including liquid phase and heat transfer.

Christopher Brophy: There is one advise for injection modeling for CFD and some experiments. I think the aerodynamics near the injection system in the engine is important". The injection is a dynamic chocking system and the dynamic/static conditions upstream and downstream and the corresponding mass rate are changed dynamically. When you do the experimental works, putting the diagnostics near the injection is extremely important. We can understand what happen actually for the engine as capturing these dynamic changes upstream and downstream.

Chiang Juay Teo: Any comments for those of CFD?

Kailas Kailasanath: What we need is a good mixing model but mixing model is one biggest challenge.

Jian-Ping Wang: In this Workshop, we have seen great progress. There are some different ongoing projects, such as fundamental research, CFD, experiment and application. However, there are a lot of unknown fundamentals. For example, from the view of the CFD, we don't have significant experimental results for comparison. Now most studies apply non-uniform injection and we indeed have more thrust and specific impulse than conventional engines, but shall get higher benefits. We may need to make more efforts to get the engine to work more efficiently.

Boo Cheong Khoo: In the CFD part, the mixing part is quite important. The other part is that even for the simulation we need a fairly long time to simulate. We hope to see more; I mean on the use of model order reduction (MOR) to cut down the number of computations. One big challenge is on the shock front to reduce the number of modes as in the MOR to represent. For the combustion wave front, the model reduction is quite good means to cut down the numbers of calculations. However, in the detonation, the biggest challenge is the required MOR modes to represent effectively the shock front. This is something for the CFD community to focus on.

Jeong-Yeol Choi: This is a chance for us to have a good discussion. The workshop pushes us to move on and make us work better than previously. Direct comparison between constant pressure combustion and RDE in numerical and experimental methods is very important. An example is Wolanski's attempt to replace/substitute the constant pressure combustor of the RDE in a gas turbine. If his test is successful, we can get the direct comparison at the system level.

Piotr Wolanski: I will encourage everybody to make the contribution as also agreed in the previously meeting that you would like to publish. I will also encourage people to write the paper and submit it to Jimmy (Jiun-Ming Li). Now we have to move on to the next workshop. We already heard from the Pierre. He volunteers to organize the next meeting.

Pierre Vidal: The organization community will be established soon. The meeting would likely to be taken place in September. The precise date will be decided as soon as we can.

Boo Cheong Khoo: Once again thank you for taking time to attend this workshop. Definitely one very important challenge ahead is to get the RDE working much better to attract commercial interest.

GPSR Compliance

The European Union's (EU) General Product Safety Regulation (GPSR) is a set of rules that requires consumer products to be safe and our obligations to ensure this.

If you have any concerns about our products, you can contact us on

ProductSafety@springernature.com

In case Publisher is established outside the EU, the EU authorized representative is:

Springer Nature Customer Service Center GmbH
Europaplatz 3
69115 Heidelberg, Germany